Signals and Control Systems

Series Editor
Maurice Charbit

Signals and Control Systems

Application for Home Health Monitoring

Smain Femmam

iSTE

WILEY

First published 2016 in Great Britain and the United States by ISTE Ltd and John Wiley & Sons, Inc.

ISTE Ltd
27-37 St George's Road
London SW19 4EU
UK

www.iste.co.uk

John Wiley & Sons, Inc.
111 River Street
Hoboken, NJ 07030
USA

www.wiley.com

Library of Congress Control Number: 2016957652

British Library Cataloguing-in-Publication Data
A CIP record for this book is available from the British Library
ISBN 978-1-78630-127-7

FSC
www.fsc.org
MIX
Paper from responsible sources
FSC® C013604

Contents

Preface

This preface gives an overview of signals and systems, generalities and notions of process control and regulation.

Introduction

The main objective of these two volumes is the analysis and the study of linear, time-invariant, causal signals and deterministic systems of finite dimensions. We will focus our efforts on defining a set of tools useful to automatic control and signal processing, after which we will discuss methods for the representation of linear dynamic systems for the analysis of their behavior. Finally, the identification and the synthesis of control laws will be addressed for the purpose of stabilization and regulation in systems control. Chapter 6 of the other volume [FEM 16] will be dedicated to the use of the Nao robot for a specific application; in this case, home care service.

Signals and systems: generalities

Whether in the analog or digital field the study of the characteristic properties of signals and systems, the knowledge of mathematical tools, processing and analysis methods are constantly evolving and of late are increasingly significant. The reason is that the current state of technology, in particular of electronics and computer science, makes possible the implementation of very advanced processing systems, which are effective and increasingly less expensive in spite of their complexity. Aims and requirements generally depend on applications. Figure 1 presents the connections between the various disciplines, the scientific and technological resources for their operation with the aim of processing signals or automatic control for the operation or the development of current applications[1].

In all areas of physics, for study, analysis and understanding of natural phenomena, a stage for modeling and for the study of the structure of the physical process is necessary. This has led to the development of techniques for modeling, representation and analysis of systems using a fairly general terminology. This terminology is difficult to introduce in a clear manner but the concepts which it relies upon will be defined in detail in the following chapters.

Signal processing concerns the various operations carried out on analog or digital physical quantities with the purpose of analyzing, interpreting and extracting information. These operations are illustrated in Figure 2.

The mastery and the implementation of signal processing techniques require the knowledge of a number of theoretical tools. The objective of this book is to establish the basic concepts of the theoretical study and clarify common processing methods.

A physical process is divided into several components or parts forming what is called a system. This is the case, for example, of an engine that consists of an amplifier, power supply, an electromagnetic part and a position and/or speed sensor. The input to the system is the voltage applied to the amplifier and the output is either the position or the speed of the rotation of the motor shaft.

Among the objectives of the control engineer is the modeling, behavior analysis and the regulation or control of a system, aiming for the dynamic optimization of its behavior. The operation of the system or control is designed to ensure that the variables or system outputs follow a desired trajectory (curve over time in general) or have dynamics defined by the specifications document. For temperature regulation of a speaker to a reference value, one of the following diagrams can be used. Details of vehicle operation is shown in Figure 3.

Notions of process and operation control

The objective of automatic control is to design control and operation systems that are able to assign to a dynamic process (physical, chemical physical, biological, economical, etc.) a behavior defined in advance by the operator based on the requirements specifications. For example, we can consider speed regulation of a car that gives the process (the car) a previously determined speed, regardless of the disturbances that may occur (variation of the slope, etc.). Other examples include a

1 The author advises to avoid the book Traitement du signal Asservissement linéaires by I. Jelinsky in the series Vuibert Technologie in order to avoid giving readers any incorrect interpretation and confusion with the concepts presented in this book.

radar antenna alignment system for the monitoring of the trajectory of an airplane or a satellite, and an air conditioner designed to stabilize the temperature at a constant value fixed in advance.

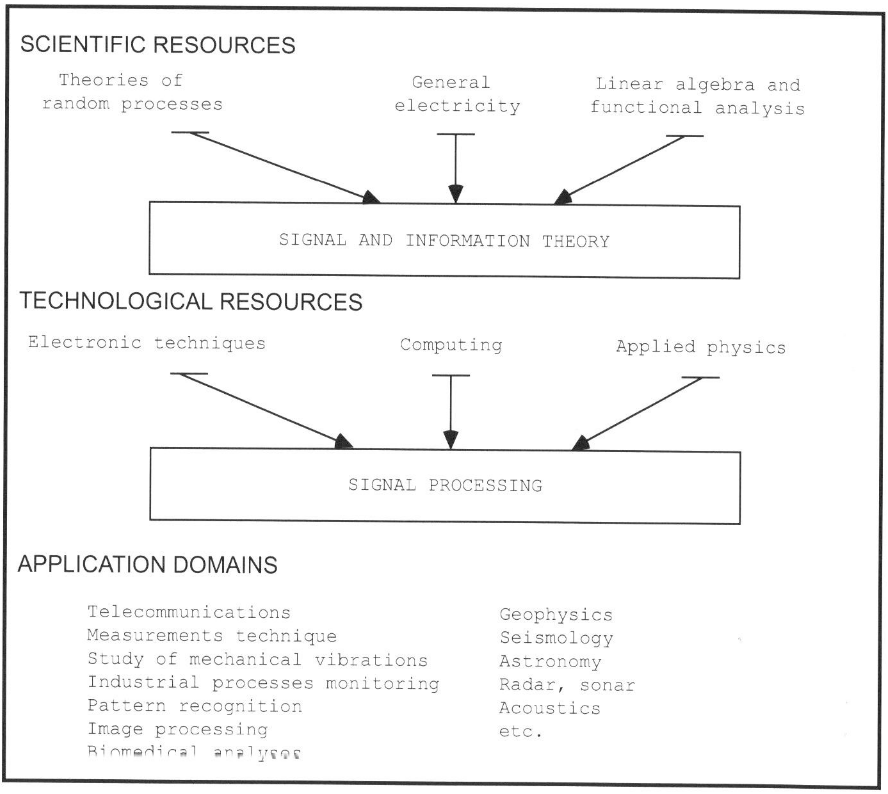

Figure 1. *Applied scientific and technological resources*

A process can be defined by establishing relationships between input and output quantities (this will be formally defined in different ways throughout this book. It is represented in Figure 4.

In the example of the car, the output is the speed and inputs may be the position of the accelerator pedal, the slope of the road and/or any other physical quantities that have an influence on the speed (the output of the system). Inputs consist of variables that can be manipulated (position of the pedal) and upon which no action is possible (the slope of the road). The latter are called disturbance inputs, they may be measurable or not accessible, random or deterministic. The variables that can be manipulated can be used as control inputs.

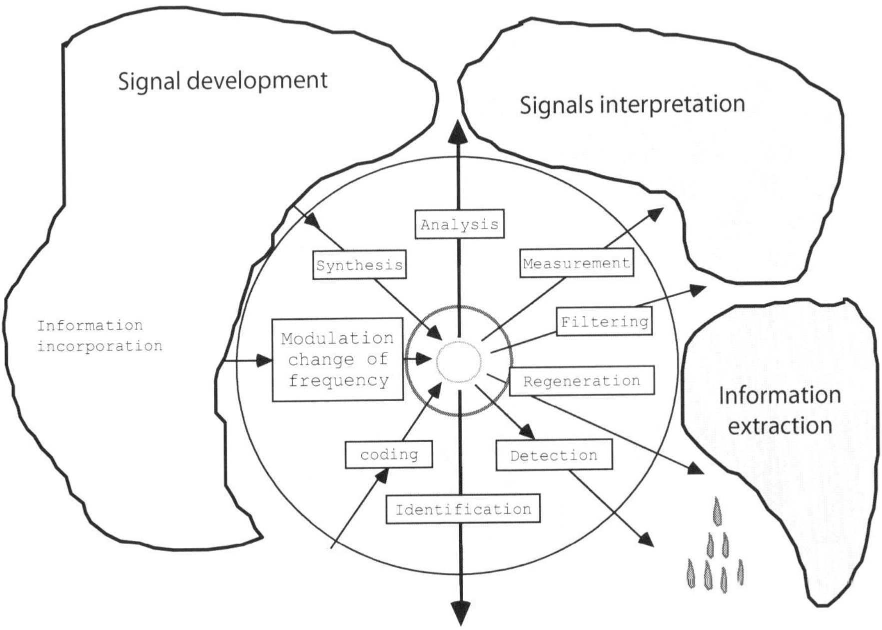

Figure 2. *Basic concepts of theoretical study and processing methods*

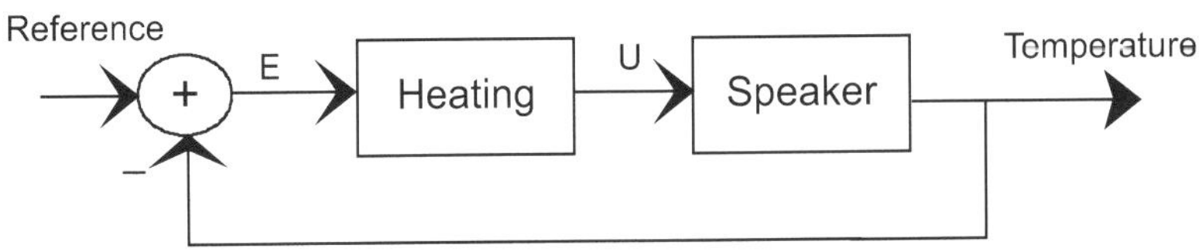

Figure 3. *Diagram of the model for vehicle operation*

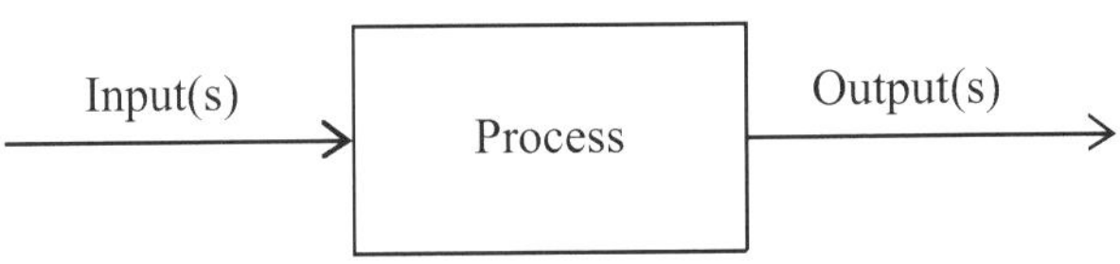

Figure 4. *Process with multiple input and output quantities*

In order to maintain the constant speed of a vehicle, a mathematical model of the process must be developed, in which the vehicle speed is linked to the position of the accelerator pedal, and then by inverting this model, the necessary input to obtain a specified speed can be derived. As a result, if the system output is not taken into consideration, an open-loop control is carried out (see Figure P.5).

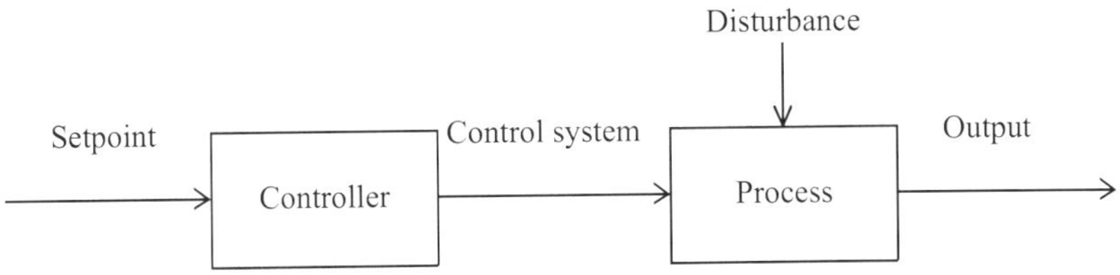

Figure 5. *Open-loop process*

This diagram shows that the control system does not account for disturbances; it cannot function properly. For example, if the vehicle is confronted with different slopes, the slope is considered a disturbance input for the process. It will be required that the model take the slope into account and thus a measurement system of the slope (which would result in a compensation of the measurable disturbances).

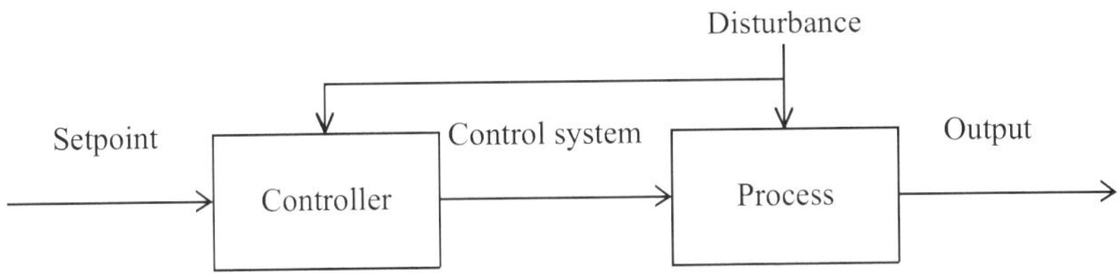

Figure 6. *Open-loop process with disturbances*

To improve the behavior, a control system can be defined that calculates, based on the desired speed obtained speed difference, the necessary action on the pedal to regulate (stabilize) the speed at a value specified by the operator. We will thus obtain an automatic control system of the speed of the vehicle. A sensor measuring the speed obtained is necessary. This system automatically performs what the driver does: it compares the target speed with the actual speed (displayed by the dashboard (sensor)) and acts upon the accelerator to reduce the speed difference to zero. The result will then be a *control system* or *loop system* or servo (servo system). The block diagram (functional) of the principle of a servo system is shown in Figure 7:

– y^d: setpoint is an electrical quantity that represents the desired value of the output of the system;

– ε: error signal between the setpoint and the actual output of the system;

– u: control signal generated by the control system;

– y: is a physical quantity that represents the system output.

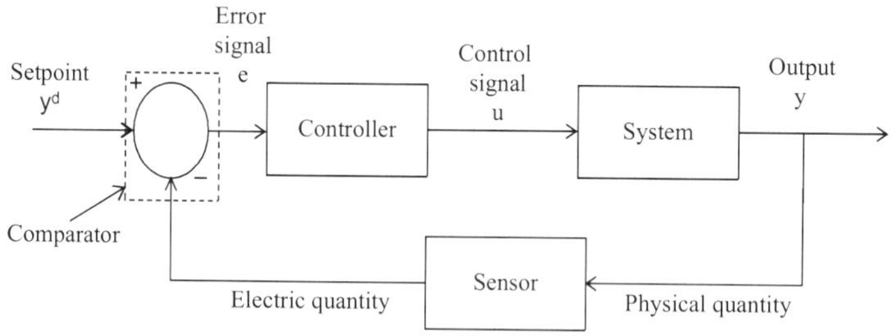

Figure 7. *Functional block diagram of a servo system*

The physical quantity y is measured with a sensor that translates it into an electrical quantity. This electric quantity is compared to the setpoint using a comparator.

One of the great advantages of a looped system compared to an open-loop system lies in the fact that the loop system automatically rejects disturbances. In control systems, when the setpoint (reference) is constant, it is referred to *control* (for example oven temperature control and speed control of a motor); in the case where the reference is not constant, this is referred to as *tracking* (for example target tracking by an antenna).

An additional input (measure of the slope), with respect to the equipment existing in the previous example, would complete this diagram with an *anticipation* about the effect of the disturbance (due to the slope variation).

A system is said to be controlled when there is a loop between the output and the input or when the variable to be adjusted is the setpoint input of the system. For example, for the heating system of a house or of an enclosure, the input is the temperature setpoint and the output is the temperature in the enclosure. An open-loop heating system is a system that does not show any loopback, taking into account the effective temperature of the enclosure. Thus, it is sensitive to external shocks, a rise in the external temperature would cause an excess of heating.

Examples:

The examples are as follows: temperature system control of an oven, fluid flow servo control, speed system control of the trajectory of a vehicle. When the desired path is reduced to a point, this is referred to as regulation and not as system control because the aim here is to stabilize the output of the system at a point. A control system can be qualified by its degree of stability, accuracy, response speed, sensitivity to disturbances acting on the system, robustness with respect to disturbances on measures and errors or variations of the characteristic parameters of the system. The accuracy of a control system can be characterized by the maximum amplitude of the position error.

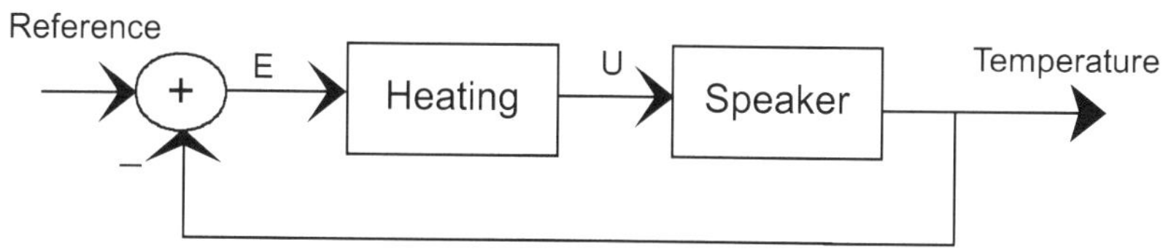

Figure 8. *Speed regulation of a motor*

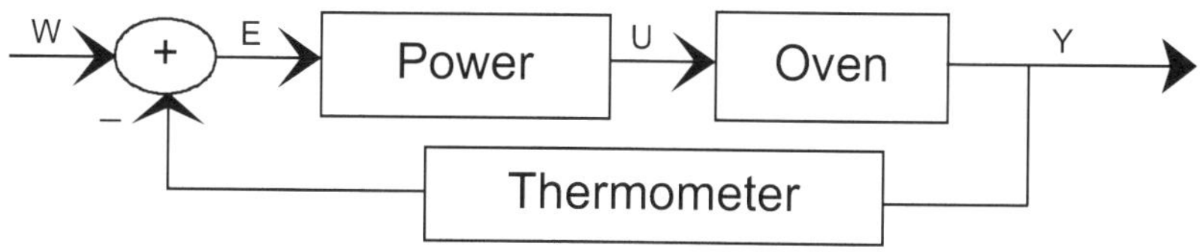

Figure 9. *Temperature regulation of an oven*

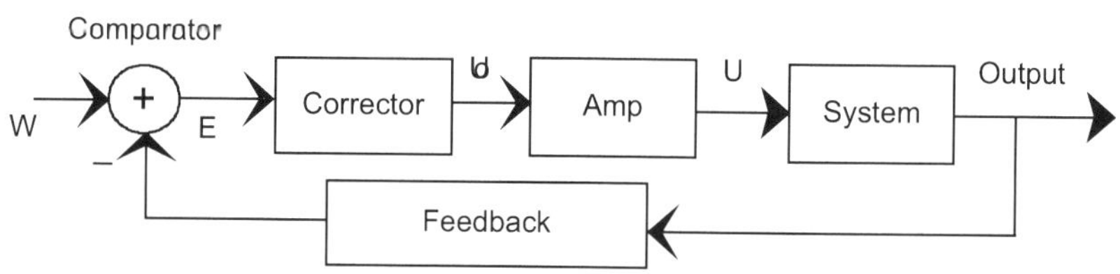

Figure 10. *Control system with correction*

In the definition of a control system, we will write transfer functions as follows: $H(p)$ transfer of the system to be controlled, p is the Laplace operator; $R(p)$ transfer of the sensor or measurement unit, $C(p)$ transfer of the corrector or controller. The setpoint is $\omega(t)$ and the output to be controlled $y(t)$. The direct chain consists of $C(p)$ and $H(p)$. Block $R(p)$ constitutes the feedback chain. $e(t)$ is the difference between output and setpoint also called control error or trajectory tracking error. In order to simplify study, we consider a unity feedback scheme in which $R(p) = 1$.

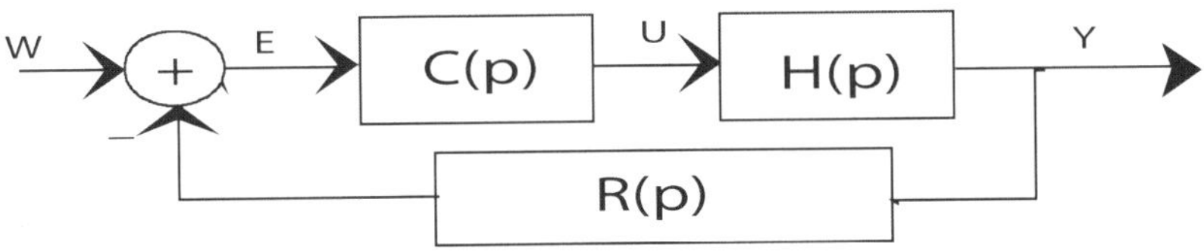

Figure 11. *Controlled transfer function with feedback*

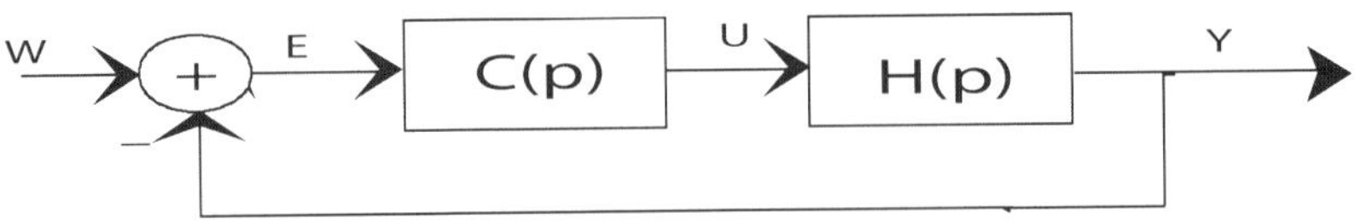

Figure 12. *Controlled transfer function without feedback*

In general, transfers $H(p)$ and $R(p)$ are known, estimated or can be obtained and the goal is the determination of a corrector $C(p)$ that can satisfy the required performances for the closed-loop system (transfer from w to y).

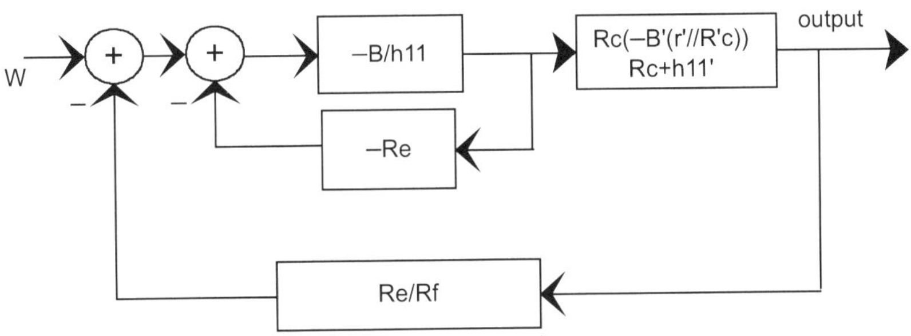

Figure 13. *Control system with closed-loop system controller*

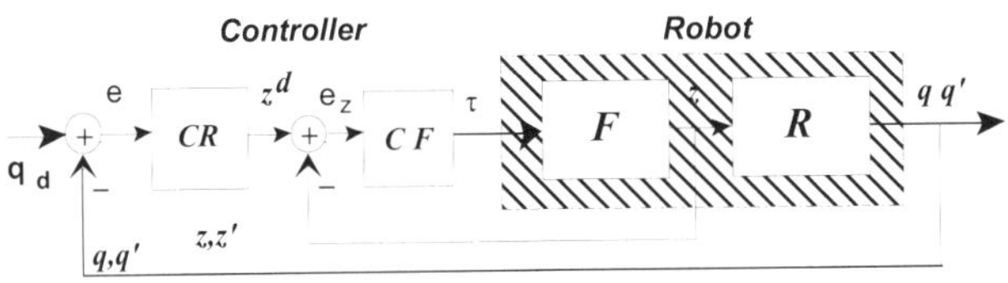

Figure 14. *System control with correction of a robotized system*

Several types of systems can be distinguished:

– continuous systems for which all measured quantities are continuous;

– discrete systems for which all measured quantities are only measured at very specific times (discontinuous or discrete); these are referred to as sampled data or digital systems;

– linear systems (they can be described by linear differential equations);

– nonlinear systems (described by nonlinear differential equation). It is possible, often in the first approximation, to linearize nonlinear systems based on an operation point (equilibrium), considering small variations around this point;

– time-invariant systems (described by differential equations with constant coefficients) and time-variable systems (described by differential equations with time-variable coefficients).

In this book, we consider time-invariant linear, continuous and sampled-data systems.

NOTATION 1.– *Consider a continuous r-input system denoted u and m outputs y,*

$$u \in U \subset R, y \in Y \subset R \qquad \text{[P.1]}$$

$$u \in U \to . \to y \in Y \qquad \text{[P.2]}$$

A minimal state representation of this system will be written as:

$$\dot{X} - A.X + Bu \qquad \text{[P.3]}$$

$$y = C.X + D.u \qquad \text{[P.4]}$$

Its transfer function

$$G(p) = C(pI - A)^{-1}B + D \qquad \text{[P.5]}$$

is denoted as:

$$G(p) = \begin{bmatrix} A & B \\ C & D \end{bmatrix} = [A, B, C, D] \qquad \text{[P.6]}$$

A is stable: the eigenvalues of A have a real part < 0.

$G(p)$ is stable: the poles of $G(p)$ are in $Re(p) < 0$.

$G(p)$ is instable: the poles of $G(p)$ are in $Re(p) > 0$.

$\tilde{G}(p) = G(-p)$

$G(p)$ is an eigen transfer function if $G(\infty)$ is finite.

$G(p)$ is a strictly eigen transfer function if $G(\infty) = 0$.

A^T is the transpose of matrix

$$A = [a_{ij}] = \begin{bmatrix} a_{11} & a_{12} \\ a_{21} & a_{22} \end{bmatrix} \quad [P.7]$$

$$A^T = \begin{bmatrix} a_{11} & a_{21} \\ a_{12} & a_{22} \end{bmatrix} \quad [P.8]$$

A^{-T} is the inverse transpose matrix of A.

A^* is the conjugate transpose matrix of $A = [a_{ij}]$ (or Hermitian transpose matrix of A). $A^* = [a^*_{ij}]$

$\lambda_i(A)$ is the ith eigenvalue of A.

$\sigma_i(A)$ is the ith singular value of A.

$\underline{\sigma}(A)$ and $\bar{\sigma}(A)$ are the minimal and maximal singular values of A.

$Diag(a_i)$ is the diagonal matrix whose diagonal elements are the a_i.

C_- is the set of complex numbers with a negative real part.

C_+ is the set of complex numbers with a positive real part.

$\boldsymbol{C}^n$ is the set of complex vectors with elements in C.

$\boldsymbol{C}^{nxm}$ is the set of complex matrices of dimensions (nxm) with elements in C.

$< x, y >$ is the scalar product of x and y.

$h \circledast u$ is the convolution product of $h(t)$ and $u(t)$.

$\mathcal{F}$ is the Fourier transform operator.

θ_h convolution operator by $h(t)$, $H(p)$ should denote the Laplace transform of $h(t)$.

$\Lambda_{Hg} = H(p).g(p)$ *is the Laurent operator or multiplication in the frequency domain.*

$\oplus$ *Direct sum of two spaces.*

$H_2^{\perp}$ *Orthogonal space at* H_2*;*

with the set H_2=$\{H(p)$ analytic matrix function in $Re(p) > 0\}$

Π_1 is the orthogonal projection on H_2 and Π_2 is the orthogonal projection on $H_2^{\perp}$.

NOTATION 2.– $\mathcal{B}(x, r) \equiv$ *Globe of radius* r*, centered in* x *of the space; example* $\mathcal{B}(0, 1)L_2 = \mathcal{B}L_2$ *refers to the Globe of unit radius of the space* L_2*.* L_2 *is the set of square-integrable functions.*

In the literature, a real rational transfer function refers to a rational transfer function with real coefficients.

Nao robot: application for home care

In view of these considerations and following previous and present works, the purpose of this section is the perception for the joint understanding of space and surroundings of an autonomous, cognitive and sociable personal robot. This Nao robot will typically act according to a planned action scheme, ensuring their viability, the consistency of current models and the presence of entities opposed to its initial plan, while interacting with the physical world through perception. The selected approach, which is clearly aimed at integrating perceptual functions on robotic platforms, relies on probabilistic modeling to consider multiple and uncertain percepts. At the sensory level, these percepts will be mainly originating from the vision embedded in the robot. These perceptual functions are to be derived from autonomy, cognition and sociability capabilities outlined for our Nao robot.

An increasingly more worrying situation due to the aging of the population is the increase in the number of elderly people living alone, far away from their loved ones, sometimes physically or mentally vulnerable. Studies recommend that, at the social level as well as at the economic level, people stay at home, eventually resorting to help (cleaning, care, gardening, etc.), to preserve as much as possible the references and the social network of the person. However, with the cost of home care being relatively burdensome, assistance times are limited and are not always enough to fight solitude. In addition, when the elderly person presents risks of accidents (fall, stroke, etc.), or when undergoing medical treatment, monitoring solely based on domestic help has become risky, or even insufficient (admission to hospital then being preferable).

To fight more effectively against isolation and the dependence of elderly people, the presented system aims to implement a prototype for an automated and robotic system for home care. Without claiming that it is capable of recreating or replacing real human contact, the proposed system, based on the integration of several technologies, will create a permanent, friendly and reassuring, presence at home. Due to the use of modern means of audio/video communication (speech recognition, speech synthesis, etc.), by the mobile humanoid Nao robot from Aldebaran Robotics and that of fixed cameras allowing the environment to be perceived, the objective is to provide a set of services that makes it possible to assist senior citizens in their daily tasks (engaging with relatives, remote medical consultation, looking for objects, medication reminders, etc.), entertain them (games, physical exercises, friendly presence, etc.) and detect dangerous or abnormal situations (falls, prolonged inactivity, unusual behavior, etc.) to, in the end, alert the doctor and relatives.

1

Control, Servo-mechanisms and System Regulation

This chapter explores compensator servo-mechanisms and control, correction and proportional control.

1.1. Introduction

1.1.1. *Generalities and definitions*

In all areas of physics, for the research, analysis and understanding of natural phenomena, a stage for modeling and the study of the structure of the physical process is necessary. This has led to the development of modeling, representation and analysis techniques of systems using a fairly general terminology. This terminology is difficult to introduce in a clear manner but the concepts, which it relies upon, will be defined in detail in the following chapters.

A physical process is divided into several components or parts forming a system. For example, this is the case of an engine that consists of an amplifier, power supply, an electromagnetic part and a position and/or speed sensor. The system input is the voltage applied to the amplifier and the output is either the position or the speed of rotation of the motor shaft.

Among the objectives of the control engineer, we can identify modeling, behavior analysis and the regulation or control with the aim of dynamically optimizing the behavior of the system. It should be noted that one preliminary and very important step is the *configuration* of the system before its control. During this step, the automation expert must define sensible choices of sensors, actuators and their placement in the system to optimize the control (control means verification of the good functioning of all sensors, actuators, system and corrector or control law). It is only after this stage that control synthesis finds its place, which might simply be reflected by the use of a conventional controller (proportional, proportional, integral and derivative (PID), phase advance or phase delay or other).

Driving the system or control serves the purpose of ensuring that the variables to be adjusted or system outputs follow a desired trajectory (curve with respect to time in general) or have dynamics defined by the specification requirements, for example temperature control of an oven, fluid flow control or speed and trajectory control of a moving object. When the desired trajectory is reduced to a point, this is referred to as regulation and not as control because the main purpose here is to stabilize the output of the system in a point. The role of control is to allow or to improve the resulting performance of a system, using actuators and sensors available for information acquisition and enabling reaction based on behavior. In general, this can be done using a negative-feedback loop (return or feedback loop) and sometimes a compensation or anticipation chain of the dynamic effects of the system (feedforward or (pre or post) compensation). The operation of a vehicle is according to the block diagram shown in Figure 1.1.

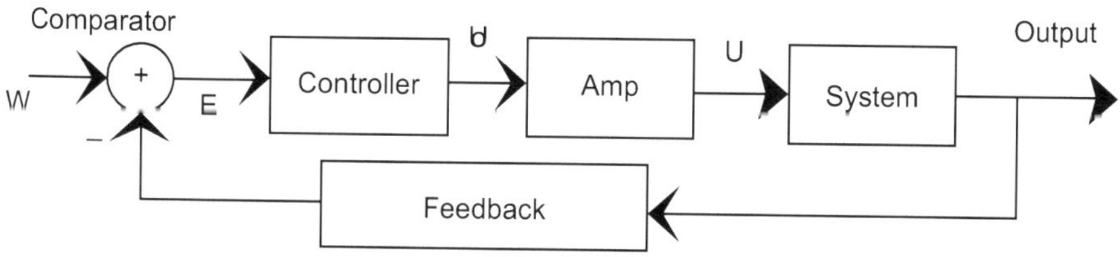

Figure 1.1. *Schematic diagram of a controlled system with compensation and feedback sequence*

In the definition of a control system, we will express transfer functions as follows:

– $H(p)$ transfer of the system to be controlled, p is the Laplace operator;

– $R(p)$ transfer of the sensor or measuring device;

– $C(p)$ transfer of the corrector or servo controller element.

The setpoint is $w(t)$ and the output to control is $y(t)$. The direct chain consists of $C(p)$ and $H(p)$ and $R(p)$ constitute the feedback chain. The difference between output and setpoint is $e(t)$ and is also called control error or trajectory tracking error. In order to simplify the study, we are considering a unity feedback scheme in which $R(p) = 1$.

In general, transfers $H(p)$ and $R(p)$ are known or can be obtained and the objective is to obtain a corrector $C(p)$ that is able to satisfy the performances required for the closed-loop system (transfer from w to y).

For the regulation of the temperature of a speaker to a reference value, it is possible to use one of the following block diagrams.

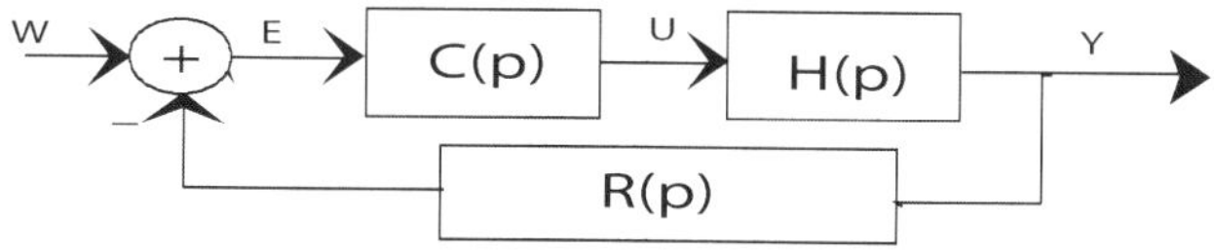

Figure 1.2. *Schematic diagram of a feedback system* $R(p) = 1$

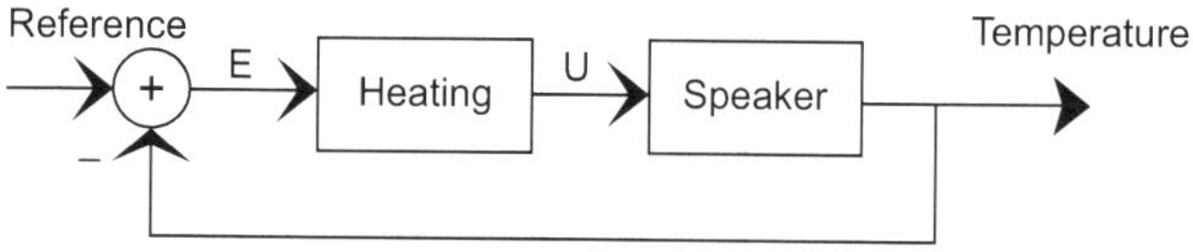

Figure 1.3. *Speed regulation of a motor*

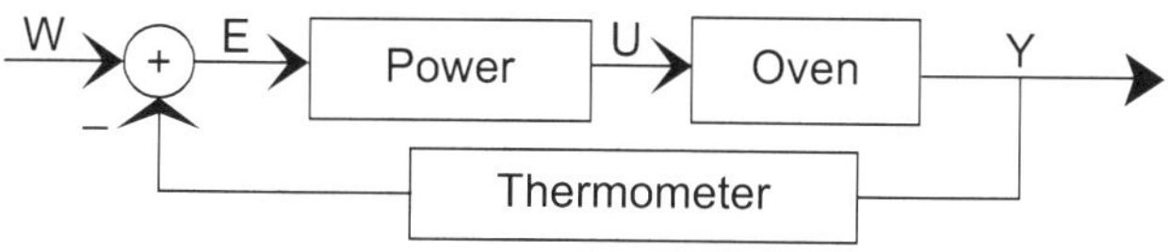

Figure 1.4. *Temperature regulation of an oven*

Vehicle operation follows the principle of the diagram shown in Figure 1.5.

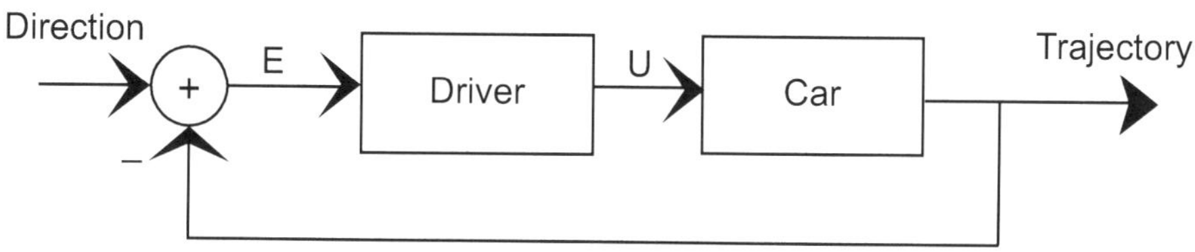

Figure 1.5. *Schematic diagram of the model for vehicle operation*

In this preface, we are going to cover some conventional methods for the design of a control system. This study will serve the purpose of finding a control structure allowing a servo system to be given dynamic characteristics or performances established *a priori* in the definition of the requirements, either in terms of temporal response or in terms of frequency response. In general, the latter is defined to ensure:

– the stability of the controlled system (loop system);

– the smallest possible permanent errors;

– a suitable dynamic behavior: a response quickly reaching its asymptote, the lowest overshoot possible, etc.

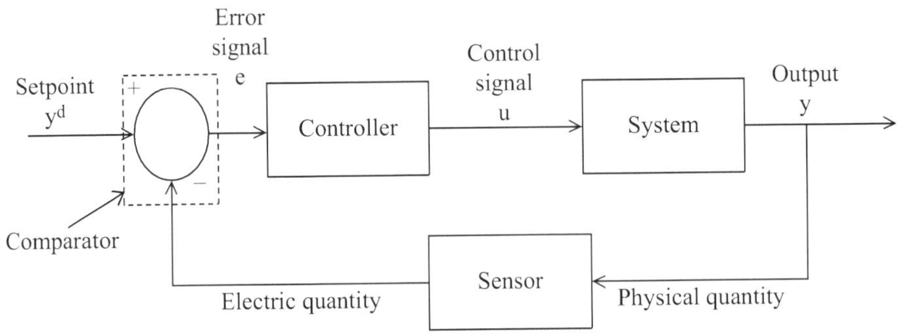

Figure 1.6. *Servo system*

The conventional operation of a servo system is shown in Figure 1.6:

– y^d: the setpoint is an electrical quantity that represents the desired output value of the system;

– ε: the error signal between the setpoint and the actual output of the system;

– u: the control signal generated by the controller;

– y: a physical quantity that represents the system output.

The physical quantity y is measured with a sensor that translates it into an electrical quantity. By means of the comparer, this electric quantity is compared to the setpoint, which is an electric quantity.

A model describing the dynamic behavior (physic) of the open-loop (OL) system is necessary for control synthesis. In general, the accuracy required for modeling is dependent of the finality of the control and the required performance. It should be noted that there are several types of models.

The *simulation model* is useful for the study of behavior and the response of the system to different excitations. It allows that the laws of control be tested and that performance be evaluated before application to the actual system. It has to be as accurate as possible (including disturbance, noises, nonlinearity and all the parts able to be modeled etc.).

The *control model* is usually simpler, sometimes linear, somewhat reduced compared to the simulation model. It is used to infer the appropriate control law so as to minimize complexity (reduction of computation times, ease of implementation, etc.). Consequently, the resulting control law is verified with the simulation model to measure the impact of the dynamic terms neglected in the synthesis stage. If it proves insufficient, either a more complete model is retained or compensators are added.

An ensuing model of the physical system may be empirical, be the result of physical modeling or derived from a process of identification based on information about the observation of the system after excitation. When a representation of the system is available, this is a function of some parameters. The estimation of these parameters from experimental data is the *identification* step.

In linear systems control, modeling is a very important phase. In order to properly control a system, a good model thereof must be known. For example, in order to drive a car, the better its dynamic behavior or model is known (by training), the better it can be controlled at high speed and therefore the better it will be driven. As a result, it will achieve the best performance. The dynamic model is acquired by learning or by system identification.

During the development of an application for automation purposes, we generally follow the following steps:

1) modeling;

2) identification;

3) behavior analysis;

4) controller synthesis;

5) control implementation;

6) analysis and study of the system in closed loop;

7) verification of the performance and eventually repetition of steps (2), (3) or (4).

The modeling stage becomes crucial when the requirements are strict with respect to performances and when the control implemented proves to be complex. In order to introduce the different types of modeling, we will study some examples.

1.1.2. *Control law synthesis*

1.1.2.1. *Specifications and configuration*

Control should enable the closed-loop system to ensure that a certain number of constraints called *specifications* be satisfied. Among the specifications, we can distinguish:

– stability;

– performance;

– robustness.

A *servo-mechanism* can be qualified by its degree of stability, accuracy, response speed, sensitivity to disturbances acting on the system, robustness with regard to disturbances on measures and errors or variations of the characteristic parameters of the system. The accuracy of a control system can be characterized by the maximal amplitude of the position error.

1.1.2.2. *Performances: regulation, disturbance rejection and anticipation*

Disturbance rejection: the process is often subjected to certain inputs considered as being disturbances. The latter must have a minimal effect on the behavior of the system when it is controlled. The *regulation* is the ability of the system to mitigate or even absorb the effects of disturbances.

Trajectory tracking: the loop system must be fast enough, must not present significant overshooting or oscillations in order to correctly follow a *desired trajectory* or setpoint varying in time.

1.1.2.3. *Robustness and parametric uncertainties*

A loop system is said to be robust if its characteristics do not vary much or do not appear too degraded when changing the parameters of the physical system to be controlled or the neglected dynamics during modeling or when disturbances occur. These changes may originate either from the change in characteristics of the system or from the difference between physical system and control model.

Some examples:

– variation in mass of a satellite after fuel consumption;

– aging of a mechanical structure and change in frequency of the natural modes;

– reduced model for the control neglecting the high-frequency dynamics of the physical process;

– external disturbances such as those conveyed by electrical networks and noises in sensors;

– failure occurring in systems that alters their dynamics.

1.1.2.4. *Constraints on control: control system input energy*

Control u is the output of a dynamic system called *controller or control law*, and it may be subjected to constraints (amplitude limits and speed variations, actuators limit, structure limit, etc.). Constraints are sometimes:

– the use of time-invariant linear correction or a simple proportional feedback;

– a control calculated in the discrete domain by a processor using integer or fixed-point representation;

– computation time constraint, limitation of the order of the controller, trajectories continuity and their derivatives up to some order.

Controls admissibility: the amplitudes of signals and control structure must not be too large compared to those physically feasible.

EXAMPLE 1.1.– *Direct current motor with tachometric feedback.*

1.1.3. *Comprehension and application exercises*

1.1.3.1. *Study of a servo-mechanism for the attitude of a satellite*

The aim is to control the attitude of a satellite such to orientate an antenna connected to the satellite with regard to a given axis. The output variable of the system is therefore the attitude $\theta(t)$. For the satellite to start rotate, a thrust $u(t)$ is applied through a nozzle, which produces a couple $\gamma(t) = Lu(t)$ acting on the satellite, where L refers to the distance of the thrust point to the axis of rotation of the satellite. We want to impose direction $\theta^d(t)$ by acting upon $u(t)$. The variable J designates the moment of inertia of the satellite; the dynamic equation is written as:

$$\gamma(t) = J\ddot{\theta}(t) = Lu(t). \qquad [1.1]$$

Hence the transfer function between the input $u(t)$ and output $\theta(t)$,

$$H_o(p) = \frac{\Theta(p)}{U(p)} = \frac{L}{Jp^2}. \qquad [1.2]$$

The system behaves as a double integrator. When a short impulse is given to the system, it will begin to rotate indefinitely (the impulse is integrated twice). Control is achieved using the difference between the desired attitude (setpoint) and the actual attitude (output) to calculate the control $u(t)$ to apply to orientate the antenna. The diagram of the control is shown in Figure 1.7.

We must determine a controller $C(p)$ that connects the error $\varepsilon(p)$ to the control signal $U(p)$. As a first step, we propose a regulation proportional to the error correction ($u(t) = K\varepsilon(t)$), therefore we will write $C(p) = K$, in which K is constant. This control is known as proportional control. The transfer function of the now loop system is given by:

$$H(p) = \frac{\Theta(p)}{\Theta^d(p)} = \frac{C(p)H_o(p)}{1 + C(p)H_o(p)} = \frac{KH_o(p)}{1 + KH_o(p)} = \frac{K\frac{L}{Jp^2}}{1 + K\frac{L}{Jp^2}}$$

$$= \frac{K\frac{L}{J}}{p^2 + K\frac{L}{J}}. \qquad [1.3]$$

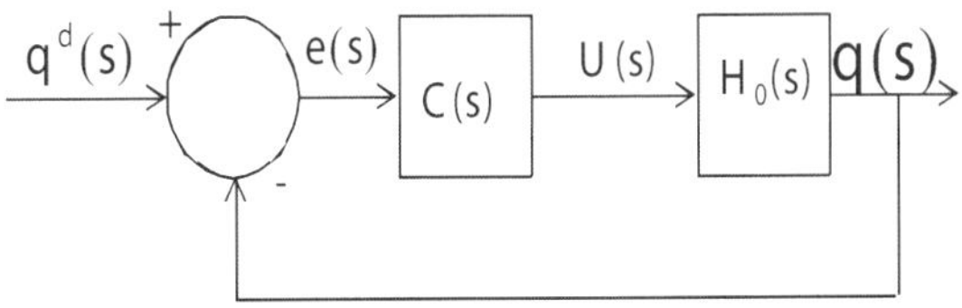

Figure 1.7. *Control diagram*

Suppose that the attitude is initially of 0, and that it is desirable that the satellite assume an attitude of setpoint θ_0. It can be said that the setpoint signal is a Heaviside function of amplitude θ_0, wherefrom

$$\Theta^d(p) = \frac{\theta_0}{p}. \qquad [1.4]$$

which gives as output:

$$\Theta(p) = \frac{K\frac{L}{J}}{p(p^2 + K\frac{L}{J})}\theta_0. \qquad [1.5]$$

By dividing into simple elements, we get:

$$\theta(t) = \theta_0(1 - \cos(\omega_0 t)) \qquad \text{with} \qquad \omega_0 = \sqrt{K\frac{L}{J}}. \qquad [1.6]$$

It can be noted that the attitude of the satellite oscillates around the desired attitude. The result is thus not satisfactory; it is necessary to reconsider the controller to improve the performance of the closed-loop system. The problem comes from the fact that when we assume the value is0, the rotation of the satellite should be slowed down, whereas it is at this moment that the control is zero, since it is proportional to the error. However, it can be observed that when the error is zero, its derivative is maximal (in absolute value). Consequently, the idea is to introduce the error and its derivative in the correction. We then choose a proportional correction and derivative ($u(t) = K_p\varepsilon(t) + K_v\dot{\varepsilon}(t)$). It can be written in a simplified way:

$$C(p) = 1 + Tp. \qquad [1.7]$$

The transfer function of the closed-loop system is therefore given by

$$\begin{aligned} H(p) &= \frac{\Theta(p)}{\Theta^d(p)} = \frac{C(p)H_o(p)}{1 + C(p)G(p)} = \frac{(1+Tp)\frac{L}{Jp^2}}{1 + (1+Tp)\frac{L}{Jp^2}} \\ &= \frac{(1+Tp)\frac{L}{J}}{p^2 + T\frac{L}{J}p + \frac{L}{J}}. \end{aligned} \qquad [1.8]$$

REMARK 1.1.– *The system using proportional and derivative (PD) control is not physically feasible since the degree of the numerator is greater than the degree of the denominator. On the other hand, a good approximation is always possible to achieve.*

Consider the same regulation conditions (i.e. step response). The output of the system is thus given by

$$\Theta(p) = \frac{(1+Tp)\frac{L}{J}\theta_0}{p(p^2 + T\frac{L}{J}p + \frac{L}{J})}. \qquad [1.9]$$

The shape depends on the roots of the following characteristic equation:

$$p^2 + T\frac{L}{J}p + \frac{L}{J} = 0. \qquad [1.10]$$

For example, we take $\frac{L}{J} = 10^{-2}$. If $T > 20$, the solutions are real and negative,

$$p_{1,2} = \frac{-10^{-2}T \pm 10^{-1}\sqrt{10^{-2}T - 4}}{2} \qquad [1.11]$$

and the response is shown in Figure 1.8.

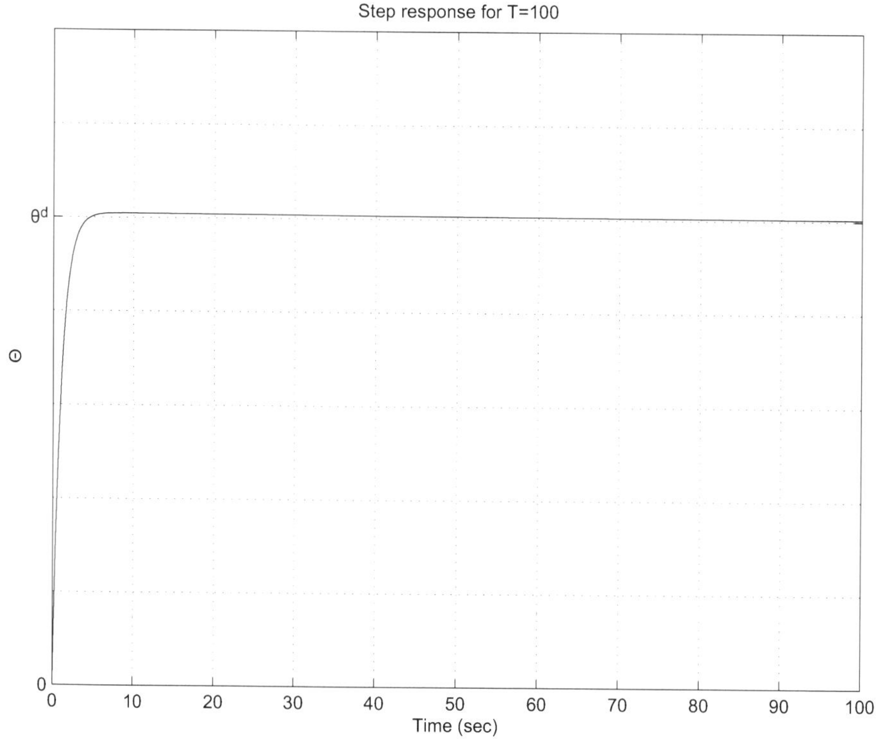

Figure 1.8. *Step response*

For $T = 100$, the roots of the denominator are approximately -1 and -0.01, therefore the dominant term in the response should be the second root (larger time constant 100 s). In fact, it is observed that this is not true, the apparent time constant is of 1 s. The reason is that the numerator of the transfer function has a root equal to $-1/T = -0.01$, which compensates for the effect of the pole for -0.01. We are here confronted with a system that does not have dominant poles.

If we take $T < 20$, solutions are complex conjugate and as a result the response shows damped oscillations. For $T = 10$, we have the response as shown in Figure 1.9.

REMARK 1.2.– *With increasingly smaller values of T, we tend towards an oscillating solution that corresponds to the case of proportional control ($T = 0$).*

It is thus seen that the shape of the response depends completely on the roots of the characteristic equation (poles of $H(s)$) and sometimes depends on the roots of the numerator of the transfer function (zeroes of $H(s)$).

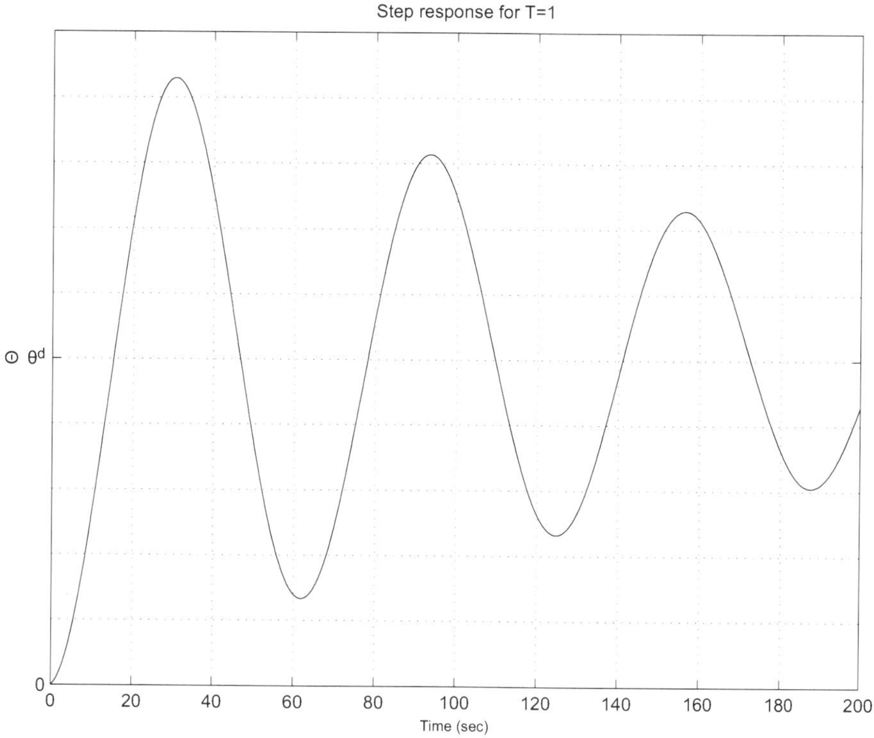

Figure 1.9. *Response with damped oscillations*

In this example, we have highlighted several characteristics of a control:

– the notion of loop that allows a process to be controlled, which in the case being considered could not be controlled in the OL system;

– the notion of control system (controller) that can be more or less adapted to the process to be controlled;

– the influence of poles and zeroes of the transfer function.

1.2. Process control

1.2.1. *Correction in the frequency domain*

As a first step, we are going to focus only on looping a system with a cascade controller (regulation). The use of an anticipation chain and compensators will be addressed farther.

In this section, we are going to cover process control using conventional methods for simple control. These controllers that make use of simple actions are regulated by an approximate study in the frequency domain. Consider a system whose frequency response has a phase margin $\Delta\Phi = \Phi_0$ and a gain margin ΔG. Suppose that these characteristics are not sufficient to provide the desired performance. For a simple system and using Bode, Nyquist or Black–Nichols representations, the observation of its frequency response makes it possible to observe that for improving the performance of a system, it is necessary to make sure that the frequency response of the corrected system passes far away from the critical point $((0dB, -\pi Rad)$, $(0dB, -180°)$ $-1)$. For this purpose, the Bode diagram may inspire two types of corrective actions, one shifting the phase curve upward in the neighborhood of the critical point (phase advance control, to increase the phase margin), the other offsetting the gain downward (phase delay control to increase the gain margin). The two corrections can be achieved using transfer functions of simple controllers and may be combined, but their effectiveness is limited as soon as the order of the system is greater than 2 or 3. Despite the possibility that this type of action can be multiplied, it is preferable to use other synthesis methods, more flexible and more efficient for more complex systems.

1.2.2. *Phase advance controller and PD controller*

The operation principle is that this controller increases the phase of the direct chain to increase the phase margin of the system. The transfer function of a PD controller is written as: $C(p) = K(1 + T_d.p)$. The derivative action is not physically feasible, it must be approximated by $T_d.p \simeq \frac{T_d p}{1+\tau p}$ with $\tau \ll T_d$, which gives us

$$C(p) = K(1 + \frac{T_d p}{1 + \tau p}) = K(\frac{1 + (T_d + \tau)p}{1 + \tau p}). \qquad [1.12]$$

The transfer function of a phase advance controller is defined as follows:

$$C(p) = \frac{1 + aTp}{1 + Tp} \qquad (a > 1). \qquad [1.13]$$

The Bode plot of the phase advance controller is shown in Figure 1.11.

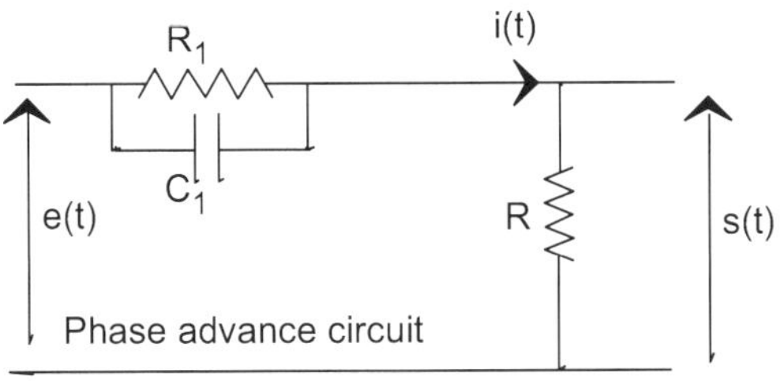

Figure 1.10. *Advance phase control circuit*

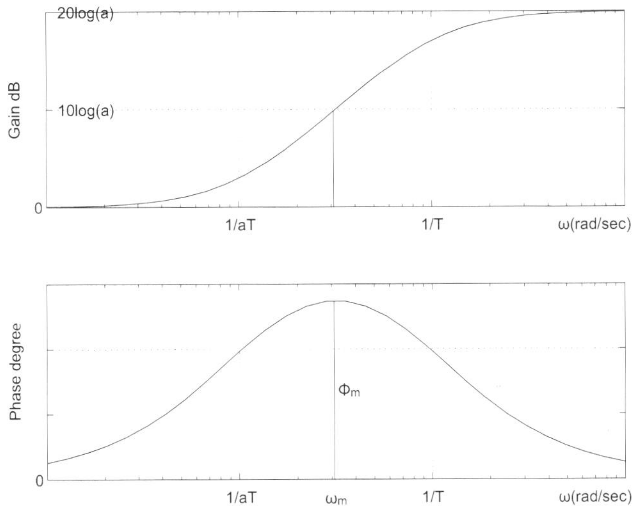

Figure 1.11. *Bode plot of the phase advance controller*

The maximum phase Φ_m of the phase advance controller is obtained for $\omega = \omega_m$, with:

$$\omega_m = \frac{1}{T\sqrt{a}} \qquad [1.14]$$
$$\sin(\Phi_m) = \frac{1-a}{1+a}.$$

We have a phase margin of Φ_0, which means that the controller should add a phase of $\Phi_m = 50° - \Phi_0$. The modulus of the controller is equal to $10\log_{10}(a)$ to $\omega = \omega_m$. As a result, if the controller is calculated to get Φ_m at ω_c, the cutoff pulse of the system corresponding to 0 dB, the new crossing point at 0 dB would be moved to the right of the starting point and therefore the phase margin would be different from the expected margin. To overcome this problem, ω_m is chosen at the point where the modulus of the system is equal to $-10\log_{10}(a)$, which makes it so that after correction the modulus of the controlled system will cross 0 dB at $\omega = \omega_m$. The phase margin of the controlled system will be equal to $\Phi_m + 180° - \Phi_{\text{read}}|_{G=-10\log(a)}$.

To determine the coefficients of the controller, the calculated phase margin is overestimated by 5° to take into account the fact that we use the asymptotic diagram: $\Phi_m = \Phi_{m(\text{calculated})} + 5°$. After having defined Φ_m, we can derive a by the formula

$$a = \frac{1+\sin(\Phi_m)}{1-\sin(\Phi_m)}. \quad [1.15]$$

Then, since this phase must be placed in $\omega_m = \omega_c = \frac{1}{T\sqrt{a}}$ corresponding to the modulus of the transfer function in the OL system,

$$G = -10\log_{10}(a). \quad [1.16]$$

This allows us to calculate T,

$$T = \frac{1}{\omega_m\sqrt{a}}. \quad [1.17]$$

REMARK 1.3.– *The phase advance control increases the bandwidth of the system and as a result the system becomes faster.*

Since the determination of the controller coefficients uses approximations, we must verify the results obtained by printing the Bode plot of $C(p)H_o(p)$. *If the phase margin after correction does not match the expected result, this may be caused by too quick a variation of the system phase around the critical point. This variation results in a fall of phase that largely exceeds the* 5° *of margin.*

The phase advance controller is not suitable in case of systems having too quick phase variations.

1.2.3. *Phase delay controller and integrator compensator*

The operation principle is that this controller decreases the gain of the direct chain to pulses corresponding to a dephasing shift of the system close to $-\pi$ rad. The transfer function of a proportional and integral (PI) controller is written as:

$C(p) = K(1 + \frac{1}{T_i.p}) = K\frac{1+T_i.p}{T_i.p}$. The integral action is often approximated by $\frac{1}{T_i.p} \simeq \frac{1}{\frac{1}{\alpha}+T_i.p}$, which gives us

$$C(p) = K(1 + \frac{1}{\frac{1}{\alpha} + T_i.p}) = K\alpha\frac{(\frac{1}{\alpha}+1)+T_i.p}{1+\alpha T_i.p}. \qquad [1.18]$$

The transfer function of a phase advance controller (integral compensator) is defined as follows:

$$C(p) = \frac{1+aTp}{1+Tp} \qquad (a < 1). \qquad [1.19]$$

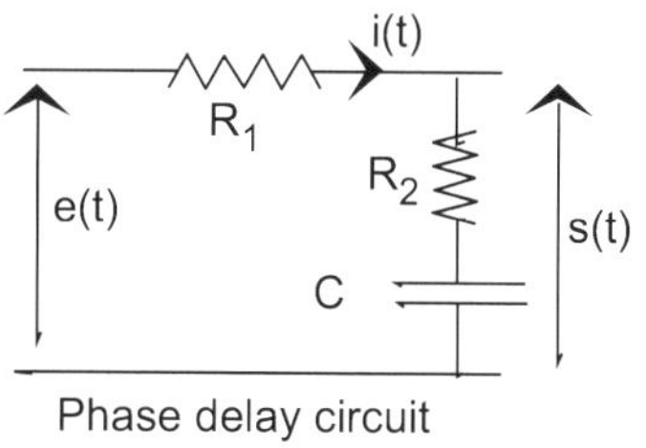

Figure 1.12. *Phase delay controller circuit*

Its Bode plot is given by Figure 1.13.

It can be observed that the phase of the controller is negative and consequently it will delay the phase of the system.

To obtain a desired phase margin of $50°$, we will act this time not upon the phase but upon the modulus so as it passes through $0\,\mathrm{db}$ at pulse ω_c that corresponds to a system phase that is equal to ($\Phi_c = -180° + 50° = -130°$). As the modulus is cancelled out for ω_c ($\Phi_c = -130°$), then the phase margin is therefore $\Delta\Phi = 180° - 130° = 50°$. To offset the effect of the phase introduced by the controller, we overestimate by $5°$ or $10°$ the margin, that is to say, instead of taking $\Phi_c = -130°$, we will take $\Phi_c = -125°$.

The value of a is calculated by measuring the modulus d at pulse ω_c corresponding to a system phase equal to $\Phi_c = -125°$. Thus, by imposing this pulse to the gain of the direct chain $|C(\omega_c)H_o(\omega_c)| = 1$, we therefore obtain the value of a,

$$20\log(a) = -d \Longrightarrow a = 10^{-d/20}. \qquad [1.20]$$

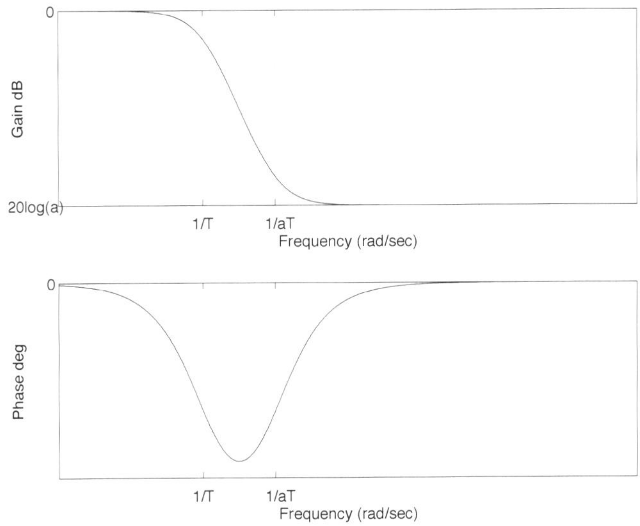

Figure 1.13. *Bode plot of the phase delay controller*

For the other parameter, we choose T in order to not affect the phase around ω_c. To this end, at least a decade is placed between $1/aT$ and the new crossing point ω_c of the modulus at 0 db after control, which gives

$$\frac{1}{aT} \approx \frac{\omega_c}{10} \Longrightarrow T = \frac{10}{a\omega_c}. \qquad [1.21]$$

REMARK 1.4.– *The main disadvantage of the integral compensator is that it reduces the bandwidth of the system, which makes the system slower.*

It is possible to combine the advantages of the two phase delay and advance controllers by implementing a PID or phase delay and phase advance controller, combining actions: the phase delay part having the purpose to stabilize the system and the phase advance part being designed to accelerate the response (make the system quick).

1.2.4. *Proportional, integral and derivative (PID) control*

The PID controller is a special case of phase advance and phase delay controller or with combined action. It is widely used in the industry. The transfer function of a PID controller is given by:

$$C(p) = K_p(1 + \frac{1}{T_i p} + T_d p). \quad [1.22]$$

The problem in designing a PID controller is therefore that of determining parameters K_p, T_i and T_d. To illustrate the influence of the choice of each of the parameters, we will study an example.

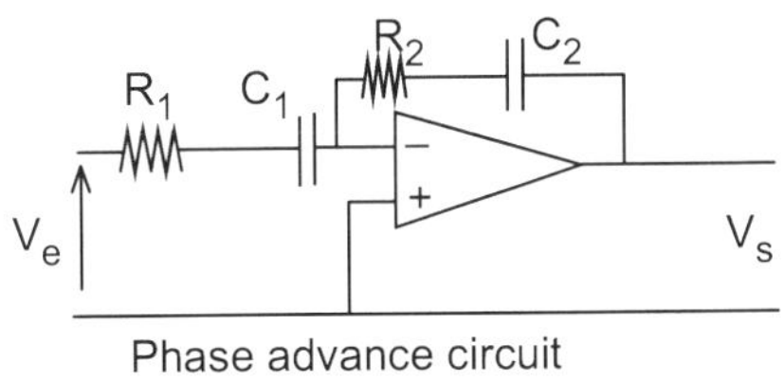

Figure 1.14. *Phase advance circuit*

EXAMPLE 1.2.– *Consider the position control of a direct current motor, whose transfer function is given by*

$$H_o(p) = \frac{100}{p(p+50)}. \quad [1.23]$$

1.2.4.1. *PD control*

The transfer function of the controller is expressed by

$$C(p) = K_p(1 + T_d p). \quad [1.24]$$

REMARK 1.5.– *Such a transfer function is not feasible, since the degree of the numerator is smaller than that of the denominator; on the other hand, what we can achieve is a function of the type:*

$$C(p) = K_p(\frac{1 + T_d p}{1 + \tau p}) \quad [1.25]$$

where τ is small enough such that the influence of the pole $-1/\tau$ is negligible.

The transfer function of a non-loop controlled system is written as:

$$H(p) = H_o(p)C(p) = \frac{100K_p(1 + T_d p)}{p(p + 50)}. \qquad [1.26]$$

We have therefore added a zero to the transfer function $H_o(p)$.

First, consider the proportional controller only ($T_d = 0$). The denominator of the transfer function of the loop system (characteristic polynomial) equation is given by

$$P(p) = p^2 + 50p + 100K_p. \qquad [1.27]$$

We have:

$$\omega_n^2 = 100K_p; \quad 2\xi\omega_n = 50 \quad \Longrightarrow \quad \omega_n = 10\sqrt{K_p}; \quad \xi = \frac{50}{20\sqrt{K_p}}. \qquad [1.28]$$

Therefore, if K_p is increased, ω_n also increases and as a result the speed of the response of the loop system, but the amplitude of the oscillations increases as well (small ξ). For $K_p = 12.5$, we get damping $\xi = \frac{\sqrt{2}}{2}$, but a slow response ($\omega_n = 35.35\,\mathrm{rad/s}$); for $K_p = 100$ the response is fast ($\omega_n = 100\,\mathrm{rad/s}$) but very oscillating because $\xi = 0.25$. It can also be seen that the static error in the velocity is equal to $\frac{50}{100K_p}$; it is improved by increasing the gain. The static error in position does not depend on the controller since the process contains an integration.

The introduction of the term involving a derivative allows for an additional degree of freedom. In effect, the denominator of the transfer function of the loop system becomes

$$P(p) = p^2 + (50 + 100K_pT_d)p + 100K_p, \qquad [1.29]$$

that is

$$\omega_n = 10\sqrt{K_p}; \quad \xi = \frac{2.5 + 5K_pT_d}{\sqrt{K_p}}. \qquad [1.30]$$

The error of velocity remains equal to $\frac{50}{100K_p}$; the derivative term does not affect the static behavior in speed. By introducing this additional degree of freedom in the controller, it is possible to ensure both large ω_n and ξ.

By taking $K_p = 100$, the same static error can be obtained in velocity as previously, a natural frequency of $\omega_n = 100\,\text{rad/s}$ but also a damping coefficient $\xi = 1$ by choosing $T_d = 0.015\,\text{s}$.

The step response of the controlled loop system is given for different values of K_p and T_d in Figure 1.15.

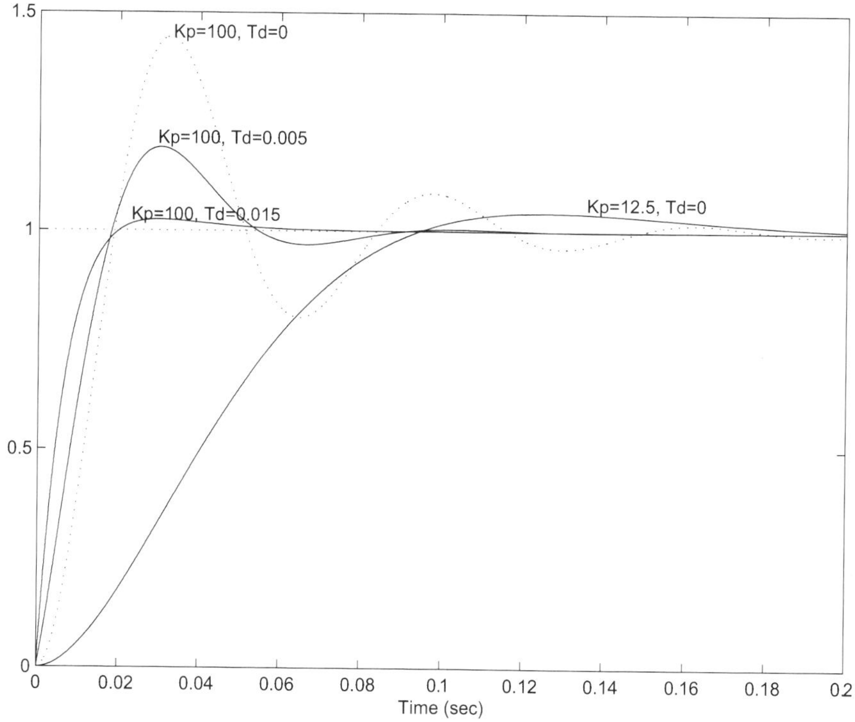

Figure 1.15. *Step response of the corrected loop system*

1.2.4.2. *PI control*

Now consider a controller of the form:

$$C(p) = K_p(1 + \frac{1}{T_i p}) \qquad [1.31]$$

which can be written as:

$$C(p) = \frac{K_p}{T_i} \cdot \frac{1 + T_i p}{p}. \qquad [1.32]$$

Therefore, a zero and a pole in 0 can be added to the system. The addition of the integration reduces the static error. In the case of the PD controller, the static error in velocity imposed the choice of T_d; this is no longer the case here because we have added a pole at the origin, which cancels out the static error in velocity. Thus, the choice of controller parameters will be primarily done based on criteria related to stability and the transient response.

We use the Routh criterion to analyze stability according to the parameters of the corrector. The characteristic equation of the loop system is given by

$$p^3 + 50p^2 + 100K_p p + 100\frac{K_p}{T_i} = 0. \qquad [1.33]$$

The results are presented in Table 1.1.

p^3	1	$100K_p$
p^2	50	$100\frac{K_p}{T_i}$
p^1	$100K_p - 2\frac{K_p}{T_i}$	0
p^0	$100\frac{K_p}{T_i}$	0

Table 1.1. *Routh table results*

We have stability if and only if $\frac{K_p}{T_i} > 0$ and $\frac{1}{T_i} < 50$. Based on this, we need to take $\frac{1}{T_i}$ as small as possible. In effect, the OL transfer function of the controlled system is given by

$$H(p) = C(p)H_o(p) = \frac{100K_p(p + \frac{1}{T_i})}{p^2(p + 50)}. \qquad [1.34]$$

The term in p^2 in the denominator ensures a zero static error in velocity and the fact of choosing the zero as small as possible makes it possible to find approximately the response of the system before integral correction. The responses corresponding to $K_p = 10$, $\frac{1}{T_i} = 0$ ($T_i = \infty$) and $K_p = 10$, $\frac{1}{T_i} = 0.01$ are identical, but the second situation has the advantage of ensuring a permanent zero error in the case of a ramp. The step response of the controlled system is given for different values of K_p and T_i in Figure 1.16.

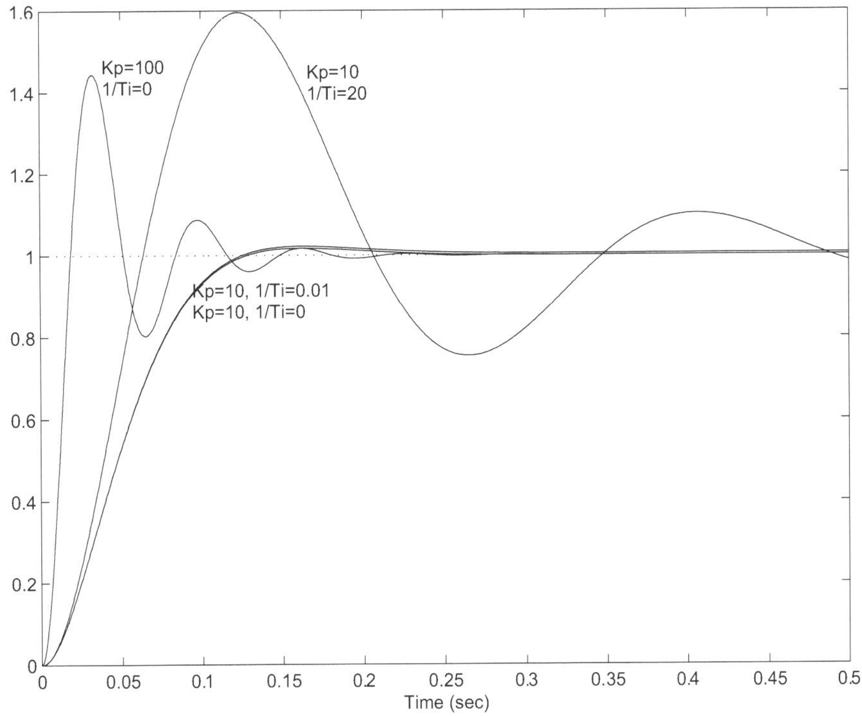

Figure 1.16. *Step response of the controlled system*

REMARK 1.6.– *The response of the system controlled by the PI* ($K_p = 10$, $\frac{1}{T_i} = 0.01$) *is slower than with the PD control* ($K_p = 100$, $T_d = 0.015$).

1.2.4.3. *PID control*

The transfer function of the controller is expressed by

$$C(p) = K_p(1 + \frac{1}{T_i p} + T_d p). \quad [1.35]$$

This controller allows an approximated implementation using approximations.

There is no general method that can find the best combination of the three actions. There are methods that make it possible to find a first approximation (Ziegler–Nichols method) that then has to be refined according to the data of the problem and to trial and error.

In the case of the correction of the previous system by this type of controller, a good solution consists of recovering $K_p = 100$, $T_d = 0.015$ and adding in the integral controller with $1/T_i = 0.01$ ($T_i = 100$). This combines the speed obtained by the PD with the zero velocity error obtained by the PI.

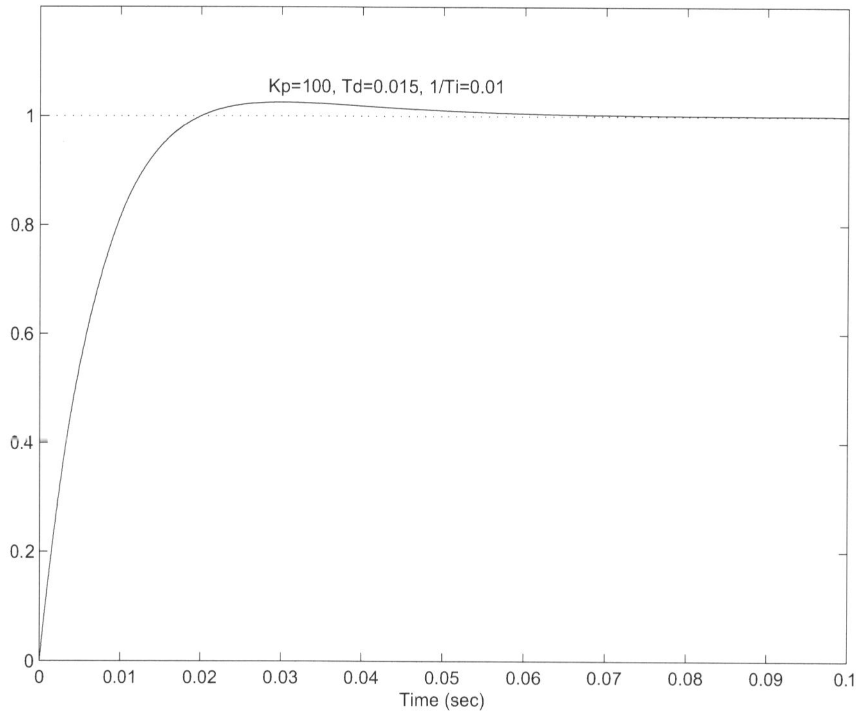

Figure 1.17. *Step response of the controlled system with correction*

In this section we have introduced a conventional method of regulation. This method is based on the knowledge of the frequency response of the system in OL and the determination of the controller consists of improving the gain margin and phase margin relatively to the system looped only by a unity feedback. It thus ensures a robustness margin (at the expense of performance) if the parameters of the transfer function were to change. We have made no assumptions about these possible variations and the knowledge of the transfer function is supposed to be acquired. It can be obtained by identification by using the methods proposed in Chapter 6 or other methods. In the following, we are addressing an example with identification based on the frequency response.

1.3. Some application exercises

1.3.1. *Identification of the transfer function and control*

The transfer function of a system can be determined from its Bode plot. The plots of the modulus and of the phase provide information about whether the system is of minimal or phase non-minimum, which allows us to propose a form of transfer function.

For the Bode plot given by Figure 1.18, we propose the following minimal phase transfer function:

$$H_o(p) = \frac{K}{p(1+\tau_1 p)(1+\tau_2 p)}. \qquad [1.36]$$

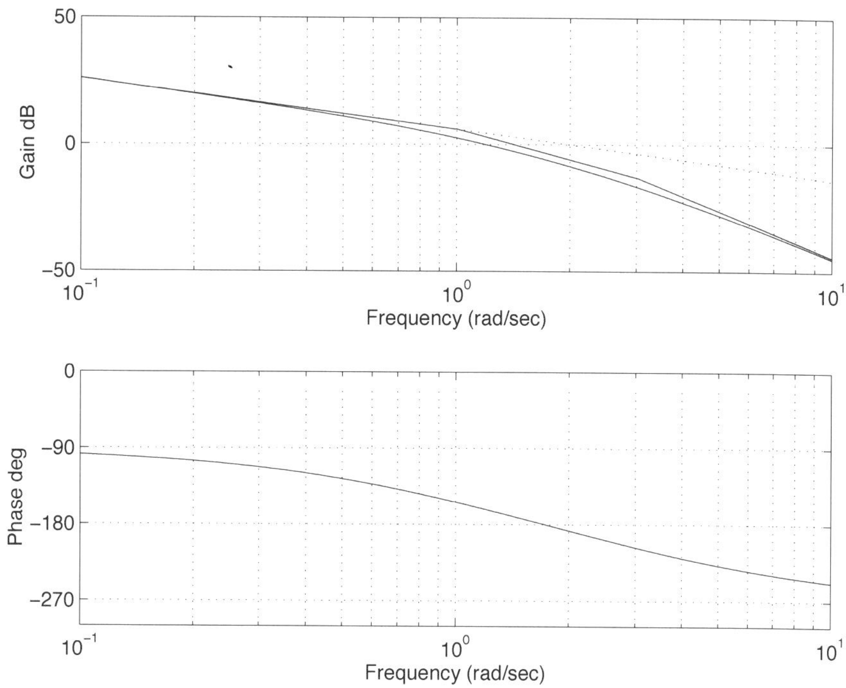

Figure 1.18. *Bode plots of a system*

The integration in the transfer function is justified by the fact that the phase starts from $-90°$ and that the very low frequency modulus follows an asymptote of -20 db/dec. The gain K can be identified by extending the asymptote due to integration. The value of K can be directly obtained at the intersection point of this asymptote with the axis 0 db. The two time constants τ_1 and τ_2 are identified from the pulsations of cutoffs ω_{c_1} and ω_{c_2} corresponding to the intersection points of the asymptotes. It is always possible to verify the results derived by the modulus plot by using the phase plot. For example, it can be verified that the phase starts from $-90°$ (integration) and tends toward $-270° = 3 \times 90°$ (two time constants).

The identified parameters of the transfer function are written as:

$$\left.\begin{array}{lll} K = 2 & & K = 2 \\ \omega_{c_1} = 1 & \Longrightarrow & \tau_1 = 1 \\ \omega_{c_2} = 3 & & \tau_2 = 1/3 \end{array}\right\} \Longrightarrow H_o(p) = \frac{2}{p(1+p)(1+p/3)}. \qquad [1.37]$$

1.3.1.1. *Calculation of static and dynamic errors*

The static error ε_p of the closed-loop system is zero because the system has an integration in the direct chain ($\varepsilon_p = 0$). The static error in the velocity or dynamic error is calculated in the following manner:

$$\varepsilon_v = \varepsilon(t = +\infty)\big|_{y^d(t)=tu(t)} = \lim_{p\to 0} p\varepsilon(p)\big|_{Y^d(p)=1/p^2} = \frac{1}{2}. \qquad [1.38]$$

1.3.1.2. *Stability study*

To study the stability, we calculate the gain and phase margins of the system from the Bode plot:

– the gain margin is $\Delta K = 6$;

– the phase margin is $\Delta\Phi = 18°$.

The behavior of the system in the closed-loop system is not satisfactory, because it shows a very low phase margin and as a result it is very poorly damped. The goal is thus to correct it so as to improve its damping and make it faster (increasing the bandwidth).

1.3.1.3. *Servo-mechanism by phase advance controller*

It is desirable to correct the system to bring the phase margin to $50°$. To this end, we will use two types of controller, phase advance controller and integral compensator (phase delay).

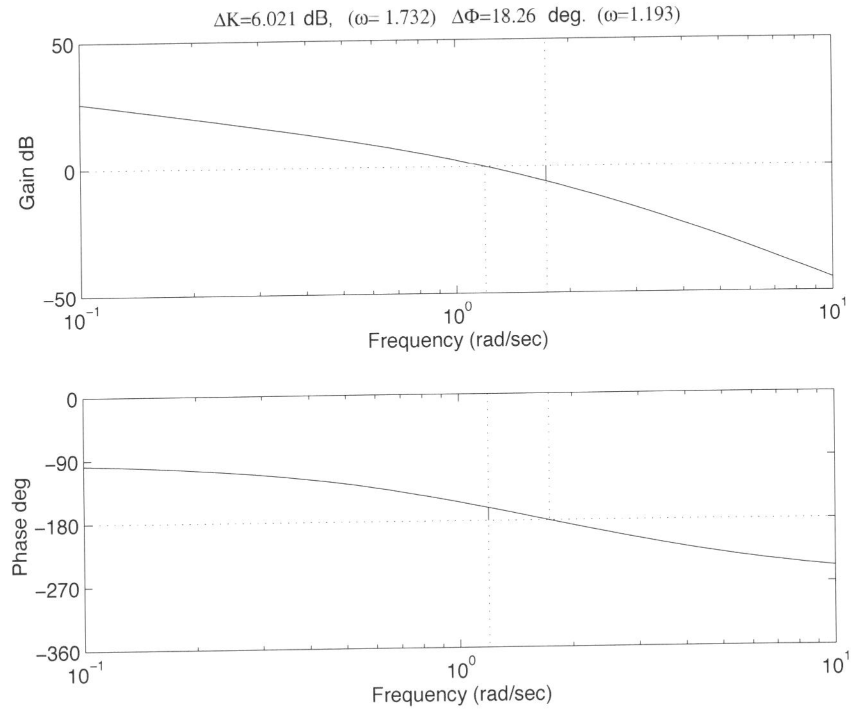

Figure 1.19. *Bode plots of a system with correction*

We have a phase margin of $18°$, which means that the controller should add a phase of $32°$. The transfer function of a phase advance controller is given as follows:

$$C(p) = \frac{1 + aTp}{1 + Tp} \qquad (a > 1). \tag{1.39}$$

The Bode plot of the phase advance controller is shown in Figure 1.20.

The maximum phase Φ_m of the phase advance controller is obtained for $\omega = \omega_m$, with

$$\omega_m = \frac{1}{T\sqrt{a}} \quad \sin(\Phi_m) = \frac{1-a}{1+a}. \tag{1.40}$$

The modulus of the controller is equal to $10 \log_{10}(a)$ for $\omega = \omega_m$. As a result, if we calculate the controller to get Φ_m with ω_c corresponding to $0\,\text{db}$, the new crossing point in 0 dB would be moved to the right of the starting point and as a result the phase margin would be different from the desired phase margin. To overcome this problem, we choose ω_m at the point where the modulus is equal to$-10\log_{10}(a)$, which makes it so that after correction the modulus of the controlled system will cross 0 db for $\omega = \omega_m$. However, the gain margin of the controlled system will be equal to $\Phi_m + 180° - \Phi_{\text{read}}|_{G=-10\log(a)}$. The phase margin will be overestimated by $5°$. The calculated Φ_m is $32°$, and therefore we are going to take as new Φ_m $32° + 5° = 37°$. Having set Φ_m, we calculate a:

$$a = \frac{1+\sin(\Phi_m)}{1-\sin(\Phi_m)} = \frac{1+\sin(37°)}{1-\sin(37°)} \Longrightarrow a = 4. \qquad [1.41]$$

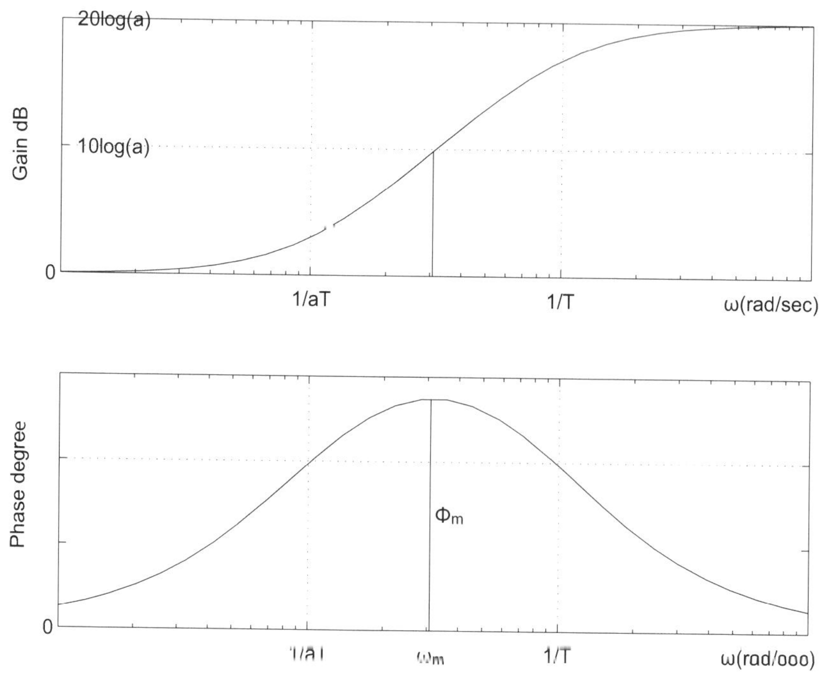

Figure 1.20. *Bode plot of the phase advance controller*

By placing ω_m at ω corresponding to the system modulus that is equal to

$$-10\log_{10}(a) = -10\log_{10}(4) = -6\,\text{db} \Longrightarrow \omega_m = 1.73\text{rd/p}. \qquad [1.42]$$

This allows us to calculate T,

$$T = \frac{1}{\omega_m \sqrt{a}} = \frac{1}{1.73\sqrt{4}} = 0.29\,\mathrm{s}. \qquad [1.43]$$

Hence, the following controller:

$$C(p) = \frac{1 + 1.16p}{1 + 0.29p}. \qquad [1.44]$$

The Bode plot of the system after correction is shown in Figure 1.21.

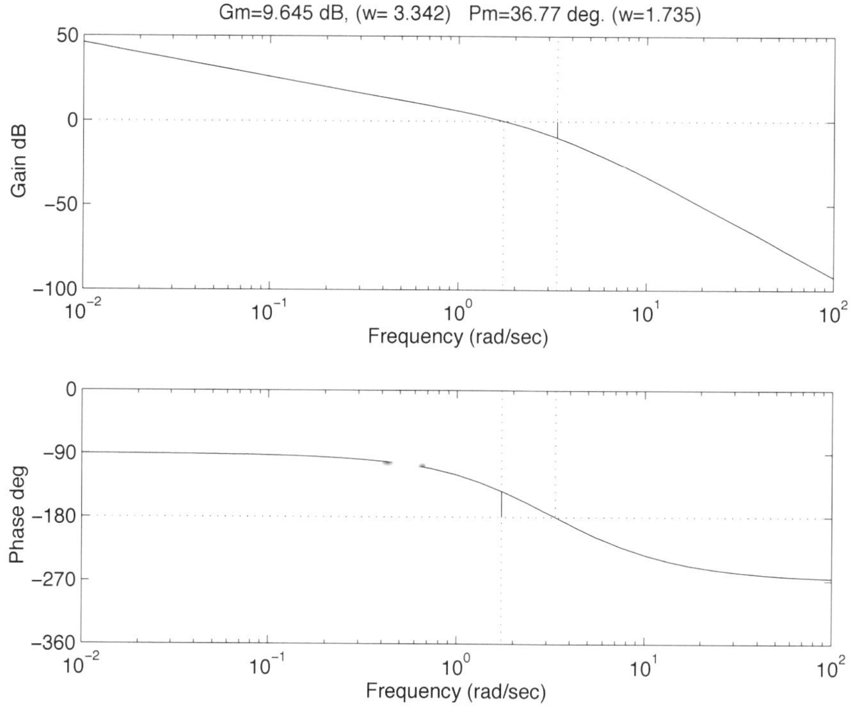

Figure 1.21. *Bode plot of the system after correction*

We can observe that the phase margin after correction does not correspond to the expected result; this is due to the variation of the phase around the critical point that is too fast. This fast variation of the phase causes a phase drop that largely exceeds the 5° (in reality $\simeq 18^\circ$ are lost). The advance phase controller is not suitable in the case of excessively fast phase variations.

REMARK 1.7.– *The results in the Bode plot of the system must always be verified after correction.*

1.3.1.4. *Integral compensator control (phase delay controller)*

In order to obtain a desired phase margin of $50°$, this time we are going to act not upon the phase but upon the modulus such that it passes through $0\,\text{db}$ at pulse ω_c that corresponds to a system phase that is equal to $(\Phi_c = -180° + 50° = -130°)$. Since the modulus cancels out for ω_c $(\Phi_c = -130°)$, then the phase margin is therefore $\Delta\Phi = 180° - 130° = 50°$. The integral compensator has the transfer function

$$C(p) = \frac{1 + aTp}{1 + Tp} \qquad (a < 1). \qquad [1.45]$$

Its Bode plot diagram is shown in Figure 1.22.

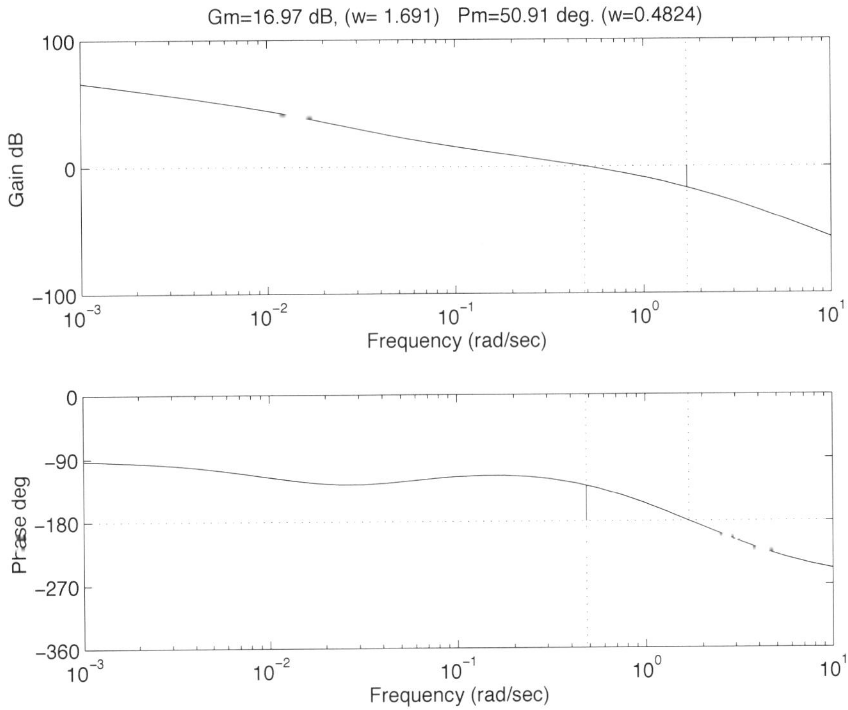

Figure 1.22. *Bode plot of the system after correction*

It can be observed that the phase of the controller is negative and consequently it will delay the phase of the system; as a result, the calculated phase margin will be affected. To compensate for this effect, we add a margin of 5°, that is to say, instead of taking $\Phi_c = -130^\circ$, we will choose $\Phi_c = -125^\circ$ and by placing ω_m we will manage to not lose more than these 5° because of the controller.

The value of a is calculated by measuring the modulus d at ω_c ($\Phi_c = -125^\circ$) and by making

$$20\log(a) = -d \Longrightarrow a = 10^{-d/20}. \qquad [1.46]$$

We choose T so as to not affect the phase around ω_c. For this purpose, we put a decade between $1/aT$ and ω_c, the impulse of the modulus crossing $0\,\text{db}$ after control, which makes

$$\frac{1}{aT} = \frac{\omega_c}{10} \Longrightarrow T = \frac{10}{a\omega_c}. \qquad [1.47]$$

The phase $\Phi = \Phi_c = -125^\circ$ is obtained for $\omega = 0.486$, which yields a modulus $d = 11.30$. Hence, the value of a is given by:

$$a = 10^{-11.30/20} = 0.27. \qquad [1.48]$$

The value of T is given by

$$T = \frac{10}{a\omega_c} = \frac{10}{0.27 \times 0.486} = 76.20\,\text{s}. \qquad [1.49]$$

Hence the following controller is obtained:

$$C(p) = \frac{1 + 20.57p}{1 + 76.20p}. \qquad [1.50]$$

The Bode plot of the system after control is shown in Figure 1.23.

The phase margin is maintained, and it is correctly written as $\Delta\Phi \simeq 50^\circ$. The advantage of this controller is its simplicity, but its main drawback is that it reduces the bandwidth of the system.

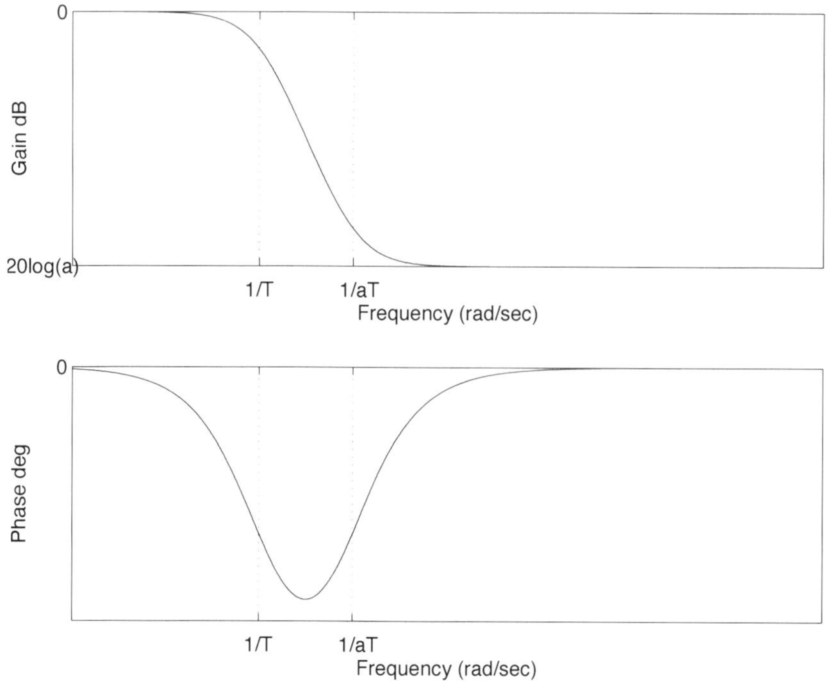

Figure 1.23. *Bode plot with integral compensator*

1.3.2. *PI control*

For the system represented by its transfer function,

$$H_o(p) = \frac{2}{(1+0.5p)^2}. \quad [1.51]$$

The following requirements must be satisfied:

– zero static position error;

– bandwidth $\omega_c \geq 4\,\text{rd/s}$ (cutoff impulse);

– phase margin $\simeq 50°$.

The Bode plot of this system is shown in Figure 1.24.

The cutoff pulse $\omega_c = 2\,\text{rd/s}$ ($|H(j\omega_c)| = 0\,\text{db}$), and the phase margin $\Delta\Phi = 90°$.

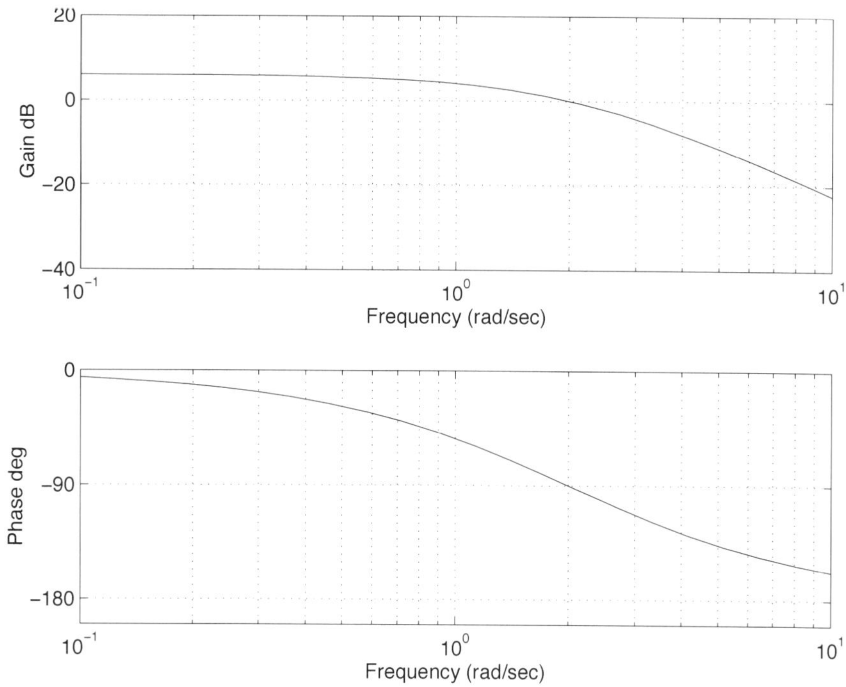

Figure 1.24. *Bode plot of the system without PI control*

The static position error is non-zero; to cancel, it is necessary to introduce an integration in the controller. The controller that we thus propose is a PI controller:

$$C(p) = K(1 + \frac{1}{T_i p}) = K\left(\frac{1 + T_i p}{T_i p}\right) = \frac{K}{T_i} \cdot \frac{1 + T_i p}{p}. \qquad [1.52]$$

The choice of $\frac{1}{T_i}$ is made so as to compensate the effect on the integration phase. The phase around ω_c should be unchanged ($\approx$ the system phase). We will therefore place $\frac{1}{T_i}$ a decade further than ω_c: $\frac{1}{T_i} = \frac{\omega_c}{10} \Longrightarrow T_i = \frac{10}{\omega_c} = \frac{10}{2} = 5\,\text{s}$. The Bode diagram of the system controlled by means of the following controller ($\frac{K}{T_i} = 1$) is shown in Figure 1.25:

$$C_1(p) = \frac{1 + T_i p}{p}. \qquad [1.53]$$

The position error is zero because there is integration in the direct chain; we are now going to determine the parameters of the controller in order to meet the specifications. The choice of T_i has been done so as to offset the phase effect of the integration, it then just suffices to calculate the second parameter K to ensure a phase margin of $\simeq 50°$ and a bandwidth of at least $4\,\mathrm{rd/s}$ ($\omega_c \geq 4\,\mathrm{rd/s}$).

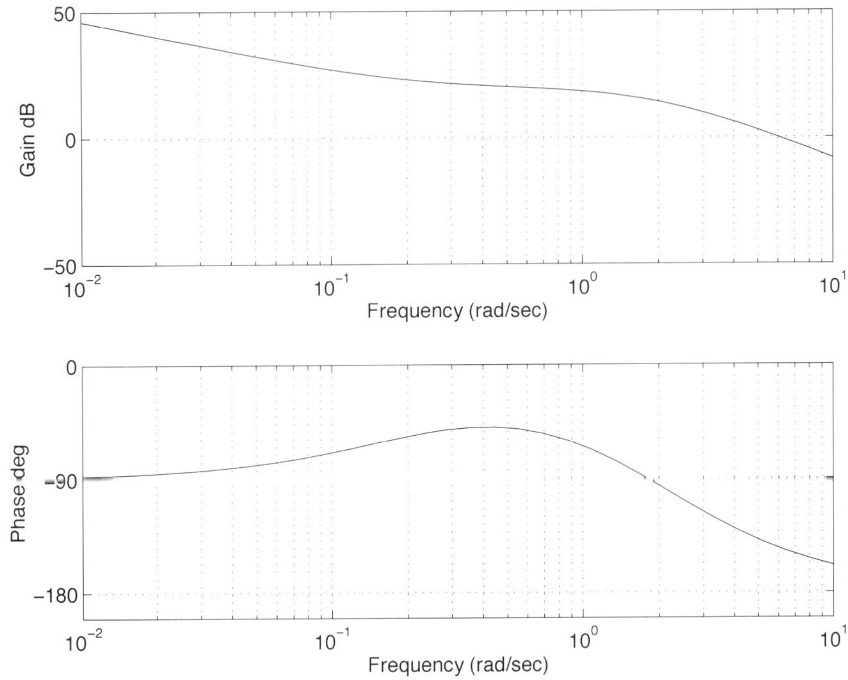

Figure 1.25. *Bode plot of the system with control*

The quantity $\frac{K}{T_i}$ is calculated by measuring the modulus d of $|C_1(j\omega_c')H_o(j\omega_c')|$ with ω_c' the new cut-off pulse corresponding to a phase $\Phi_c = -130°$ (phase margin $\approx 50°$) and by imposing:

$$20\log(\frac{K}{T_i}) = -d \Longrightarrow K = T_i \cdot 10^{-d/20}. \qquad [1.54]$$

The phase $\Phi = \Phi_c = -130°$ is obtained for $\omega = \omega_c = 4.02\,\mathrm{rd/s}$, which yields a modulus $d = 5.94$. Hence, the value of K: $K = 5 \cdot 10^{-5.94/20} = 2.5$ and the expression of the controller

$$C(p) = 2.5(1 + \frac{1}{5p}). \qquad [1.55]$$

The Bode plot of the system after control is shown in Figure 1.26. We correctly verify on the plot that the bandwidth is of $4\,\mathrm{rd/s}$ and that the phase margin is $50°$. The position static error is zero due to the integration in the direct chain. The requirements of the specifications are thus properly satisfied.

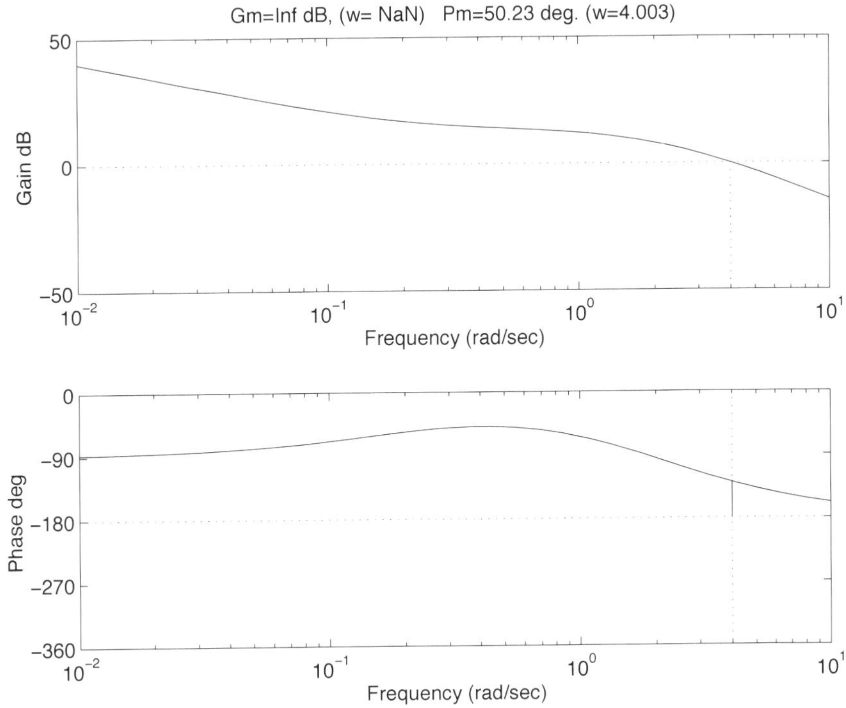

Figure 1.26. *Bode plot of the system with control*

1.3.3. *Phase advance control*

For the system represented by its transfer function

$$H_o(p) = \frac{K(1+p)}{p(1+0.2p)(1+0.05p)}, \qquad [1.56]$$

The following requirements must be satisfied:

– zero static error;

– a gain of about 30 db for $\omega = 2\,\mathrm{rd/s}$;

– phase margin $\simeq 50°$.

We want a gain of 30 for $\omega = 2\,\mathrm{rd/s}$. This allows us to define the value of K. We write:

$$\begin{aligned} 20\log_{10}\left(|H(j\omega)|_{\omega=2}\right) &= 20\log_{10}\left(\left.\frac{K\sqrt{1+\omega^2}}{\omega\sqrt{1+0.04\omega^2}\sqrt{1+0.0025\omega^2}}\right|_{\omega=2}\right) \\ &= 30\,\mathrm{db}. \end{aligned} \quad [1.57]$$

This equation can be rewritten by replacing ω by its value

$$20\log_{10}(1.033K) = 30 \Longrightarrow K = 30.6 \qquad \text{(that is } K = 30\text{)}. \quad [1.58]$$

The new transfer function of the system is

$$H(p) = \frac{30(1+p)}{p(1+0.2p)(1+0.05p)}. \quad [1.59]$$

The Bode plot of this system is shown in Figure 1.27.

The controller that we propose is a phase advance corrector because the dynamic accuracy is imposed by the value of K (gain of 30 db at $\omega = 2\,\mathrm{rd/s}$). Consequently, we cannot use an integral compensator (phase delay). The transfer function of the controller is as follows:

$$C(p) = \frac{1+aTp}{1+Tp} \qquad (a > 1). \quad [1.60]$$

The phase margin is $\Delta\Phi = 25°$, hence the phase to be added is $\Phi_m - 25° + 5° - 30°$. The value of a is given by

$$a = \frac{1+\sin(\Phi_m)}{1-\sin(\Phi_m)} = \frac{1+\sin(30°)}{1-\sin(30°)} = 3 \qquad \text{(that is } a = 4\text{)}. \quad [1.61]$$

By placing ω_m at ω corresponding to the system modulus that is equal to

$$-10\log_{10}(a) = -10\log_{10}(4) = -6\,\mathrm{db} \Longrightarrow \omega_m = 76\,\mathrm{rd/s}. \quad [1.62]$$

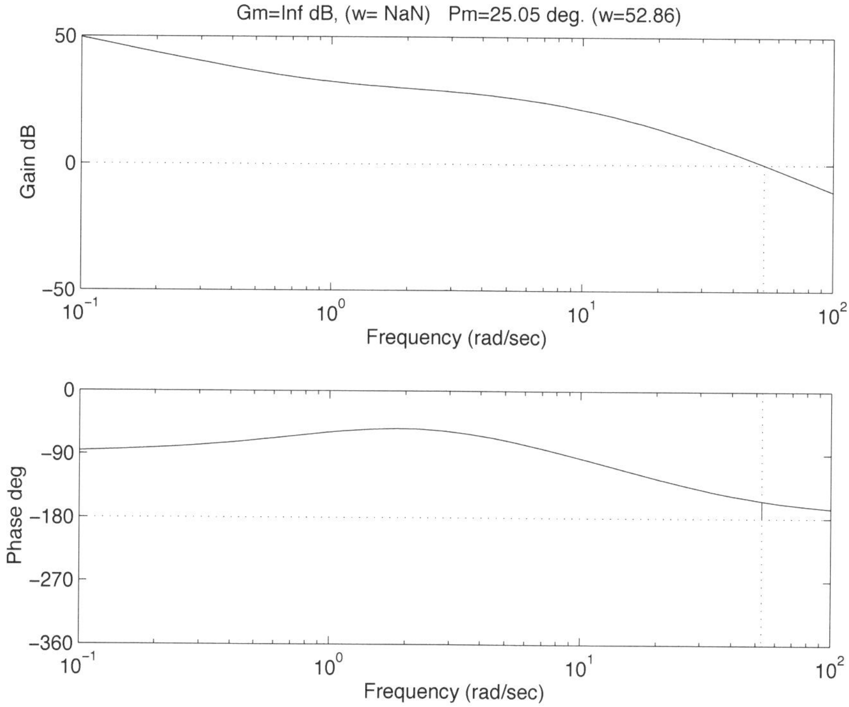

Figure 1.27. *Outline of the Bode plot of the system without phase advance control*

This allows us to calculate T,

$$T = \frac{1}{\omega_m \sqrt{a}} = \frac{1}{76\sqrt{4}} = 0.0066\,\text{s}. \qquad [1.63]$$

Hence, the following controller:

$$C(p) = \frac{1 + 0.0263p}{1 + 0.0066p}. \qquad [1.64]$$

The outline of the Bode plot of the system after being controlled is shown in Figure 1.28 in which we verify that the specifications are properly satisfied.

The cases of nonlinear systems will be addressed in the following chapters and examples of nonlinear systems using linearization methods will be given.

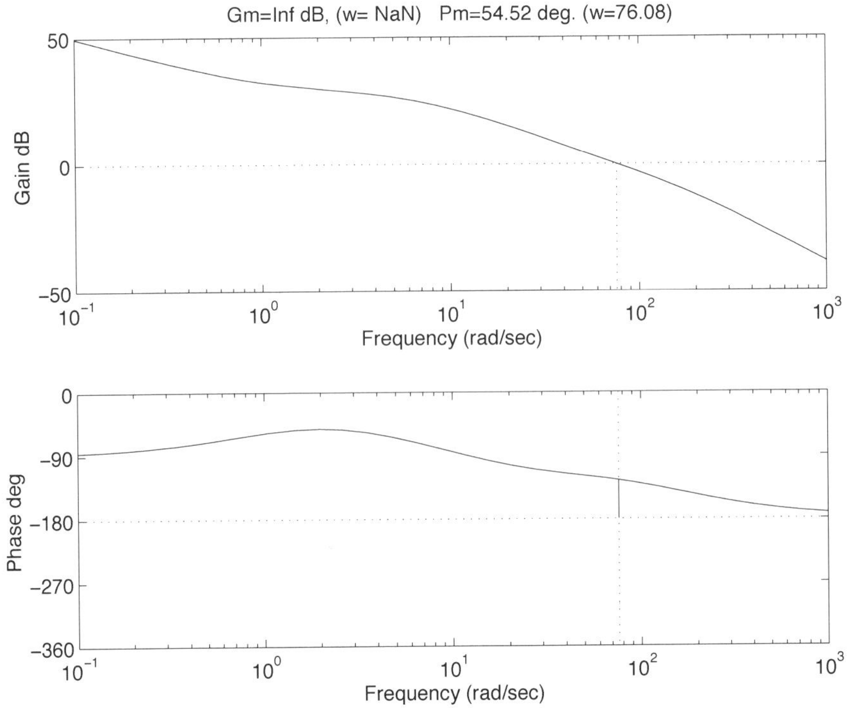

Figure 1.28. *Outline of the Bode plot of the system with phase advance control*

1.4. Some application exercises

EXERCISE 1.– A device, whose transfer function is as follows, is controlled by a controller placed in cascade with the system with a unity feedback loop. $H_0(p) = \frac{1}{p(p+1)}$:

1) Let $C(p) = K$.

a) Express the parameters of the transfer function of the system in the closed-loop system.

b) For $K = 1$, what can be said of its step response $y(t)$ to unit step function, and about permanent errors of positions ε_p and velocity ε_v?

c) For $K = 9$, what can be said of $y(t)$, of ε_p and of ϵ_v? Specify the effects due to the increase in K. Following Bode's method, determine the gain and phase margins. What can be observed from these values?

2) We want to find a permanent velocity error $\epsilon_v < 10\%$ and a phasc margin of about $50°$.

a) Determine the parameters of a controller to be inserted.

b) Verify the result using the Nyquist method by plotting the frequency response with and without control.

EXERCISE 2.– Figure 6.21 shows the frequency response of a system plotted according to Bode's method:

1) Determine the transfer function $H_o(p)$ of the system.

2) What are the position ϵ_p and velocity errors ϵ_v?

3) What are the gain ΔG and phase margins $\Delta\Phi$? The behavior of the system is not considered satisfactory. Why?

4) We want to bring the phase margin to $50°$. Study the serial compensation using phase advance and phase delay controllers. Compare both methods.

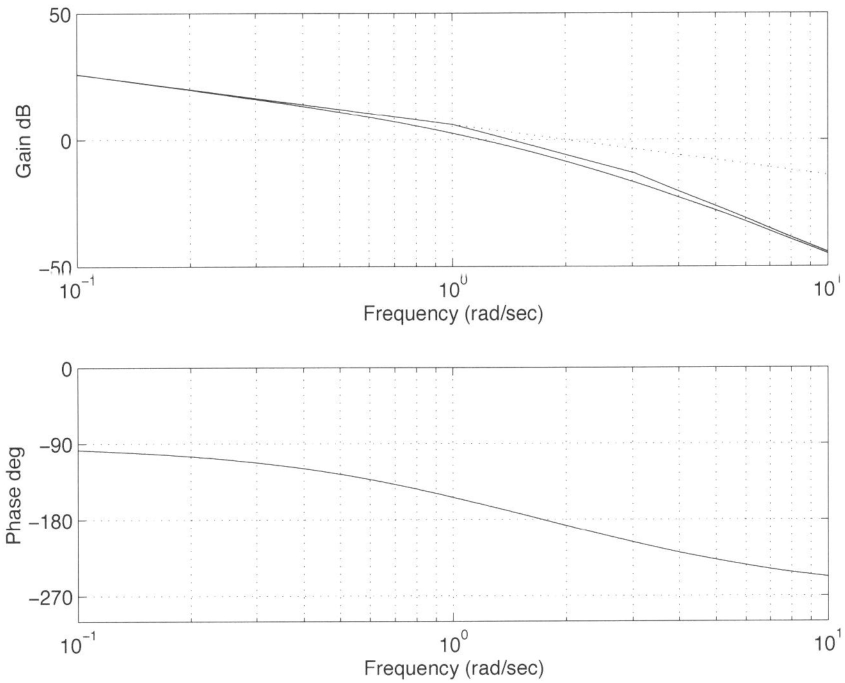

Figure 1.29. *Frequency response of the unknown system*

EXERCISE 3.– An OL system is described by Bode, see Figure 1.30.

$$H_o(p) = \frac{2}{(1+0.5p)^2}. \quad [1.65]$$

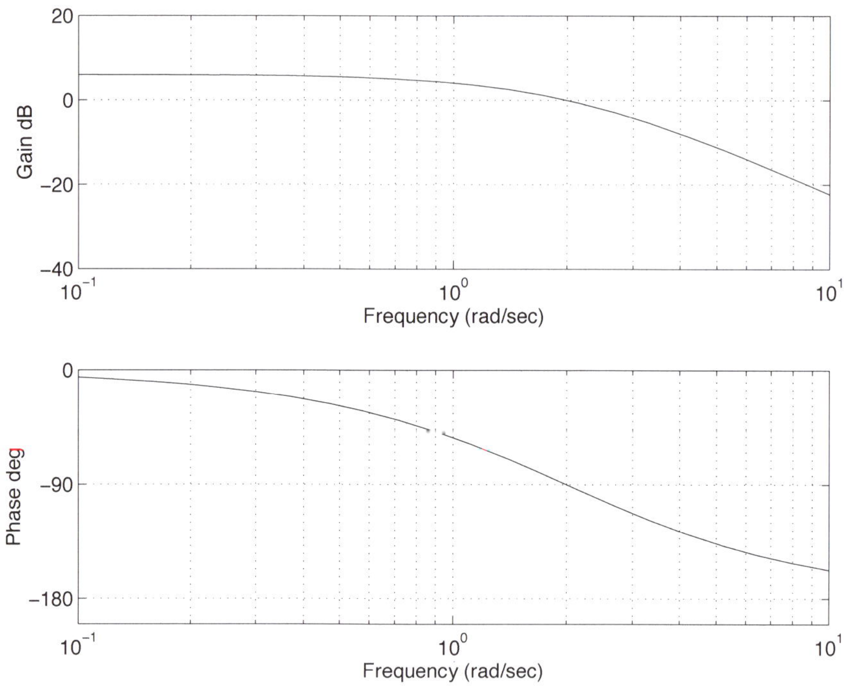

Figure 1.30. *Bode plot of* $H(p) = \frac{2}{(1+0.5p)^2}$

Determine the parameters of a controller to be inserted in series to obtain the following performance:

1) zero static (position) error;

2) bandwidth $\omega_c \geq 4$ rad/s;

3) phase margin $\simeq 50^{\circ}$.

EXERCISE 4.– Let a system be described by its transfer function in OL (Bode, Figure 1.31):

$$H_o(p) = \frac{K(1+p)}{p(1+0.2p)(1+0.05p)}. \quad [1.66]$$

This process is inserted in a control chain with unity back. The desired performances are:

1) zero static (position) error;

2) a gain of about 30 db for $\omega = 2\,\mathrm{rd/s}$;

3) phase margin $\simeq 50^o$.

Determine the parameters of a controller to be inserted in the sequence to satisfy these conditions.

Bode plot of

$$H(s) = \frac{K(1+s)}{s(1+0.2s)(1+0.05s)}. \quad [1.67]$$

1.5. Application 1: stabilization of a rigid robot with pneumatic actuator

The model of a robot for manipulation with two degrees of freedom is defined by

$$\tau = M(q)\ddot{q} + C(q,\dot{q})\dot{q} + F_v\dot{q} + G(q). \quad [1.68]$$

The system variables are vectors of dimension 2, respectively, representing positions, velocities and accelerations, $q, \dot{q}, \ddot{q}$. We define the following parameters for this robot:

– $M(q)$ is the matrix of inertia of dimension 2×2;

– $G(q)$ is the vector of gravity effects;

– $C(q,\dot{q})\dot{q}$ represents the centrifugal and Coriolis forces;

– F_v is the coefficient of viscous friction at the axis level;

– τ represents the couples applied at the axis level of the robot.

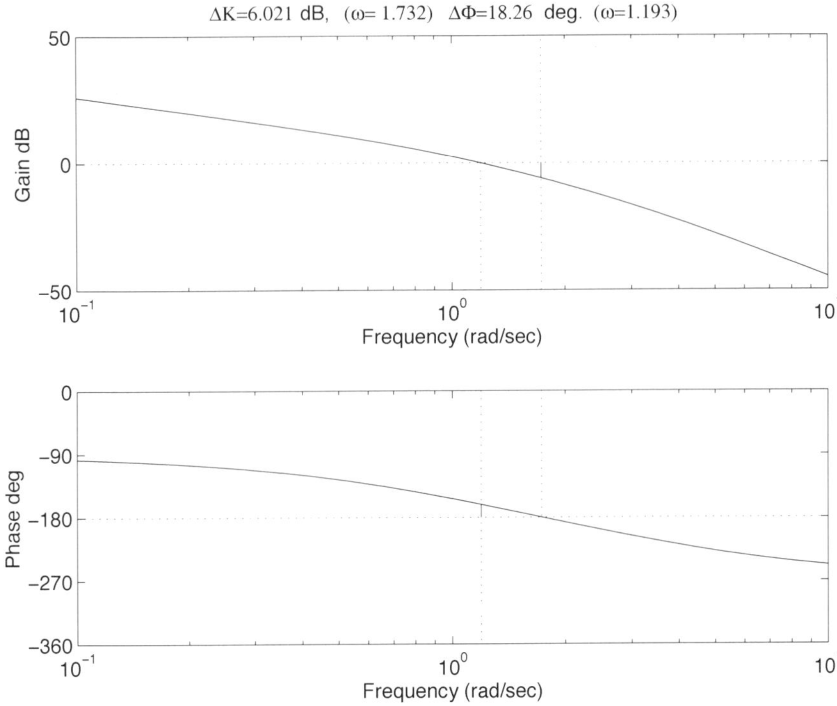

Figure 1.31. *Bode plot of* $\frac{H(p)}{K} = \frac{(1+p)}{p(1+0.2p)(1+0.05p)}$

Pneumatic actuators used for this robot can be represented by the differential equation linking the couple τ with the control voltage u and the velocity of the axes $\dot{q}$ as it follows:

$$\dot{\tau} + B\tau + E\dot{q} = Ju. \qquad [1.69]$$

First, to simplify the study, it will be assumed that the essential terms of the dynamic model: $M(q) = M$, $C(q, \dot{q})\dot{q} = C_o\dot{q}$, $G(q) = G_o$ are constant. The first part of the study concerns the first axis of the robot. In other words, variables $q, \dot{q}, \ddot{q}$ and u can be regarded as scalars.

The parameters of the system (considered linear invariant) are the following:

$g = 9.81$; $l1 = 0.11$; $l2 = 0.15$; $I_1 = 0.07\,\text{kgm}^2$; $I_2 = 0.025\,\text{kgm}^2$; $m_1 = 0.6\,\text{kg}$; $m_2 = 0.4\,\text{kg}$;

$E = 5I_{d2}$; $B = 10I_{d2}$; $J = 100I_{d2}$, I_{d2} is the identity matrix of dimension 2.

$s = \sin(q_2)$; $c = \cos(q_2)$;

$mll = 2m_2 l_1 l_2$; $mlc = 2m_2 l_1 l_2 c$; $mls = 2m_2 l_1 l_2 s$;

$c_{12} = \cos(q_1 + q_2)$; $s_{12} = \sin(q_1 + q_2)$;

Components of the inertia matrix: M_{ij}: let $A = 4m_2 l_1^2 + I_1 + I_2$;

$M_{11} = A + 4m_2 l_1 l_2 c + m_1 l_1^2 + m_2 l_2^2$; $M_{12} = I_2 + 2m_2 l_1 l_2 c + m_2 l_2^2$;

$M_{21} = I_2 + 2m_2 l_1 l_2 c + m_2 l_2^2$; $M_{22} = I_2 + m_2 l_2^2$;

Components of the matrix C: $C_{11} = -mls\dot{q}_2$; $C_{12} = -mls(\dot{q}_1 + \dot{q}_2)$; $C_{21} = mls\dot{q}_1$; $C_{22} = 0$;

Componants of vector G: $G_1 = (m_1 + 2m_2)gl_1 sin(q_1) + m_2 g l_2 s_{12}$; $G_2 = m_2 g l_2 s_{12}$.

To simplify the preliminary study of the control of this robot, we will focus on the first axis only and as a first step, couplings and nonlinearities, which can be considered as disturbance inputs, will be neglected. Next, we will be able to consider as a nominal model the one obtained when functioning around angular positions $q_1 = q_2 = 0$ and movements of small amplitudes.

$$M = \begin{pmatrix} 0.1570 & 0.0472 \\ 0.0472 & .0340 \end{pmatrix}; C_o = 0; \text{ and } \quad G_o = 0; \; F_v = 0. \qquad [1.70]$$

$$J = \begin{pmatrix} 100 & 0 \\ 0 & 100 \end{pmatrix}; \; B = \begin{pmatrix} 10 & 0 \\ 0 & 10 \end{pmatrix}; \; E = \begin{pmatrix} 5 & 0 \\ 0 & 5 \end{pmatrix}. \qquad [1.71]$$

1.5.1. *Conventional approach*

For the study below, the system equation will be taken as: $\tau = M_o\ddot{q}$ and $\dot{\tau} + B\tau + E\dot{q} = Ju$, with $M_o = 0.157$, $B = 10$, $J = 100$, and $E = 5$.

1) *Write in the form of a single differential equation the model of the first axis of the robot with its actuator.*

We shall express the model of the first axis of the robot with its actuator in the form of a single differential equation. The system equations: $\tau = M_o\ddot{q}$ and $\dot{\tau} + B\tau + E\dot{q} = Ju$, with $M_o = 0.157$, $B = 10$, $J = 100$, and $E = 5$ can be written describing $\tau = M_o\ddot{q}$ and substituting in the other equation. This gives us: $Ju = M_o\dddot{q} + BM_o\ddot{q} + E\dot{q} = 0.157\dddot{q} + 1.57\ddot{q} + 5\dot{q} = 100u$.

2) *Express the system transfer functions for the velocity* $H_1(p) = \frac{V(p)}{U(p)}$ *and for position* $H_2(p) = \frac{q(p)}{U(p)}$, *with* $v(t) = \frac{dq(t)}{dt}$ *the rotation velocity of the axis. Determine the poles and zeros of these two transfer functions.*

From the above equation, the transfer functions of the system are derived using Laplace transformation and considering zero initial conditions,

$$H_1(p) = \frac{V(p)}{U(p)} = \frac{100}{0.157p^2 + 1.57p + 5} \qquad [1.72]$$

and

$$H_2(p) = \frac{q(p)}{U(p)} = \frac{100}{p(0.157p^2 + 1.57p + 5)}. \qquad [1.73]$$

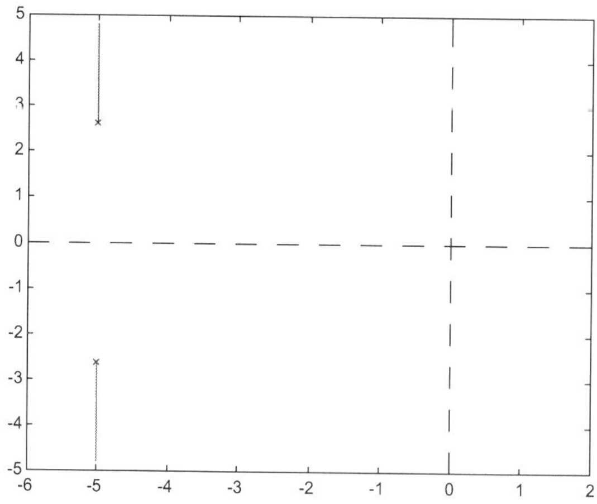

Figure 1.32. *Locus of the roots of* $H_1(p)$

3) *Plot the Nyquist locus of the transfer function* $H_o(p) = K_p.\ H_2(p) = K_p\frac{q(p)}{U(p)}$ *and analyze the stability of the system with a unity feedback loop for this position control.*

Nyquist locus of the transfer function

$$H_o(p) = K.H_2(p) = K\frac{q(p)}{U(p)} = \frac{100K}{p(0.157p^2 + 1.57p + 5)}. \qquad [1.74]$$

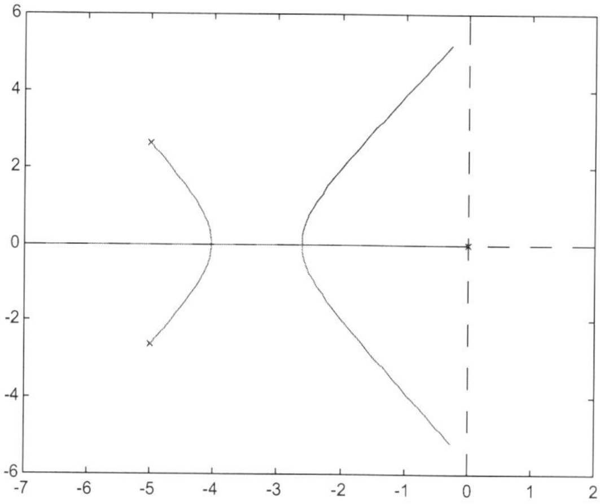

Figure 1.33. *Locus of the roots of* $H_2(p)$

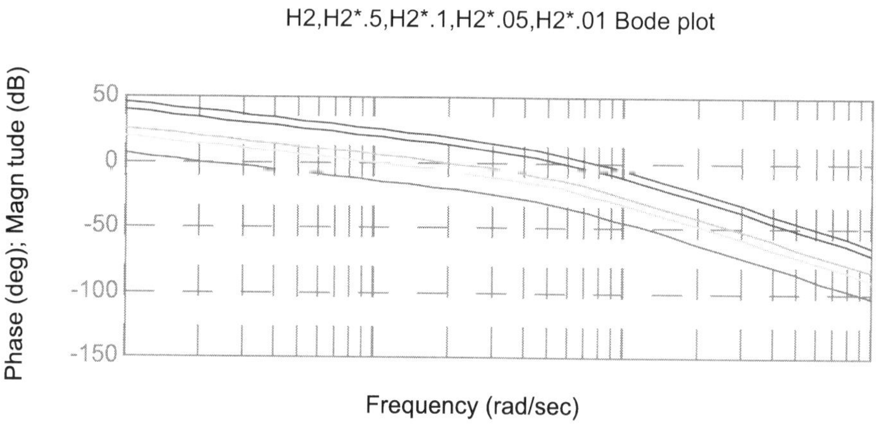

Figure 1.34. *Bode plot of* $K.H_2(p)$*. For a color version of this figure, see www.iste.co.uk/femmam/signals.zip*

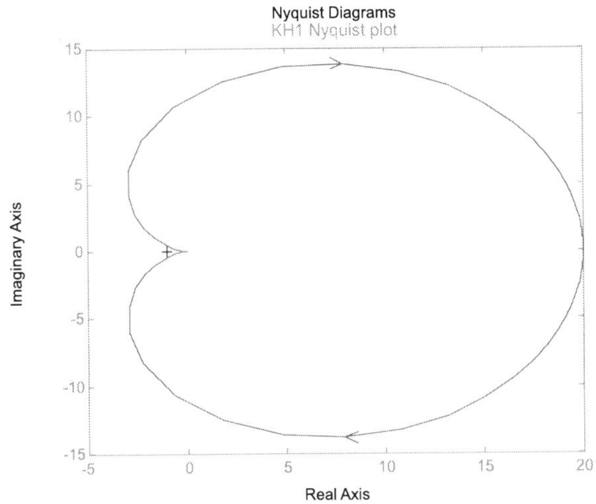

Figure 1.35. *Nyquist locus for* $K = 1(H_1)$

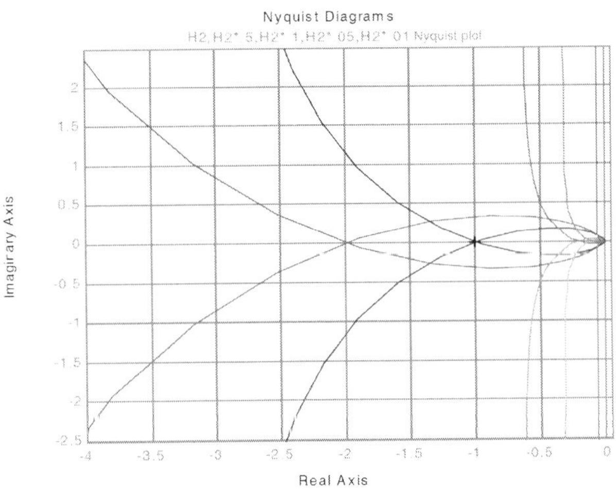

Figure 1.36. *Nyquist loci for* $H_2(p) * Ki$. *For a color version of this figure, see www.iste.co.uk/femmam/signals.zip*

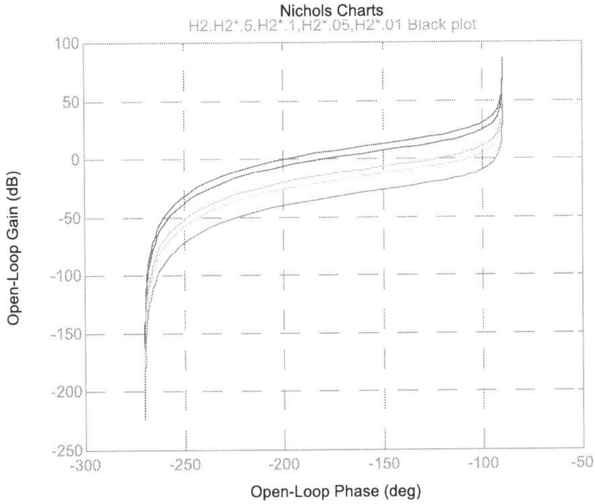

Figure 1.37. *Black plot for $K.H_2(p)$. For a color version of this figure, see www.iste.co.uk/femmam/signals.zip*

4) The stability of the system with a unity feedback loop is guaranteed if $K < 0.5$.

5) *First, we want to control the velocity of this axis; the system is then considered as defined by* $H_1(p) = \frac{V(p)}{U(p)}$.

a) *By applying the Routh criterion, analyze the stability of the system, whose velocity is looped with a proportional controller of gain* K_v ($C(p) = K_v$) with a unity feedback. Determine the conditions on the gain of a proportional feedback ensuring the stabilization of the system.

Velocity control of $H_1(p) = \frac{V(p)}{U(p)} = \frac{100}{(0.157p^2+1.57p+5)}$.

The system looped with a proportional controller of gain K ($C(p) = K$) with a unity feedback has a transfer

$$G_1(p) = \frac{100K}{0.157p^2 + 1.57p + 5 + 100K}. \quad [1.75]$$

We apply the Routh criterion to

$$0.157p^2 + 1.57p + 5 + 100K. \quad [1.76]$$

6) a) The condition on gain K ensuring the stability of the system in the closed-loop system is: $K > -0.05$.

line p^2	0.157	$5+100K$
line p^1	1.57	0
line p^0	$5+100K$	0

Table 1.2. *Routh table results*

b) *Plot the Bode graph and recall the definitions of phase margin of gain margin and static gain.*

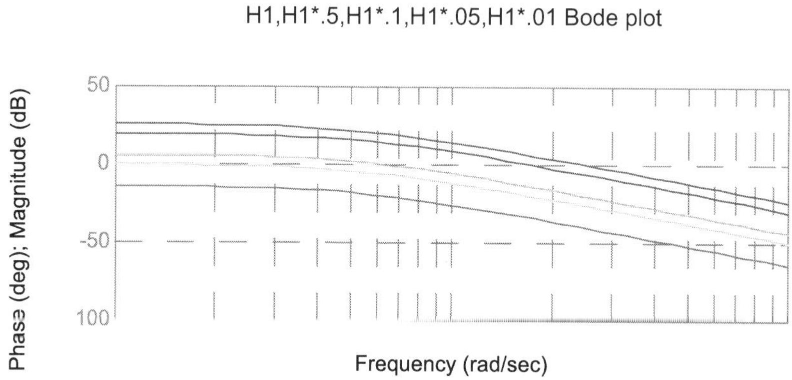

Figure 1.38. *Bode plot of $K.H_1(p)$. For a color version of this figure, see www.iste.co.uk/femmam/signals.zip*

7) a) *Plot the Nyquist locus of the transfer function $H(p) = K_v.H_1(p) = K_v\frac{V(p)}{U(p)}$ and verify the previous results.*

Nyquist locus of the transfer function

$$H(p) = K.H_1(p) = K\frac{V(p)}{U(p)} = \frac{100K}{0.157p^2 + 1.57p + 5}. \qquad [1.77]$$

b) *Can a margin be obtained with phase of 45°? What is the order of magnitude of the gain margin? Justify the answers.*

If the gain is correctly chosen, a phase margin greater than 45° can be obtained as well as an infinite gain margin regardless of the order of magnitude of K.

c) *What can be said about the static error of position ε_p and about the system controlled in velocity ε_v?*

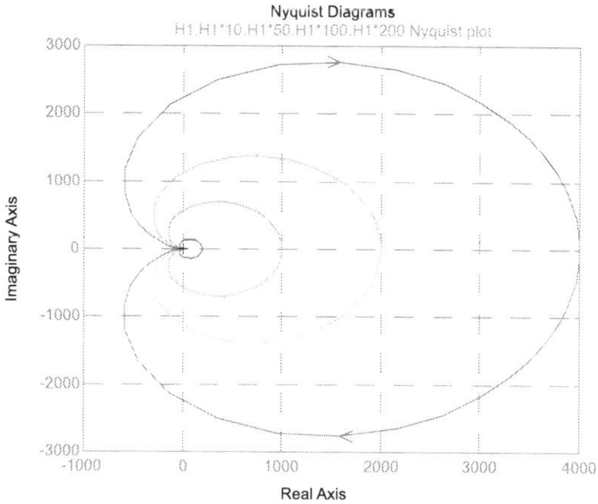

Figure 1.39. *Nyquist locus for velocity $K.H_1(p)$. For a color version of this figure, see www.iste.co.uk/femmam/signals.zip*

The static error in position $\varepsilon_p = \lim_{p\to 0} \frac{1}{0.157p^2+1.57p+5+100K} = \frac{1}{5+100K}$ and the static error in velocity $\varepsilon_v = \lim_{p\to 0} \frac{p}{0.157p^2+1.57p+5+100K} = 0$.

Some observations: in order to stabilize the system, it is more interesting to implement a first loop for the velocity feedback using a control $u = K_v(v-\dot{q})$ and then to consider the system having as input v and as output angular position q. The transfer function becomes: $H_o(p) = \frac{100K_v}{0.157p^2+1.57p+5+100K_v}\frac{1}{p}$. Here, the Nyquist locus for such a system is represented for $K_v = 1$, then for $K_v = 10$. Note the difference with the result of Question 3. Calculate the gain margin and the phase margin in these two cases and conclude on the difference and the significance of the velocity feedback (differential term).

8) The objective is to complete this servo with a position loop, after the velocity feedback giving a phase margin of 45° *(this allows us to set the value of K_v)*.

a) Express the transfer function $H_3(p)$ of the system that results therefrom (considering as the output the position q).

$$H_3(p) = \frac{100K_v}{0.157p^2 + 1.57p + 5 + 100K_v}\frac{1}{p}. \qquad [1.78]$$

b) By applying the Routh criterion, analyze the stability of the system, in which position is looped back with a proportional controller of gain K_p with a unity feedback. Determine the conditions of the gain K_p that ensure the stabilization of the system.

c) Plot the Bode chart and the Nyquist locus of the transfer function $H_o(p) = K_p.H_3(p)$. Conclude on the stability of the system using the Nyquist criterion.

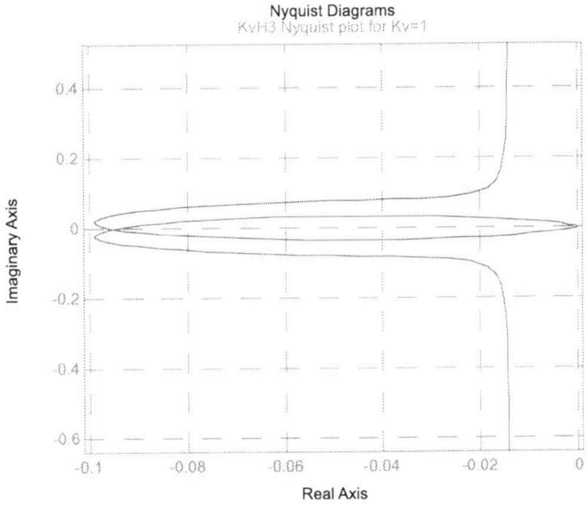

Figure 1.40. *Nyquist locus for the system position after velocity feedback $K_v = 1$ and $K = 1$*

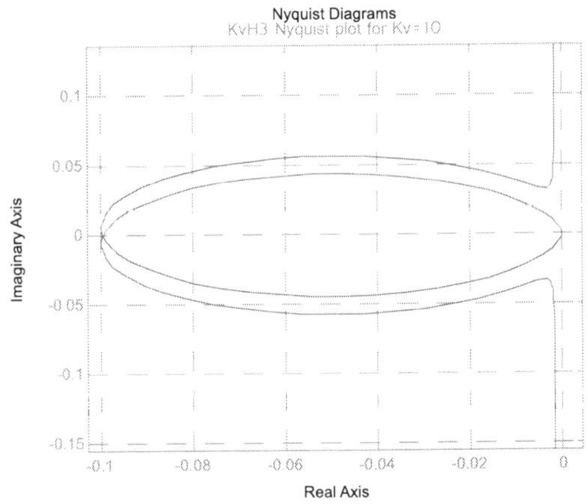

Figure 1.41. *Nyquist locus for the system position after velocity feedback $K_v = 10$ and $K = 1$*

d) Determine the optimal values that can be obtained for the phase margin and the gain margin of the system looped in this manner.

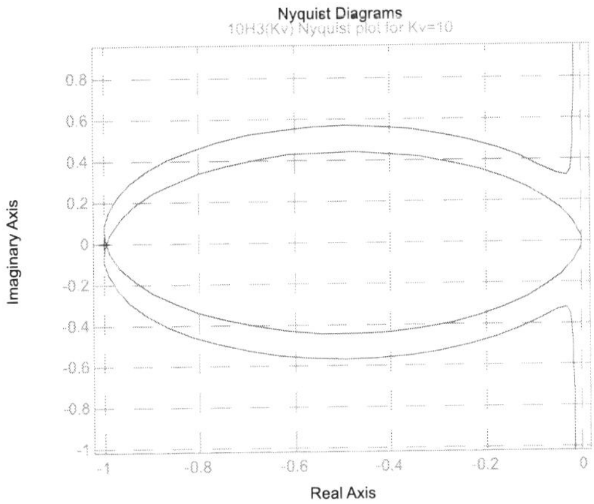

Figure 1.42. *Nyquist locus for the system position after velocity feedback* $K_v = 10$ *and* $K = 10$

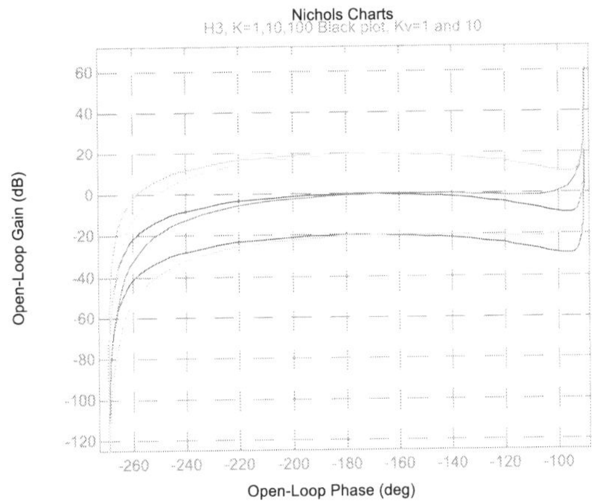

Figure 1.43. *Black plot for* $Kv = 1$ *and* 10 *with* $K = 1, 10$ *and* 100

e) Express static position ε_p and velocity ε_v errors of the system controlled in position.

f) For the position control of the system, is an acceleration feedback loop necessary and what would its contribution be in this case?

9) Compare and discuss the results of Questions 3–5 in the cases that follow:

a) what would be the effect due to a velocity sensor of transfer function $H_c(p) = \frac{1}{1+Tp}$?

b) G_o the effect of gravitation is materialized by a non-zero constant;

c) the coefficient of frictions F_v was not zero;

d) C_o originating from the Coriolis effect and centrifugal is non-zero and variable (see definition of C_{11} at the beginning of the text);

e) M_o varies according to the angular position;

f) G_o the effect of gravitation is not null and variable as well as C_o and F_v;

10) *In this section, the study of the control of the system is achieved in the state space*:

a) Give two state–space representations for the model of this axis, one of which under a controllable canonical form. Voltage u will be considered as input and as output the angular position q on the one hand and the angular velocity $\dot{q}$ on the other hand.

b) Study of the system in the state space: state–space representations for the model of this axis. In the case where the output is the angular velocity $\dot{q}$,

$$H_1(p) = \frac{100}{0.157p^2 + 1.57p + 5} = \frac{100/.157}{p^2 + 10p + 5/.157}$$
$$= \frac{636.94}{p^2 + 10.0p + 31.847}. \quad [1.79]$$

$$\dot{X} = \begin{bmatrix} -10 & -31.847 \\ 1 & 0 \end{bmatrix} X + \begin{bmatrix} 636.94 \\ 0 \end{bmatrix} u \quad \text{and} \quad y = \begin{bmatrix} 0 & 100 \end{bmatrix} X \quad [1.80]$$

$$\dot{X} = \begin{bmatrix} 0 & -10 \\ 1 & -31.847 \end{bmatrix} X + \begin{bmatrix} 1 \\ 0 \end{bmatrix} u \quad \text{and} \quad y = \begin{bmatrix} \beta_1 & \beta_2 \end{bmatrix} X \quad [1.81]$$

For the case where the output is the angular position q,

$$H_2(p) = \frac{100}{p(0.157p^2 + 1.57p + 5)} = \frac{636.94}{p^3 + 10p^2 + 31.847p}. \quad [1.82]$$

$$\dot{X} = \begin{bmatrix} -10 & 31.847 & 0 \\ 1 & 0 & 0 \\ 0 & 1 & 0 \end{bmatrix} X + \begin{bmatrix} 1 \\ 0 \\ 0 \end{bmatrix} u \text{ and } y = \begin{bmatrix} 0 & 0 & 636.94 \end{bmatrix} X \quad [1.83]$$

$$\dot{X} = \begin{bmatrix} 0 & 0 & -10 \\ 1 & 0 & -31.847 \\ 0 & 1 & 0 \end{bmatrix} X + \begin{bmatrix} 1 \\ 0 \\ 0 \end{bmatrix} u \text{ and } y = \begin{bmatrix} \beta_1 & \beta_2 & \beta_3 \end{bmatrix} X \qquad [1.84]$$

c) Give the state–space representation of the system having the state vector $x = \begin{pmatrix} q \\ \dot{q} \\ \ddot{q} \end{pmatrix}$.

$$0.157\dddot{q} + 1.57\ddot{q} + 5\dot{q} = 100u \Rightarrow \dddot{q} = -10\ddot{q} - 31.847\dot{q} + 636.94u \qquad [1.85]$$

$$\dot{X} = \begin{bmatrix} 0 & 1 & 0 \\ 0 & 0 & 1 \\ 0 & -31.847 & -10 \end{bmatrix} X + \begin{bmatrix} 1 \\ 0 \\ 0 \end{bmatrix} u \text{ and } y = \begin{bmatrix} 1 & 0 & 0 \end{bmatrix} X \qquad [1.86]$$

d) Express the characteristic equation of the system.

e) Velocity control: the system is velocity looped, with a state feedback $u = -L_1 x$, determine the gains l_1 and l_2 of vector L_1, *in order for the closed-loop system to have a natural frequency* $\omega_o = 10\, rad/s$ and a damping $\xi = 1$ (corresponding to the characteristic equation $p^2 + 2\omega_o p + \omega_o^2$).

f) The system (having the angular position on output) is position looped on a state feedback $u = -Lx$, determine the gains l_1 and l_2 of vector L, such that the closed-loop system has a natural frequency $\omega_o = 10\, rad/s$ and a damping $\xi = 1$ (corresponding to the characteristic equation $(p + 10)(p^2 + 2\omega_o p + \omega_o^2)$).

11) Compare both approaches of the position control of the system, considering disturbances, couplings and variations of the parameters.

12) Start the study again considering this time both mobile axes simultaneously and non-diagonal matrices and with variable coefficients.

1.6. Application 2: temperature control of an oven

The study consists of two parts: modeling and identification on the one hand and control on the other hand.

1.6.1. *Modeling and identification study*

In the case of thermal processes, the most often applied modeling and identification technique consists of finding a model describing the behavior of a system from experimental measurements. Most often, the measure chosen is the reading of the system response to a step setpoint (in the case of the figure in multiples of 500°). The step response of the system as a function of time, if the step function is applied at date $t = 1$ s, is shown in Figure 1.44 for the case of an empty oven, half-loaded and with a full load. It is desired to derive its transfer function when empty (half-loaded and in full load). Modeling is the most important step in the study of an automated system. In order to control or regulate a given system, it is first necessary to have its model in order to study it in simulation. Once simulation results are very satisfactory, we will apply the control laws proposed in simulation on the real system. On the other hand, if the results with the real system are not acceptable, the modeling will imperatively have to be reviewed. After modeling, the second step consists of analyzing the behavior of the system using its model. This behavior analysis allows us to elaborate a control strategy for the system taking into account the performance restrictions imposed by the specifications and of the physical limitations of the real system. To illustrate this approach, we propose the following organization chart (Figure 1.45).

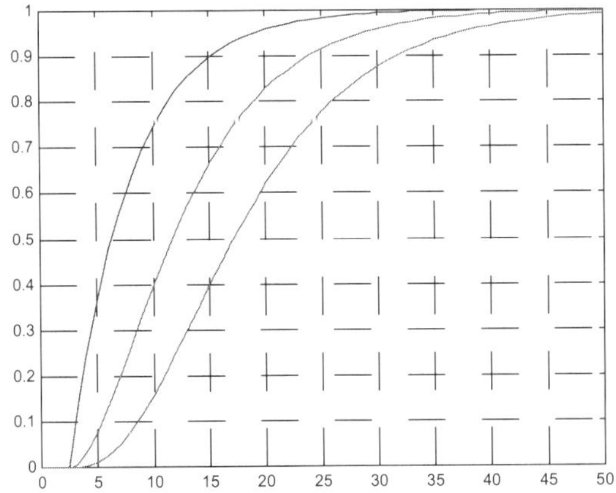

Figure 1.44. *Model describing the behavior of a system*

The modeling is achieved by writing the physical equations that describe the behavior of the system. In the continuous case, these equations are differential equations and in the discrete case recurrence equation. To these equations we apply transformations (Laplace transform: continuous; Z-transform: discrete) to shift from the time domain to the frequency domain where the analysis of the behavior of the system is more interesting. In the case where the system cannot be described by physical equations, it is always possible that an approximate model be proposed, through identification. This model should best describe the behavior of the system. It can be obtained from the identification by analogy to known systems, using the responses to test signals or by using an identification that includes the optimization of the error criterion. In this case, the algorithm uses data that correspond to the input and output signals of the system. These data must be rich enough in excitation to cover the entire spectrum of the system. The block diagram is shown in Figure 1.46.

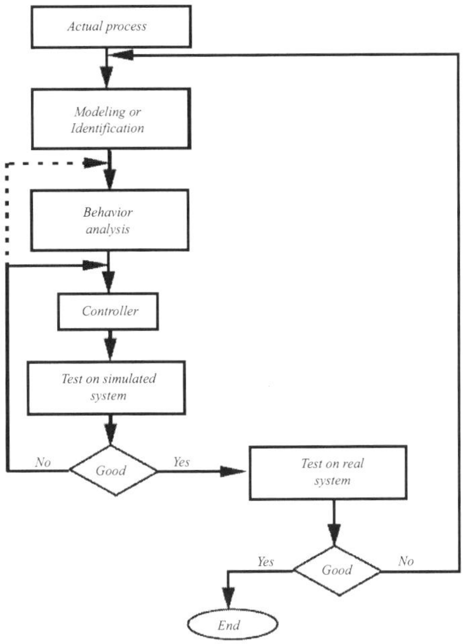

Figure 1.45. *Organigram of the modeling stages for the study of an automated system*

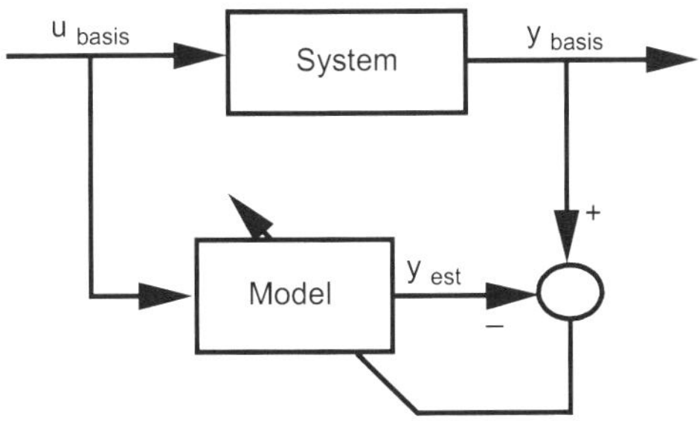

Figure 1.46. *Approximate model for identification*

Representation choice

The delay

The time constant(s)

$$H_o(p) = \frac{Ke^{-\tau}}{(1 + Tp)}. \quad [1.87]$$

$$H_o(p) = \frac{Ke^{-\tau}}{(1 + Tp)^n}. \quad [1.88]$$

Parametrization choice

Transfer function

State–space representation

Discrete representations

Identification method choice

Comparative method

Strejc method

Broida method

Least squares method

and at the end of this application as practical work we should consider control and performance improvement.

2

System Process Control

This chapter explores process control in the state space.

2.1. Introduction

The system (in the open-loop system) to be controlled has a control input u, an auxiliary input w, an output (controlled) y and an output z. Control u is a function (static of the state in general) of the output of the system y. Block C is called controller, compensator or control law. This pattern is used in control to meet most of the requirements but a number of variants do exist.

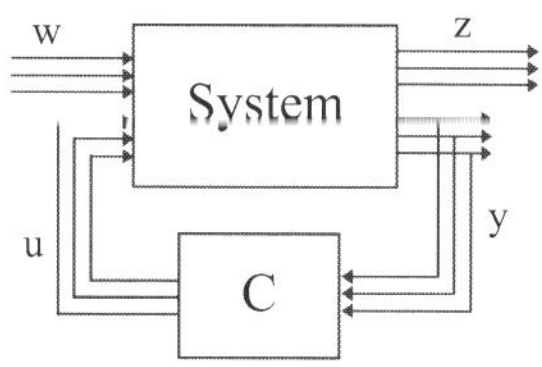

Figure 2.1. *System with controller to be automated*

2.2. Modeling

2.2.1. *Introduction*

The following topics will be detailed in this chapter: precompensator control, postcompensator control, the observability problem, static feedback and dynamic feedback.

2.3. Governability, controllability and observability

In this section, the focus is on the study of regions of the state–space that can be obtained by using a control (*governability* or *controllability*) and on those in which the state of the system can be derived through the observation of these inputs and outputs (*observability*).

2.3.1. *Characteristic polynomial, minimal polynomial and Cayley–Hamilton theorem*

Let A be a square matrix $n \times n$; the characteristic polynomial of A is defined by:

$$P(\lambda) = \det(\lambda I - A) = (\lambda - \lambda_1)(\lambda - \lambda_2)...(\lambda - \lambda_n).$$

The characteristic equation of matrix A is $P(\lambda) = 0$.

The roots λ_i of the characteristic polynomial of A are the eigenvalues $\lambda_i(A), i = 1...n$ of matrix A.

THEOREM 2.1.– The Hamilton–Cayley theorem ensures that the matrix verifies the characteristic equation $P(A) = 0$.

We call minimal polynomial of A the monic polynomial $Q(\lambda)$ of the lowest degree in λ such that $Q(A) = 0$.

The minimal polynomial of A divides its characteristic polynomial and any eigenvalue of A is a root of the minimal polynomial.

EXERCISE.– Show that for a diagonal matrix $\Lambda = \text{diag}(\lambda_i)$ of dimension n, $P(\Lambda) = 0$ if $P(\lambda)$ is its characteristic polynomial. Deduct that for any matrix of the form $A = T\Lambda T^{-1}$ we also have $P(A) = 0$. Give their minimal polynomial, consider the case of multiple eigenvalues with Λ, a Jordan-based matrix.

2.3.2. *Governability or controllability*

2.3.2.1. *Definition*

Consider the system defined by its state–space representation and the initial state $x(0)$:

$$\dot{x} = A.x + Bu(t)\,, \quad x(0) = 0. \qquad [2.1]$$

DEFINITION 2.1.– *The attainable subspace denoted,* $\mathcal{A}$*, is the set of states* x *for which there is a control* $u(t)$ *that drives the state of the system from* $x(0)$ *to* x *in finite time:*

$$\mathcal{A} = \{x | \exists T \geq 0, \exists u(t), \text{ such that } x = x(T),$$

where $x(t)$ *is the solution of equation [2.1] with* $x(0) = 0\}$.

The set $\mathcal{A}$ is a subspace of R^n because the system is linear. It should also be noted that the set of states attainable from a nonzero state $x(0)$ can be derived from $\mathcal{A}$ by translation. A system that has as an attainable subspace (or controllable or governable) the integer space R^n is called a control system.

DEFINITION 2.2.– *A system is controllable if for any initial condition* $x(t_o)$ *and any final state* $x(t_f)$*, there is an input* $u(t)$ *defined in* $[t_o, t_f]$ *that transfers the system state from* $x(t_o)$ *to* $x(t_f)$.

DEFINITION 2.3.– *The system is controllable if there is a finite control* $u(t)$ *that can bring the system from an initial state* x_0 *to a final state* x_f *in finite time.*

2.3.2.2. *Kalman controllability criterion*

THEOREM 2.2 (Kalman Theorem).– The attainable subspace is the image of the controllability matrix defined by : $\mathcal{C} = [B, AB, ..., A^{n-1}B]$.

A system is governable if and only if the rank of its controllability matrix $\mathcal{C}$ is n.

Proof: Assume that $Rank(\mathcal{C}) < n$. This means that there is a vector $v \in R^n - \{0\}$ such that $\forall i, 0 \leq i \leq n-1, v^T A^i B = 0$.

Following Cayley Hamilton's theorem, it can be deduced that $\forall i \geq 0, v^T A^i B = 0$ and thus that $\forall t \geq 0, \sum_{i=0}^{\infty} v^T A^i B \frac{t^i}{i!} = v^T e^{At} B = 0$.

When it is desirable to transfer the system state from $x(0) = 0$ to $x(T)$, we have with the evolution equation $x(T) = \int_0^T e^{At} Bu(T-t)dt$ and this gives us $vx(T) = \int_0^T ve^{At} Bu(T-t)dt = 0$ for all u. This means that states that are outside the hyperplane (subspace) defined by $v^T x = 0$ cannot be reached and therefore the system is not controllable.

Conversely, consider a system that is not completely controllable. There exists then a non-zero vector v such that $v^T x = 0$ for any $x \in \mathcal{A}$. Let x be the attainable space vector obtained using the control $u(t) = B^T e^{(T-t)A^T} v$. Therefore, the answer following direction v will be:

$$0 = v^T x = \int_0^T ve^{At} BB^T e^{tA^T} vdt = \int_0^T \|u(t)\|^2 dt$$

from which it can be concluded that $ve^{At}B = 0$ and that the rank of $\mathcal{C}$ is smaller than n.

The system is controllable if the controllability matrix

$\mathcal{C} = [B, AB, \ldots, A^{n-1}B]$ is of rank n.

In the case of a system expressed in diagonal form, it is sufficient that all the elements of vector B be non-zero.

THEOREM 2.3.– If $x(0) \in \mathcal{A}$, then for any $t, x(t) \in \mathcal{A}$, there is no control that makes it possible to leave the state of the attainable subspace $\mathcal{A}$.

PROOF.– Assume $x(0) \in \mathcal{A} = Im(\mathcal{C})$. It can be written that $x(0) = \sum_{i=0}^{n-1} \alpha_i A^i B$

The Laplace transform $x(t)$ equals

$$X(p) = (pI - A)^{-1}x(0) + (pI - A)^{-1}BU(p), \qquad [2.2]$$

$$X(p) = \sum_{k=0}^{\infty} \frac{1}{p^{k+1}} A^k \left(\sum_{i=0}^{n-1} \alpha_i A^i B \right) + \sum_{k=0}^{\infty} \frac{1}{p^{k+1}} A^k BU(p), \qquad [2.3]$$

$$X(p) = z_0(p)B + \ldots + z_{n-1}(p)A^{n-1}B, \qquad [2.4]$$

where $z_i(p)$ are complex functions of p. Using the inverse Laplace transformation, we find in the time domain ($\mathcal{C}$ is the controllability matrix):

$$x(t) = z_0(t)B + \ldots + z_{n-1}(t)A^{n-1}B = \mathcal{C} \begin{bmatrix} z_0(t) \\ \vdots \\ z_{n-1}(t) \end{bmatrix}. \qquad [2.5]$$

EXAMPLE 2.1 (Non-controllable System).– *Consider a satellite, assumed as a system plane consisting of a body of mass M and inertia J, equipped with an antenna of mass m and length l, which we want to move toward a fixed point. The system control is materialized by an actuator generating a couple τ on the axis of rotation of the antenna in O. θ represents the angle of the antenna with respect to a fixed direction and Ω is the rotation velocity of mass M. The state of the system is represented by the vector $x = [\theta, \dot{\theta}, \Omega]^T$. It is possible to write the model of the system in the form*

$$\dot{x} = \begin{pmatrix} 0 & 1 & 0 \\ 0 & 0 & 0 \\ 0 & 0 & 0 \end{pmatrix} \begin{pmatrix} \theta \\ \dot{\theta} \\ \Omega \end{pmatrix} + \begin{pmatrix} 0 \\ \frac{1}{m} \\ -\frac{1}{J} \end{pmatrix} \tau(t) \qquad [2.6]$$

and

$$y(t) = [1, 0, 0]x(t) = \theta(t). \qquad [2.7]$$

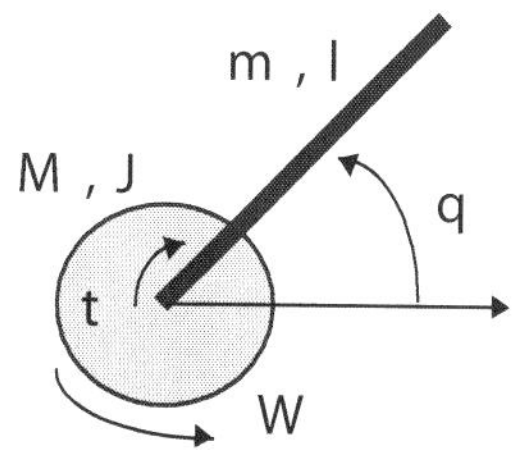

Figure 2.2. *Satellite control example*

– *The governability matrix of this system is* $\mathcal{C} = \begin{pmatrix} 0 & \frac{1}{m} & 0 \\ \frac{1}{m} & 0 & 0 \\ -\frac{1}{J} & 0 & 0 \end{pmatrix}$. *It has rank 2 and therefore the system is not completely controllable.*

EXAMPLE 2.2.– *Given the system with the state–space representation:* $\dot{X} = \begin{bmatrix} -1 & 0 \\ 1 & 1 \end{bmatrix} X + \begin{bmatrix} -2 \\ 1 \end{bmatrix} u$ *and* $y = \begin{bmatrix} 0 & 1 \end{bmatrix} X$.

The system has a controllability matrix $\mathcal{C} = [B, AB] = \begin{pmatrix} -2 & 2 \\ 1 & -1 \end{pmatrix}$. *The corresponding governability subspace is* $Im(C) = \{v$ *such that* $v = \alpha \begin{pmatrix} -2 \\ 1 \end{pmatrix}$ *and* $\alpha \in R\}$. *The system is therefore not controllable because the states that the system can attain are those such that:* $\frac{1}{2}x_1 + x_2 = 0$.

2.3.2.3. *Discrete system controllability*

A discrete univariate system has discrete state–space representation:

$$x_{k+1} = Ax_k + Bu_k. \qquad [2.8]$$

Its response to an input u at time n can be written as:

$$x_n = A^n x_0 + [B,\ AB,\ ......,\ A^{n-1}B] \begin{pmatrix} u_{n-1} \\ \vdots \\ u_0 \end{pmatrix}. \qquad [2.9]$$

Suppose that we want to bring the state to a value x_d; to this end, it requires that $\mathcal{C} = [B,\ AB,\ \ldots\ldots,\ A^{n-1}B]$ be invertible; in this case, we will have

$$\begin{pmatrix} u_{n-1} \\ \vdots \\ u_0 \end{pmatrix} = \mathcal{C}^{-1}(x_d - A^n x_0). \qquad [2.10]$$

In the multivariate case, the condition for an input u causing the desired state to exist is that the rank of $\mathcal{C}$ have to be maximal. The Kalman criterion is therefore, for the discrete case, identical to the continuous case.

2.3.2.4. *Change in basis and controllability*

THEOREM 2.4.– For any controllable system, a change in basis can be found that allows the system state–space representation to be expressed in controllable canonical or control form.

PROOF.– For the transfer function $H(p) = \frac{b_1 p^{n-1} + b_2 p^{n-2} + \ldots + b_n}{p^3 + a_1 p^2 + a_2 p + a_3}$, we show that we can obtain a state–space representation of the canonical control form:

$$\dot{X}_c = \begin{bmatrix} 0\ 0\ 0\ \ldots\ldots & -a_n \\ 1\ 0\ 0\ \ldots\ldots & -a_{n-1} \\ \vdots & \vdots \\ 0\ 0\ ..\ 0..1 & -a_1 \end{bmatrix} X_c + \begin{bmatrix} 1 \\ \vdots \\ 0 \end{bmatrix} u$$

and

$$y = \begin{bmatrix} \beta_1 & \ldots\ldots & \beta_n \end{bmatrix} X_c. \qquad [2.11]$$

Consider the system defined by its state–space representation,

$$\dot{X} = A.X + Bu(t) \qquad [2.12]$$

$$y(t) = C.X. \qquad [2.13]$$

The change in basis is achieved by $x = TX_c$ and gives $T^{-1}AT = A_c$ and $T^{-1}B = B_c$; we denote by T_i the columns of matrix T, $T = [T_1, T_2, \ldots..T_n]$. This yields $T_1 = B$, $T_2 = AB$, $T_3 = A^2B$, $T_n = A^{n-1}B$, hence $T = \mathcal{C}$. The matrix for the desired change in matrix is directly related to the controllability matrix.

It can thus be observed that the controllability matrix of the state–space representation in the canonical control form is $\mathcal{C}_c = I$, which justifies the designation of representation.

In order to obtain the controllable canonical form, the transformation for the change in state is $T = \mathcal{C}_c\mathcal{C}_{co}^{-1}$, where $\mathcal{C}_{co}$ is defined by:

$$\mathcal{C}_{co} = \begin{bmatrix} 1 \; a_1 \; a_2 \; \ldots\ldots \; a_{n-1} \\ 0 \; 1 \;\; a_1 \; \ldots\ldots \; a_{n-2} \\ \vdots \qquad\qquad\qquad \vdots \\ 0 \; 0 \;\; .. \;\; \ldots.0 \; 1 \end{bmatrix}. \qquad [2.14]$$

Decomposition of the system.

THEOREM 2.5.– For a system not completely controllable, there is an inverse transformation T such that the system can be expressed in the state form with:

$$T^{-1}AT = \begin{bmatrix} A_1 & A_2 \\ 0 & A_3 \end{bmatrix} \quad \text{and} \quad T^{-1}B = \begin{bmatrix} B_1 \\ 0 \end{bmatrix} \qquad [2.15]$$

with $A_1 \in R^r \times R^r$ and $B_1 \in R^r \times R^m$, where r is the rank of the controllability matrix and m is the number of system inputs. In this decomposition, the pair (A_1, B_1) is controllable.

PROOF.– In the state space after transformation, the system can be written as:

$$\begin{aligned} \dot{z}_1 &= A_1 z_1 + A_2 z_2 + B_1 u, \\ \dot{z}_2 &= A_3 z_2. \end{aligned} \qquad [2.16]$$

It can thus be seen that the $n - r$ states z_2 are ungovernable because they have a dynamic independent of the states z_1 (which are dependent of the inputs) and the control inputs u.

2.3.2.5. *Popov–Belevitch–Hautus controllability tests*

For the system [2.12], the characteristic polynomial is written as:

$$\begin{aligned} \det(pI - A) &= \det(pI - T^{-1}AT) = \det(pI - \begin{bmatrix} A_1 & A_2 \\ 0 & A_3 \end{bmatrix}) \\ &= \det(pI - A_1)\det(pI - A_3) \end{aligned} \qquad [2.17]$$

and the transfer function is defined by

$$\begin{aligned} H(p) &= C(pI - \begin{bmatrix} A_1 & A_2 \\ 0 & A_3 \end{bmatrix})^{-1} \begin{bmatrix} B_1 \\ 0 \end{bmatrix} \\ &= C(pI - A)^{-1}B = C(pI - A_1)^{-1}B_1. \end{aligned} \quad [2.18]$$

The transfer function does not depend on the non-controllable part of the system. As a result, we can classify the eigenvalues of A and the associated eigenvectors into two sets such that the eigenvalues associated with A_1 define the controllable modes and those associated with A_3 correspond to non-controllable modes. The Popov–Belevitch–Hautus theorem that follows allows us to make a distinction between controllable and non-controllable modes.

THEOREM 2.6.– PBH controllability test. A pair (A, B) is controllable if and only if there are no eigenvectors (on the left side) v of A such that $v^T B = 0$, that is to say that

$$\forall p \in C, \text{rank}(\begin{bmatrix} pI - A & B \end{bmatrix}) = n, \quad [2.19]$$

Otherwise, any couple eigenvalue–eigenvector (λ, v) of $C^{n \times n}$ such that

$$A^T v = \lambda v, \;\; v^T B = 0 \text{ and } \; v \neq 0 \quad [2.20]$$

is called non-controllable (ungovernable) mode of the system.

PROOF.– Let λ be an eigenvalue of matrix A of the state–space representation of the system and v a left eigenvector associated with this eigenvalue: $A^T v = \lambda v, \;\; \text{with} \;\; v^T v = 1.$

For any trajectory of the system, we can write:

$$\frac{d}{dt}(v^T x) = v^T(Ax + Bu) = \lambda(v^T x) + (v^T B)u \quad [2.21]$$

and if $v^T B = 0$, then $v^T x = e^{\lambda t}(v^T x(0)), \quad \forall u.$

This means that the component of state x that is in the direction of v cannot be controlled if $v^T B = 0$. The attainable subspace is therefore smaller than R^n.

Conversely, consider a system not completely controllable, there is then a state–space representation such that

$$T^{-1}AT = \begin{bmatrix} A_1 & A_2 \\ 0 & A_3 \end{bmatrix} \quad \text{and} \quad T^{-1}B = \begin{bmatrix} B_1 \\ 0 \end{bmatrix}. \qquad [2.22]$$

Then, consider a pair eigenvalue–eigenvector (λ, v) in $C^{n \times n}$ such that:

$$v^T A_3 = \lambda v^T \quad \text{and} \quad v \neq 0. \qquad [2.23]$$

It can be observed that the vector $w = T^{-T} \begin{bmatrix} 0 \\ v \end{bmatrix}$ verifies

$$w^T A = \lambda w^T \quad \text{and} \quad w^T B = 0. \qquad [2.24]$$

2.3.3. *Observability*

2.3.3.1. *Definitions*

DEFINITION 2.4.– *The unobservable subspace, denoted $\mathcal{I} \subset R^n$, is the set of states x_i such that the solution of the differential equation $\dot{x} = Ax$, $y = Cx$ with the initial condition $x = x_i$ yields the output $y(t) = 0$.*

In other words, the initial condition $x(0)$ can be determined from the knowledge of output $y(t)$ only, if the initial state $x(0)$ is located in the unobservable subspace. For an observable system, the unobservable subspace is reduced to $\mathcal{I} = \{0\}$.

DEFINITION 2.5.– *A system is said to be observable (completely) if for any vector x of R^n, the knowledge only of input $u(t)$ and output $y(t)$ is enough to determine the initial state $x(0)$.*

2.3.3.2. *Kalman observability criterion*

THEOREM 2.7.– The unobservable subspace is the kernel of the so-called observability matrix and defined as follows:

$$\mathcal{O} = \begin{bmatrix} C \\ CA \\ \vdots \\ CA^{n-1}. \end{bmatrix}. \qquad [2.25]$$

A system is said to be (completely) observable if and only if the rank of the observability matrix is n. The observability matrix is said to have full rank.

PROOF.– Consider the system with one zero input $u(t) = 0$. The output of the system is thus given by $y(t) = Ce^{At}x(0)$. According to Cayley–Hamilton's theorem, we have

$$\forall t \geq 0,\ Ce^{At}x(0) = \sum_{i=0}^{\infty} CA^i x(0)\frac{t^i}{i!} = 0 \quad \Leftrightarrow \quad \mathcal{O}x(0) = 0. \qquad [2.26]$$

It should be noted that the observability property is dependent only on the pair (A, C). The pair (A, C) is said to be observable if the observability matrix $\mathcal{O}$ is of full rank.

DEFINITION 2.6.– *The system is observable if regardless of the initial state $x(0)$ it is possible to determine $x(0)$ from $u(t)$ and $y(t)$ with $0 \leq t \leq \tau$.*

The system is observable if the observability matrix: $Obs = [C^T, A^T C^T, \ldots, A^{T(n-1)}C^T]$ has rank n. In the case of a system expressed in diagonal form, it is sufficient that all elements of vector C be non-zero.

PROPOSITION 2.1.– *The system is observable if and only if the symmetric matrix*

$O(T) = \int_0^T e^{sA^T} C^T C e^{As} ds$ *is invertible (or positive definite) for at least* $T > 0$.

PROOF.–

$$y(t) = Ce^{At}x_o + \int_0^t Ce^{A(t-\tau)}Bu(\tau)d\tau \qquad [2.27]$$

from which

$$Ce^{At}x_o = y(t) - \int_0^t Ce^{A(t-\tau)}Bu(\tau)d\tau \qquad [2.28]$$

by multiplying on the left hand side by $e^{tA^T}C^T$ and by integrating from 0 to $T > 0$, we get the vector

$$O(T)x_o = \int_0^T e^{sA^T} C^T C e^{As} ds.x_o = f(y,(t),u(t)), \qquad [2.29]$$

therefore if $O(T)$ is invertible, the initial state x_o can be deducted.

Conversely, if $O(T)$ is not invertible for a certain T, we can consider a vector v corresponding to the zero eigenvalue such that

$$Ce^{At}v = 0 \Rightarrow Ce^{At}(v + x_o) = Ce^{At}x_o. \qquad [2.30]$$

It simply means that states x_o and $x_o + v$ are indistinguishable and thus the system is not fully observable.

EXAMPLE 2.3.– *Let the system have the state–space representation:* $\dot{X} = \begin{bmatrix} 1 & 0 \\ -2 & -1 \end{bmatrix} X + \begin{bmatrix} 1 \\ 0 \end{bmatrix} u$ *and* $y = \begin{bmatrix} 1 & 1 \end{bmatrix} X$.

Matrix A has two eigenvalues $\lambda_1 = 1$ *and* $\lambda_2 = -1$. *For the eigenvalue* $\lambda_1 = 1$, *the corresponding eigenvector is* $v_1 = \begin{pmatrix} 1 \\ -1 \end{pmatrix}$. *The system has a controllability matrix* $\mathcal{O} = \begin{pmatrix} C \\ CA \end{pmatrix} = \begin{pmatrix} 1 & 1 \\ -2 & -2 \end{pmatrix}$, *which is a matrix of rank* 1. *The unobservability subspace corresponding to it is* $\mathcal{I} = \{v$ *such that* $v = \alpha \begin{pmatrix} 1 \\ -1 \end{pmatrix}$ *and* $\alpha \in R\}$. *The system is therefore not completely observable. If we consider the initial state* $x(0) = \begin{pmatrix} 1 \\ -1 \end{pmatrix}$ *and zero input* $u(t) = 0$, *we thus have* $x(t) = e^{\lambda_1 t}\begin{pmatrix} 1 \\ -1 \end{pmatrix} = e^t\begin{pmatrix} 1 \\ -1 \end{pmatrix}$ *and the output remains zero all the time because* $y(t) = x_1(t) + x_2(t) = 0$. *The corresponding mode* e^t *is a hidden mode because it is unobservable on output.*

2.3.3.3. *Discrete system observability*

Consider a discrete univariate system defined by its state–space representation

$$x_{k+1} = Ax_k + Bu_k, y_k = Cx_k. \qquad [2.31]$$

We define an observer for this system to rebuild the state using the following equations:

$$\begin{aligned} y_0 &= Cx_0, \\ y_1 &= Cx_1 = CAx_0 + CBu_0, \\ y_2 &= Cx_2 = CAx_1 + CBu_1 = CA^2x_0 + CABu_0 + CBu_1. \end{aligned} \qquad [2.32]$$

We obtain

$$\begin{pmatrix} y_0 \\ y_1 \\ y_2 \end{pmatrix} = \begin{pmatrix} C \\ CA \\ CA^2 \end{pmatrix} x_0 + \Gamma \begin{pmatrix} u_0 \\ u_1 \\ u_2 \end{pmatrix} \qquad [2.33]$$

with

$$\Gamma = \begin{bmatrix} 0 & 0 & 0 \\ CB & 0 & 0 \\ CAB & CB & 0 \end{bmatrix}. \qquad [2.34]$$

If the observability matrix $\mathcal{O} = \begin{pmatrix} C \\ CA \\ CA^2 \end{pmatrix}$ is invertible, then we can derive from the initial state of the previous equation:

$$x_0 = \mathcal{O}^{-1} \left(\begin{pmatrix} y_0 \\ y_1 \\ y_2 \end{pmatrix} - \Gamma \begin{pmatrix} u_0 \\ u_1 \\ u_2 \end{pmatrix} \right). \qquad [2.35]$$

For any time k, it can be written as:

$$x_{k+1} = A^k x_0 + \sum_{i=0}^{k-1} A^{k-i-1} B u_i, \qquad [2.36]$$

$$x_{k+1} = A^k \mathcal{O}^{-1} \left(\begin{pmatrix} y_0 \\ y_1 \\ y_2 \end{pmatrix} - \Gamma \begin{pmatrix} u_0 \\ u_1 \\ u_2 \end{pmatrix} \right) + \sum_{i=0}^{k-1} A^{k-i-1} B u_i. \qquad [2.37]$$

2.3.3.4. *Change in basis and observability*

THEOREM 2.8.– For any observable system, there is a change in basis allowing the system to be expressed in observable canonical form or observability.

This theorem is obtained by duality of the case of a controllable system.

Decomposition of the system and extraction of the unobservable part.

THEOREM 2.9.– For a system (C, A) not completely observable, there is an inverse transformation T such that the system can be expressed in state form with:

$$T^{-1}AT = \begin{bmatrix} A_1 & A_2 \\ 0 & A_3 \end{bmatrix} \quad \text{and} \quad CT = [0 \;\; C_2] \qquad [2.38]$$

with $A_1 \in R^r \times R^r$ and $C_2 \in R^p \times R^r$, where r is the rank of the controllability matrix and p is the number of outputs of the system. In this decomposition, the pair (A_3, C_2) is observable.

In the state space after transformation, the system can be written as:

$$\begin{aligned} \dot{z}_1 &= A_1 z_1 + A_2 z_2 + Bu \\ \dot{z}_2 &= A_3 z_2 \\ y &= C_2 z_2. \end{aligned} \qquad [2.39]$$

It can thus be observed that the r states z_1 are unobservable because they do not influence the output and are independent of the states z_2 (which are the only contributions to the output).

2.3.3.5. *Popov–Belevitch–Hautus observability tests*

After change in state defined by [2.38], for the system [2.1], the characteristic polynomial can be written as:

$$\begin{aligned} \det(pI - A) &= \det(pI - T^{-1}AT) = \det(pI - \begin{bmatrix} A_1 & A_2 \\ 0 & A_3 \end{bmatrix}) \\ &= \det(pI - A_1)\det(pI - A_3) \end{aligned} \qquad [2.40]$$

and the transfer function is defined by

$$H(p) = [0 \;\; C_2]\,(pI - \begin{bmatrix} A_1 & A_2 \\ 0 & A_3 \end{bmatrix})^{-1}B = C_2(pI - A_3)^{-1}B. \qquad [2.41]$$

The transfer function does not depend on the non-observable part of the system. We can thus classify the eigenvalues of A and the associated eigenvectors into two subsets such that the eigenvalues associated with A_1 define the unobservable modes and those associated with A_3 correspond to the observable modes. The Popov–Belevitch–Hautus theorem that follows allows for the distinction between observable and unobservable modes to be made.

THEOREM 2.10.– PBH observability test. A pair (A, C) is observable if and only if there are no (right) eigenvectors v of A such that: $Cv = 0$, that is to say that

$$\forall p \in C, \text{rank}(\begin{bmatrix} C \\ pI - A \end{bmatrix}) = n. \quad [2.42]$$

Conversely, any couple eigenvalue–eigenvector (λ, v) of $C^{n \times n}$ such that:

$$Av = \lambda v, \; Cv = 0 \text{ and } v \neq 0 \quad [2.43]$$

is called unobservable mode of the system.

2.3.4. *Observer*

Knowing the properties of observability above and acknowledging that sometimes it may prove expensive or even impossible to make use of sensors to measure the complete state of a system, it might be rewarding to know if from the knowledge of the system output or part of the state, the complete state vector can be reconstructed Such a reconstructor of the state is called observer or state estimator.

DEFINITION 2.7.– *An observer designates a dynamic system having inputs $y(t)$ and $u(t)$ and on output the system state $x(t)$. For an observable system, it allows rebuilding the entire state at each time t.*

Consider the univariate system

$$y^{(3)}(t) + a_1\ddot{y}(t) + a_2\dot{y}(t) + a_3 y(t) = b_1\ddot{u}(t) + b_2\dot{u}(t) + b_3 u(t). \quad [2.44]$$

Let (A, B, C) be a realization of this system

$$\dot{x} = Ax + Bu(t) \quad , \; y(t) - Cx(t). \quad [2.45]$$

Can the system state $x(t)$ be determined assuming functions $y(t)$ and $u(t)$ as known?

Successively deriving output $y(t) = Cx(t)$, we obtain $\dot{y}(t) = C\dot{x}(t) = CAx(t) + CBu(t)$ and $\ddot{y}(t) = CA\dot{x}(t) + CB\dot{u}(t) = CA^2x(t) + CABu(t) + CB\dot{u}(t)$, we thus obtain the equation

$$\begin{bmatrix} y(t) \\ \dot{y}(t) \\ \ddot{y}(t) \end{bmatrix} = \begin{bmatrix} C \\ CA \\ CA^2 \end{bmatrix} x(t) + \Gamma \begin{bmatrix} u(t) \\ \dot{u}(t) \\ \ddot{u}(t). \end{bmatrix} \quad [2.46]$$

where Γ is a Toeplitz matrix[1]:

$$\Gamma = \begin{bmatrix} 0 & 0 & 0 \\ CB & 0 & 0 \\ CAB & CB & 0. \end{bmatrix}. \qquad [2.47]$$

Consequently, if $\mathcal{O}$ is invertible, the system state can be reconstructed:

$$x(t) = \mathcal{O}^{-1} \left(\begin{bmatrix} y(t) \\ \dot{y}(t) \\ \ddot{y}(t) \end{bmatrix} \_\Gamma \begin{bmatrix} u(t) \\ \dot{u}(t) \\ \ddot{u}(t). \end{bmatrix} \right). \qquad [2.48]$$

The observer thus obtained is called differential observer, because it uses derivators for the reconstruction of the system state. As a result, its use in practice is avoided because the derivation amplifies noises at high frequencies.

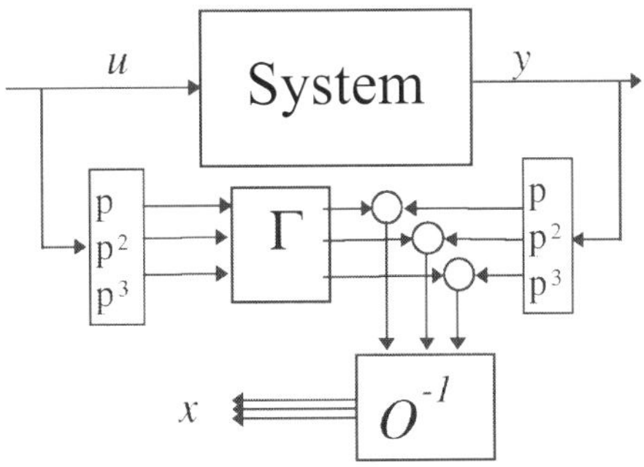

Figure 2.3. *Differential observer*

2.3.5. *Observer for state reconstruction*

DEFINITION 2.8.– *A linear asymptotic observer is a dynamic system of the form:*

$$\begin{cases} \dot{\hat{x}} = A\hat{x} + Bu + L(y - \hat{y}) = A\hat{x} + Bu + L(y - C\hat{x}) \\ \hat{y} = C\hat{x} \end{cases} \qquad [2.49]$$

where L is such that $\hat{x}$ tends to x when t tends toward infinity.

1 In a Toeplitz matrix, the terms of a same diagonal are equal.

Consider the case of a univariate observable system in canonical observability form (or companion observable):

$$\dot{X} = \begin{bmatrix} 0\ 0\ 0\ \ldots\ldots & -a_n \\ 1\ 0\ 0\ \ldots\ldots & -a_{n-1} \\ \vdots & \vdots \\ 0\ 0\ ..\ 0..1 & -a_1 \end{bmatrix} X + \begin{bmatrix} b_1 \\ \vdots \\ b_n \end{bmatrix} u$$

and

$$y = \begin{bmatrix} 0 & \ldots\ldots 0 & 1 \end{bmatrix} X_c. \quad [2.50]$$

THEOREM 2.11.– It is possible to build an observable univariate system, an asymptotic observer whose poles can be placed arbitrarily.

Justification. The state observation error is denoted as $e = x - \hat{x}$; it is derived by subtracting the two previous equations. We then obtain:

$$\dot{\hat{x}} = A\hat{x} + Bu + L(y - \hat{y}) = (A - LC)\hat{x} + Bu + Ly$$

$$\dot{x} = A.x + Bu,$$

$$\dot{e} = (\dot{x} - \dot{\hat{x}}) = (A - LC)e, \quad [2.51]$$

$$A - LC = \begin{bmatrix} 0\ 0\ 0\ \ldots\ldots & -l_n - a_n \\ 1\ 0\ 0\ \ldots\ldots & -l_{n-1} - a_{n-1} \\ \vdots & \vdots \\ 0\ 0\ ..\ 0..1 & -l_1 - a_1 \end{bmatrix} \quad [2.52]$$

The last equation shows that the choice of the correction gain of the observer L (l_m, for $1 \leq m \leq n$) allows us to arbitrarily define the poles of the observer and to obtain the asymptotic convergence of the observed state toward the actual state of the system. The eigenvalues of $(A - LC)$ can be arbitrarily chosen for e to asymptotically tend toward zero:

$$Det(\lambda I - A + LC) = P(\lambda) = 0 \quad [2.53]$$

$$P(\lambda) = Det\left(\lambda I - \begin{bmatrix} \lambda & 0\ 0\ \ldots\ldots & l_n + a_n \\ -1 & \lambda\ 0\ \ldots\ldots & l_{n-1} + a_{n-1} \\ \vdots & \lambda & \vdots \\ 0 & 0\ ..\ 0..-1 & l_1 + a_1 \end{bmatrix}\right). \quad [2.54]$$

$P(\lambda) = \lambda^n + (l_1 + a_1)\lambda^{n-1} + (l_2 + a_2)\lambda^{n-2} + ... + (l_{n-1} + a_{n-1})\lambda + (l_n + a_n)$ from which we obtain L.

In the general case, $\dot{z} = Fz + Gu$ and $y = Hz$. If the system is observable, it can be transformed ($x = T^{-1}z$) to express it in observable canonical form, for example the [2.50] form. We will have $\dot{z} = T\dot{x} = FTz + Gu \Rightarrow \dot{x} = T^{-1}FTx + T^{-1}Gu$ and $y = Hz = HTx$,

$A = T^{-1}FTx$, $B = T^{-1}G$ and $C = HT$

This allows us to define an observer for the canonical form as previously with: $\dot{e} = (\dot{x} - \dot{\hat{x}}) = (A - LC)e = T^{-1}\dot{\varepsilon}$.

Let $L = T^{-1}\bar{L}$, using transformation we can obtain an observer for the original state–space representation because it can be shown that $A - LC = T^{-1}(F - \bar{L}H)T$ and that the characteristic polynomial feature is unchanged: $Det(\lambda I - A + LC) = Det(\lambda T^{-1}T - T^{-1}(F - \bar{L}H)T) = P(\lambda)$,

$P(\lambda) = Det(T^{-1})Det(\lambda I - F + \bar{L}H)Det(T) = Det(\lambda I - F + \bar{L}H)$,

from which $\dot{\varepsilon} = (\dot{z} - \dot{\hat{z}}) = T(\dot{x} - \dot{\hat{x}}) = T(A - LC)T^{-1}\varepsilon = (TAT^{-1} - TLCT^{-1})\varepsilon = (F - TLH)\varepsilon \Rightarrow \bar{L} = TL$ is the observer gain for the original state–space representation.

EXAMPLE 2.4.– *Study of an observer. Let the system be*

$$\dot{x} = \begin{bmatrix} 1 & 2 & 0 \\ 3 & -1 & 1 \\ 0 & 2 & 0 \end{bmatrix} x + \begin{bmatrix} 2 \\ 1 \\ 1 \end{bmatrix} u \text{ and } y = \begin{bmatrix} 0 & 0 & 1 \end{bmatrix} x. \qquad [2.55]$$

Expressing the observable canonical form gives:

$$\dot{z} = \begin{bmatrix} 0 & 0 & -2 \\ 1 & 0 & 9 \\ 0 & 1 & 0 \end{bmatrix} z + \begin{bmatrix} 3 \\ 2 \\ 1 \end{bmatrix} u \text{ and } y = \begin{bmatrix} 0 & 0 & 1 \end{bmatrix} z. \qquad [2.56]$$

The transformation is defined by $T = \begin{bmatrix} 6 & -2 & -7 \\ 0 & 2 & 0 \\ 0 & 0 & 1 \end{bmatrix}$, $T^{-1} = \begin{bmatrix} \frac{1}{6} & \frac{1}{6} & \frac{7}{6} \\ 0 & \frac{1}{2} & 0 \\ 0 & 0 & 1 \end{bmatrix}$, $A = T^{-1}FTx$, *that is the characteristic polynomial* $P(\lambda) = (\lambda + 3)(\lambda + 4)(\lambda + 5) = \lambda^3 + 12\lambda^2 + 47\lambda + 60 = 0$. *In order for the dynamic of the observation error to have the eigenvalues -3, -4 and -5, we impose:* $P(\lambda) = \lambda^3 + (l_1 + a_1)\lambda^2 + (l_2 + a_2)\lambda + (l_3 + a_3) = \lambda^3 + 12\lambda^2 + 47\lambda + 60$ *where we obtain* $L^T = (58, 56, 12)$ *and thus*

$$\bar{L} = TL = \begin{bmatrix} 6 & -2 & -7 \\ 0 & 2 & 0 \\ 0 & 0 & 1 \end{bmatrix} \begin{bmatrix} 58 \\ 56 \\ 12 \end{bmatrix} = \begin{bmatrix} 152 \\ 112 \\ 12 \end{bmatrix} \qquad [2.57]$$

or still

$$\dot{\hat{x}} = \begin{bmatrix} 1 & 2 & 0 \\ 3 & -1 & 1 \\ 0 & 2 & 0 \end{bmatrix} \hat{x} + \begin{bmatrix} 2 \\ 1 \\ 1 \end{bmatrix} u + \begin{bmatrix} 152 \\ 112 \\ 12 \end{bmatrix} (y - \hat{y}) \text{ and } \hat{y} = \begin{bmatrix} 0 & 0 & 1 \end{bmatrix} \hat{x} \qquad [2.58]$$

which is equivalent by means of transformation T *to the dynamic system:*

$$\dot{\hat{z}} = \begin{bmatrix} 0 & 0 & -60 \\ 1 & 0 & -47 \\ 0 & 1 & -12 \end{bmatrix} \hat{z} + \begin{bmatrix} 58 \\ 56 \\ 12 \end{bmatrix} y + \begin{bmatrix} 3 \\ 2 \\ 1 \end{bmatrix} u \text{ and } \hat{y} = \begin{bmatrix} 0 & 0 & 1 \end{bmatrix} \hat{z}. \qquad [2.59]$$

EXAMPLE 2.5.– *Study of an observer for the actuator model*

The purpose of this observer is to rebuild the components of the state vector of the representation previously seen. Both reconstituted quantities will be the couple τ and the auxiliary variable $W = \tau + \ddot{A}\ddot{A}q$. Although the reconstitution of the couple τ be unnecessary because it is a measurable quantity, this observer will have the advantage of being able to reconstruct quantity W, and therefore quantity τ because the speed q is supposed to be measurable. For the model of the actuator, we have a representation of the type:

$$\dot{X} = AX + BU \text{ and } Y = CX. \qquad [2.60]$$

This system is said observable if its observability matrix W_0 is of full rank.

In the case of the actuator model proposed by Kircanski, we have the observability matrix being the identity matrix; it is obviously of full rank. The system is therefore completely observable. The state vector is estimated by: $Y = CX$, where e is the estimation error,

$$e = Y - \hat{Y} = Y - C\hat{X}. \qquad [2.61]$$

Equation [2.60] can then be written in the form:

$$X = (A - KC)X + BU + KY. \qquad [2.62]$$

The observer can be defined as follows: let X be the estimation error of the state vector, deducting equations [2.61] and [2.62], we get:

$$X = (A - KC)X. \qquad [2.63]$$

The last step is to find the values $k1$ and $k2$ components of vector K that cause the rejection of the observer error to be faster than the system to be observed.

To calculate the dynamics of such a system, the characteristic equation must be solved: $det(A - lI) = 0$. In our example above, the characteristic equation is written as: $l + 45.88l + 17,400 = 0$.

By identifying this equation to the second-order characteristic equation, that is $l + 2xwl + w = 0$, we get: $x = 0.1739$. For the observer, the characteristic equation is $det((A - KC) - lI) = 0$. The characteristic equation is given by: $l + (45.88 + k1)l + (17,400 + k2) = 0$. It is desirable that the previous equation be the second-order characteristic equation with a fast dynamic $(x\prime = 0.707)$ and the same natural frequency w as the actuator. We therefore identify the equation $l + 2x\prime wl + w = 0$, which gives us $k1 = 140.6k2 = -6450$.

The full equation of the observer has the couple τ and the auxiliary variable $W = \tau + \ddot{A}\ddot{A}q$ for components. Although the reconstitution of the couple τ is unnecessary because it is a measurable quantity, this observer will have the advantage of being able to rebuild the quantity W and therefore the quantity τ because the velocity q is supposed to be measurable. For the actuator model, we have a representation of the type:

$$X = AX + BU, \qquad Y = CX. \qquad [2.64]$$

EXAMPLE 2.6.– *Study of an observer for the two-stage model.*

Observer: from what has been previously seen, in order for the passive control previously exposed to be implementable, it is necessary that one be able to have access to quantities z and $\hat{z}$, which are not measurable. It is therefore necessary to design a state observer that reconstructs these quantities.

Two-stage modeling can be written in the following form:

$$\tau = JKz + BKz + Bq + Jq \qquad [2.65]$$

$$z = M(q)q + C(q, \hat{q})q + G. \qquad [2.66]$$

The following state–space representation is then chosen to represent the system, $x = q$; $\hat{x} = \hat{q}$; $\tilde{x} = z$; $\dot{x} = z + Kq$.

This gives according to equations [2.65] and [2.66]: $x = q = \hat{x}$,

$$x = q = M(q)(z - C(q, \hat{q})q - G) = Mx + f(x) \qquad [2.67]$$

$$x = z = x - Kx \qquad [2.68]$$

$$x = z = -ax - \hat{a}x + a\tau. \qquad [2.69]$$

with:

$$a = KJBK, \hat{a} = KJ \qquad [2.70]$$

and

$$f(x) = (M(x) - M)p3p - M(x)(C(x, \hat{x})x + G(x)). \qquad [2.71]$$

The system is then written as a state–space representation:

$$y = Cx. \qquad [2.72]$$

The final step consists of choosing vector C such that the system be fully observable and that the variable y be measurable. If we choose $C = (1, 0, 0, 0)$, that is to say that only the angular position q is measured, it is shown that the system is fully observable; in other words, the observability matrix (W) is of full rank. In effect, additionally $det(W) = M$.

Since the system is observable, we can write the equation of the observer:

$$x = (A - KC)x + Bu + Ky + f(x). \qquad [2.73]$$

Let x be the observation error: $x = x - \hat{x}$. According to equations [2.72] and [2.73], we can write:

$$x = (A - KC)x + f(x) - f(x). \qquad [2.74]$$

It has to be shown now that the equation of the closed-loop (CL) system is stable and, in addition, the function $h = f(x) - \hat{f}(x)$ must be bounded.

An increase in $f(x)$:

$$\begin{aligned} h &= (M(x) - M)(x - \hat{x}) - M(x)(C(x, \hat{x})x - C(x, \hat{x})) \qquad [2.75] \\ h &= (M(x) - M)(x - \hat{x}) - M(x)(2C(x, \hat{x})x + C(x, \hat{x})x) \\ h &= (M(x) - M)x - M(x).2C(x, x + \ddot{A}\ddot{A})x. \end{aligned}$$

We can then increase the norm of h:

$$\|h\|\, \acute{o}b\, \|x\| \; + 2b\, \|x\| + \ddot{A}\ddot{A}.\, \|x\| => \|h(t)\|\, \acute{o}b\, \|x\| + b\, \|x\| \,.\, \|x\| \,. \qquad [2.76]$$

However, the norm of the velocity being bounded by m, we have:

$$\|h(t)\|\, \acute{o}b\, \|x\| + bm\, \|x\| => \|h(t)\|\, \acute{o}L\, \|x\| \,. \qquad [2.77]$$

–*System stability*:

We consider the following candidate Lyapunov function: $V = xPx$ where P is the solution of the Lyapunov equation, that is, $AP + PA = -Q$ with $A = A - KC$ and Q a positive definite symmetric matrix. By deriving the candidate Lyapunov function along the trajectories of the system (equation [2.74]), we get: $V = xPx + xPx$ and $V = -xQx + 2xPh(t)$.

Or even:

$$V\acute{o} - l(Q)\, \|x\| \; + 2\, \|x\| \; \|P\|\, L\, \|x\| \,. \qquad [2.78]$$

For the Lyapunov conditions to be verified, V has to be negative, which given equation [2.78] can be written as:

$$l(Q) > 2Ll(P). \qquad [2.79]$$

If we impose $Q = 2Id$, the condition [2.79] can be expressed in the following form.

–Expression of the observable canonical form:

Gain K has to be calculated in a way to have a given dynamic of error rejection. For this purpose, the state–space representation is transformed [2.66] and [2.72] into an observable canonical form. This transformation consists of finding a matrix T such that $z = Tx$, which transforms the state–space representation into:

$$z = Az + Bu, \tag{2.80}$$

$$y = Cx. \tag{2.81}$$

With $A = TAT$, $B = TB$, $C = CT$ and $K = TK$. Matrix T is chosen so that A be of the form:

we choose A in this form because the system characteristic equation then becomes

$det(A - lI) = d$ and $(A - lI) = l + al + al + al + a$.

Once A and C are known, the new observability matrix $W = pwwýp$ can be calculated and T can be deducted because $WW = T$. The advantage of such a transformation is that in CL, we have the characteristic equation of the CL system:

$$P(l) = l + àl + àl + àl + à \text{ with } à = a - k. \tag{2.82}$$

We can then identify to this equation the characteristic equation of a system whose transfer function would be *H*(*p*) of equation [2.41]. In other words, equation [2.82] is identified: $l + 4zwl + 2(2zw + 2w)l + 4zwl + w$.

Finally, we have:

$$k = a + 4zwk = a + 4w(z + 1)k = a + 4zwk = a + w. \tag{2.83}$$

2.3.6. *Minimal state–space representation*

After the concepts of observable, unobservable, controllable and non-controllable modes, the relationship between the eigenvalues of a realization and the poles[2] of a system can be interpreted with regard to the minimality or non-minimality of the realization and stability.

2 Roots of the denominator of the transfer function of the system.

2.3.6.1. *Definition and examples*

DEFINITION 2.9.– *A realization of the system (A, B, C) is minimal if there is no realization having the same transfer function with less states*[3].

PROPOSITION 2.2.– *A triplet (A, B, C) is minimal if and only if (A, B) is controllable and (A, C) is observable.*

PROOF.– If (A, B) is not fully controllable, then we can extract the non-controllable part and thereof derive a realization with a smaller order.

Conversely, suppose that $C(pI-A)^{-1}B = \tilde{C}(pI-\tilde{A})^{-1}\tilde{B}$ with $\dim(\tilde{A}) = r < n$, by expanding in series around $p = \infty$, we get:

$$\sum_{i=1}^{\infty} \frac{CA^{i-1}B}{p^i} = \sum_{i=1}^{\infty} \frac{\tilde{C}\tilde{A}^{i-1}\tilde{B}}{p^i}, \qquad [2.84]$$

hence we deduce that $CA^{i-1}B = \tilde{C}\tilde{A}^{i-1}\tilde{B}$ for $i \geq 1$.

Therefore, we finally conclude that

$$\mathcal{O}.\mathcal{C} = \begin{bmatrix} \tilde{C} \\ \tilde{C}\tilde{A} \\ \vdots \\ \tilde{C}\tilde{A}^{n-1} \end{bmatrix} \begin{bmatrix} \tilde{B} \\ \tilde{A}\tilde{B} \\ \vdots \\ \tilde{A}^{n-1}\tilde{B} \end{bmatrix} \qquad [2.85]$$

and that rank($\mathcal{O}.\mathcal{C}$)$\leq r < n$, which implies that rank($\mathcal{O}$)$< n$ or rank($\mathcal{C}$)$< n$.

EXAMPLE 2.7.– *Consider the following state–space representation:*

$$\dot{X} = \begin{bmatrix} 0 & 1 \\ 1 & 0 \end{bmatrix} X + \begin{bmatrix} 1 \\ 0 \end{bmatrix} u \textit{ and } y = \begin{bmatrix} 1 & -1 \end{bmatrix} X. \qquad [2.86]$$

It can be shown that it corresponds to the transfer function

$$H(p) = \frac{p-1}{p^2-1} = \frac{p-1}{(p-1)(p+1)} = \frac{1}{p+1}. \qquad [2.87]$$

3 State vector of lower dimension.

This state–space representation is therefore not minimal because it is possible to find another representation having a state vector of dimension 1. *It can be verified that it is not observable.*

This same system can be represented by the form: $\dot{X} = -X + u$, $y = X$, *which is a minimal form corresponding to the same transfer function.*

2.3.6.2. *Modes stability and minimality*

THEOREM 2.12.– A system with the state–space representation (A, B, C) is bounded input bounded output (BIBO) stable if and only if the observable and controllable eigenvalues are stable.

For a minimal system, there is no difference between BIBO stability and the stability of all of these natural modes[4].

EXAMPLE 2.8.– *Let the system having the state–space representation be*

$$\dot{X} = \begin{bmatrix} 1 & 0 \\ -2 & -1 \end{bmatrix} X + \begin{bmatrix} 1 \\ 0 \end{bmatrix} u \text{ and } y = \begin{bmatrix} 1 & 1 \end{bmatrix} X. \qquad [2.88]$$

Matrix A *has two eigenvalues* $\lambda_1 = 1$ *and* $\lambda_2 = -1$. *This system has the transfer function* $H(p) = \frac{1}{p+1}$, *which has a single stable pole.*

On the other hand, the realization (A, B, C) *has an unstable natural mode* $(\lambda_1 = 1)$, $e^t \begin{bmatrix} 1 \\ -1 \end{bmatrix}$, *which do not appear on output because it is unobservable.*

One can thus distinguish:

– the states not necessary to the input–output description of the system;

– the unobservable modes and non-controllable modes of the poles of the system;

– the eigenvalues of A that are not poles of the system transfer function $H(p)$

4 Defined by the eigenvalues of the realization being considered.

2.4. State feedback, control by poles placement and stability

2.4.1. *State feedback control*

In the following, we consider the system $y = S(u)$, of order n, controllable and defined by its state–space representation (A, B, C, D) of transfer function $H(p)$ with m inputs and p outputs,

$$\dot{x} = A.x + Bu(t) \quad , \quad x(0) = 0. \qquad [2.89]$$

DEFINITION 2.10.– *State feedback regulation consists of using a proportional feedback in the state $u = -Kx$ where $K \in R^{m \times n}$ is the matrix feedback gain (m is the number of inputs of the system).*

The system in the CL system with this control is written as: $\dot{x} = (A - BK)x$.

Therefore, through the choice of the values of the components of K, we can impose the eigenvalues of matrix $A - KB$ to specify the dynamics of the state of the loop system.

DEFINITION 2.11.– *State feedback control consists of using a law of control proportional feedback in the state $u = v - Kx$ where $K \in R^{m \times n}$ is the feedback gain matrix (m is the number of inputs of the system).*

$$\dot{x} = (A - BK).x + Bv, \qquad [2.90]$$

$$y = (C + DK).x + Dv. \qquad [2.91]$$

In general, for systems that come across in practice, $D = 0$.

EXAMPLE 2.9.– *Let the unstable system defined by its state–space representation be: $\dot{x} = x + u$. We apply a feedback state $u = -kx$ to this system. We obtain in the CL system$\dot{x} = (1-k)x$. The CL system is stable if $k > 1$ and it is all the more faster than k is great. For any initial state $x(0) = 1$, its response is $x(t) = x(0)e^{(1-k)t}$ for $t > 0$. The resulting input is $u(t) = -k.x(0)e^{(1-k)t}$. Its amplitude, in the neighborhood of 0 increases with k. For large gains, the input tends toward an impulse control $u(t) = \delta(t)$ to bring the state to zero.*

The choice of the control gain k must be done such that to achieve a compromise between:

– a fast system with eigenvalues having negative real parts large in absolute value; possibly large amplitudes of certain states and the control input;

– smaller gain values; stable eigenvalues smaller in absolute values, smaller control amplitude, a slower system and less sensitive to noise.

2.4.2. *Poles placement and stabilizability*

The control law defined above and represented below makes it possible to ensure the stabilization of the CL system and to guarantee a level of performance.

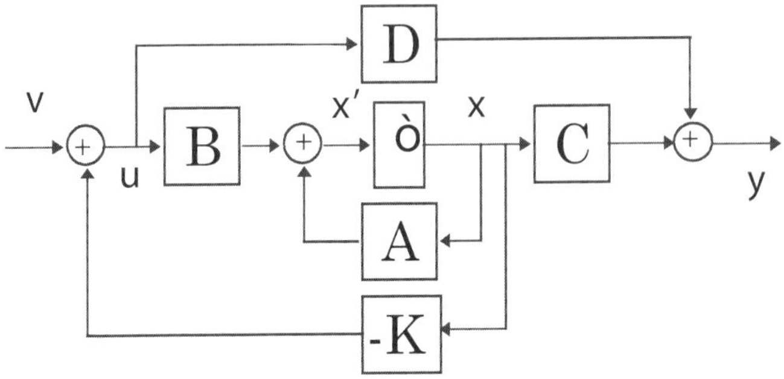

Figure 2.4. *System with state feedback*

EXAMPLE 2.10.– *Arbitrary placement of eigenvalues. Consider the unstable system defined by:*

$$\dot{x} = \begin{bmatrix} 1 & 1 \\ 1 & 2 \end{bmatrix} .x + \begin{bmatrix} 1 \\ 0 \end{bmatrix} u. \qquad [2.92]$$

Its characteristic equation is $\det(pI - A) = p^2 - 3p + 1 = (p - 2.618)(p - 0.382) = 0.$

The two poles of the system are unstable. Consider the control law

$$u = v - [k_1, k_2].x \qquad [2.93]$$

$$A - BK = \begin{bmatrix} 1 & 1 \\ 1 & 2 \end{bmatrix} - \begin{bmatrix} 1 \\ 0 \end{bmatrix} [k_1, k_2] = \begin{bmatrix} 1 - k_1 & 1 - k_2 \\ 1 & 2 \end{bmatrix}, \qquad [2.94]$$

the system in the CL system becomes:

$$\dot{x} = \begin{bmatrix} 1 - k_1 & 1 - k_2 \\ 1 & 2 \end{bmatrix} .x + \begin{bmatrix} 1 \\ 0 \end{bmatrix} v. \qquad [2.95]$$

The characteristic polynomial of the CL system is

$$\det(pI - A + BK) = p^2 - (3 - k_1)p + (1 + k_2 - 2k_1) = 0. \qquad [2.96]$$

By appropriately choosing k_1 and k_2, we can make the system stable and arbitrarily choose the eigenvalues of the system in CL. For instance, the poles of the CL system can be placed in $\lambda_1 = -2$ and $\lambda_2 = -1$, by imposing $P(p) = (p+2)(p+1) = p^2 - (3 - k_1)p + (1 + k_2 - 2k_1)$, which yields for gain values $k_1 = 6$ and $k_2 = 13$.

It is always possible to place the system eigenvalues arbitrarily as long as the λ_i be complex conjugate two-by-two (because matrix $A - BK$ is real) and the system be fully controllable.

EXAMPLE 2.11.– *Let the system be*

$$\dot{x} = \begin{bmatrix} 1 & 1 \\ 0 & 2 \end{bmatrix} .x + \begin{bmatrix} 1 \\ 0 \end{bmatrix} u. \qquad [2.97]$$

We apply the control law

$$u = v - [k_1, k_2].x \qquad [2.98]$$

$$A - BK = \begin{bmatrix} 1 & 1 \\ 0 & -2 \end{bmatrix} - \begin{bmatrix} 1 \\ 0 \end{bmatrix} [k_1, k_2] = \begin{bmatrix} 1 - k_1 & 1 - k_2 \\ 0 & 2 \end{bmatrix}, \qquad [2.99]$$

the system in CL becomes:

$$\dot{x} = \begin{bmatrix} 1 - k_1 & 1 - k_2 \\ 0 & -2 \end{bmatrix} .x + \begin{bmatrix} 1 \\ 0 \end{bmatrix} v. \qquad [2.100]$$

The characteristic polynomial of the CL system is

$$\det(pI - A + BK) = (p - 1 + k_1)(p + 2). \qquad [2.101]$$

It can be seen that in this case the pole $p = -2$ is fixed and cannot be arbitrarily displaced; it corresponds to a non-controllable mode of the system.

THEOREM 2.13.– If the pair (A, B) is fully controllable, then for any monic polynomial[5] $P_d(p)$, of degree n, there is a matrix $K \in R^{m\times n}$ such that $\det(pI - A + BK) = P_d(p)$.

THEOREM 2.14.– If the pair (A, B) is not fully controllable, then the polynomial obtained in CL with state feedback is such that $\det(pI - A + BK) = P_{nc}(p)P_c(p)$, where $P_{nc}(p)$ is the characteristic polynomial of the non-controllable (ungovernable) part of the system and $P_c(p)$ is that, arbitrary, placed for the controllable part. More particularly, there are polynomials that cannot be assigned by state feedback.

PROOF IN THE UNIVARIATE CASE.– Let $P(p) = \det(pI - A + BK) = p^n + a_1p^{n-1} + ... + a_{n-1}p + a_n$.

We assume that the pair (A, B) is fully controllable and express the system in the canonical form of the control, that is to say that there is an invertible transformation T such that

$$A_o = T^{-1}AT = \begin{bmatrix} -a_1 & -a_2 & \cdots\cdots & -a_n \\ 1 & 0 & & 0 \\ . & \vdots & \vdots & \vdots \\ 0 & \cdots\cdots & 1 & 0 \end{bmatrix}, \qquad [2.102]$$

$$T^{-1}B = b_c = \begin{bmatrix} 1 \\ \vdots \\ 0 \end{bmatrix}. \qquad [2.103]$$

Let $k_c = KT = \begin{bmatrix} k_{c1} & \cdots\cdots & k_{cn} \end{bmatrix}$, we thus have $P(p) = \det(pI - A + BK) = \det(pI - A_c + b_ck_c)$,

$$P(p) = \det \begin{bmatrix} k_{c1} - a_1 & k_{c2} - a_2 & \cdots\cdots & k_{cn} - a_n \\ 1 & 0 & & 0 \\ \vdots & . & \vdots & \vdots \\ 0 & \cdots\cdots & 1 & 0 \end{bmatrix}, \qquad [2.104]$$

$$P(p) = p^n + (a_1 - k_{c1})p^{n-1} + ... + (a_{n-1} - k_{cn-1})p + (a_n - k_{cn})).$$

This result shows that if the system is completely controllable, it is possible to arbitrarily place all the poles of the CL system by a judicious choice of the components of the gain K.

5 A polynomial is said to be monic if its coefficient (a_n) of the highest degree in p is equal to 1.

If the pair (A, B) is not controllable and $rank(\mathcal{C})$ $=rank[B, AB, ..., A^{n-1}B] = r < n$, a change in basis of the state space can be found, such that

$$\tilde{A} = T^{-1}AT = \begin{bmatrix} A_1 & A_2 \\ 0 & A_3 \end{bmatrix} \text{ and } \tilde{b} = T^{-1}B = \begin{bmatrix} b_1 \\ 0 \end{bmatrix}, \qquad [2.105]$$

with (A_1, b_1) controllable and A_1 of dimension r.

By defining $\tilde{k} = KT = [k_1, k_2]$ with k_1 of dimension r and k_2 of dimension $n - r$, we obtain $P(p) = \det(pI - A + BK) = \det(pI - \tilde{A} + \tilde{b}\tilde{k}) = \det(pI - A_1 + b_1k_1)\det(pI - A_3)$.

It can be observed that the polynomial $\det(pI - A_3)$ corresponding to the non-controllable modes is established; however, we can arbitrarily place the poles or controllable eigenvalues.

REMARK 2.1.– *In general, there are different matrices* K *such that* $\det(pI - A + BK) = P_d(p)$ *for a fully controllable system. The theorem stated above does not give these matrices.*

Non-controllable poles cannot be moved. These are not a problem if they are sufficiently stable.

This theorem shows that eigenvalues can be arbitrarily placed; for a multivariate system, it can be advantageous to also place its modes or eigenvectors (for the matrix $A - BK$*).*

The systems that can be stabilized are those whose non-controllable poles are sufficiently stable and unstable poles are controllable.

THEOREM 2.15 (PBH stabilizability test).– A pair (A, B) is stabilizable, if there exists a matrix $K \in R^{m\times n}$ such that $A - BK$ is stable. The pair (A, B) is stabilizable, if and only if all of its unstable modes are controllable, that is,

$$\forall p \in C, Re(p) \geq 0 \Longrightarrow rank\,(pI - A,\, B) = n.$$

2.4.3. *Finite-time response for a discrete system, deadbeat response*

For a fully controllable discrete system, the polynomial z^n is particularly interesting when obtained in the CL system. It corresponds to a response in n sampling steps. The Cayley–Hamilton theorem allows us to write that $(A - BK)^n = 0$. This means that the response of the free system (with zero input) in the CL system is zero at the end of n sampling periods (regardless of the value of the initial state $x(0)$). The CL system is stable and the effect of a non-zero initial condition disappears after n points, if n is the order of the system. This justifies the designation of deadbeat control law. This control is not often used in practice because it can lead to excessive gains or large amplitude control signals likely to cause a nonlinear functioning of the system.

EXAMPLE 2.12.– *Deadbeat control for a discrete system.*

Let the system of order $n = 2$ *be*

$$x_{k+1} = \begin{bmatrix} 1 & 0 \\ 2 & 2 \end{bmatrix} .x_k + \begin{bmatrix} 1 \\ 0 \end{bmatrix} u_k. \qquad [2.106]$$

The pair (A, B) *is completely controllable.*

We can find $K = [k_1, k_2]$ *such that* $(A - BK)^2 = 0$. *To this end, we have to choose* (k_1, k_2) *such that* $\det(zI - A + BK) = z^2$.

It thus yield $k_1 = -3$, $k_2 = -2$.

THEOREM 2.16.– If for a system (A, B) is completely controllable then for any K, we have $(A - BK, B)$ also completely controllable.

THEOREM 2.17.– State feedback control does not affect the zeros of a system realization.

PROOF FOR THE UNIVARIATE CASE.–

If the system is completely controllable, it is then possible to find a realization in a controllable canonical state form (A_c, B_c, C_c). For example, for an order $n = 3$ if

$$H(p) = \frac{b_1 p^2 + b_2 p + b_3}{p^3 + a_1 p^2 + a_2 p + a_0}, \qquad [2.107]$$

we have

$$\dot{X} = \begin{bmatrix} -a_1 & -a_2 & -a_3 \\ 1 & 0 & 0 \\ 0 & 1 & 0 \end{bmatrix} X + \begin{bmatrix} 1 \\ 0 \\ 0 \end{bmatrix} u \qquad [2.108]$$

and

$$y = \begin{bmatrix} b_1 & b_2 & b_3 \end{bmatrix} X. \qquad [2.109]$$

The system looped by state feedback has for realization (in controllable canonical form $(A_c - B_cK_c, B_c, C_c)$). For the latter form, the corresponding transfer function is: $G(p) = \frac{Y(p)}{V(p)} = C_c.(pI - A_c + B_cK_c)^{-1}B_c$ (or $G(z) = \frac{Y(z)}{V(z)} = C_c.(zI - A_c + B_cK_c)^{-1}B_c$ for the discrete case. The numerator of the transfer function only depends on C_c, that is to say only of the open-loop system.

If the state feedback introduces a pole zero compensation of the CL transfer function, it is shown that the state feedback does not affect the controllability, but in this case it affects the observability. There is loss of observability using pole zero simplification in the transfer function.

2.4.4. *Use of observers in control: separation principle*

2.4.4.1. *Continuous system observer and control*

In continuous system control when state is not completely measurable, we use an observer for the reconstruction of the state in order to implement state feedback control. In this case, when the system is observable and controllable, the principle of separation makes it possible to ensure the stabilization of the system by arbitrarily placing the poles of the observer and the regulator in a separate manner.

Consider the system [2.1], if the pair (A, B) is controllable and the pair (A, C) is observable, there are then matrices $K \in R^{m\times n}$ and $L \in R^{n\times p}$ such that matrices $(A-BK)$ and $(A - LC)$ have all their eigenvalues with strictly negative real parts (placed at specified values). It is therefore possible to consider the asymptotic observer:

$$\begin{aligned} \dot{\hat{x}} &= A\hat{x} + Bu + L(y - C\hat{x}) \\ \text{with} \quad \dot{e} &= (A - LC)e. \end{aligned} \qquad [2.110]$$

In state feedback control, the reconstructed state is used:

$$u = -K\hat{x} + v. \qquad [2.111]$$

Furthermore, for the system, the observer and the control, we get the following equations:

$$\begin{cases} \dot{\hat{x}} = (A - BK - LC)\hat{x} + LCx + Bv \\ \dot{x} = Ax - BK\hat{x} + Bv = (A - BK)x - BKe + Bv \\ \dot{e} = (A - LC)e \\ y = Cx \\ u = -K\hat{x} + v. \end{cases} \qquad [2.112]$$

The four last equations allow us to define the system having input v, state $\begin{pmatrix} x \\ e \end{pmatrix}$ and output $\begin{pmatrix} y \\ u \end{pmatrix}$ by:

$$\begin{pmatrix} \dot{x} \\ \dot{e} \end{pmatrix} = \begin{pmatrix} A - BK & -BK \\ 0 & A - LC \end{pmatrix} \begin{pmatrix} x \\ e \end{pmatrix} + \begin{pmatrix} B \\ 0 \end{pmatrix} v \qquad [2.113]$$

$$\begin{pmatrix} y \\ u \end{pmatrix} = \begin{pmatrix} C & 0 \\ K & K \end{pmatrix} \begin{pmatrix} x \\ e \end{pmatrix} + \begin{pmatrix} 0 \\ 1 \end{pmatrix} v. \qquad [2.114]$$

REMARK 2.2 (Separation Principle).– *The fact that the matrix of equation [2.113] be triangular indicates that the placement of the observer poles, to impose a dynamic to the observation error, is independent of the system control and the placement of the poles for the control.*

Regarding the system control, it can be observed that the dynamics of the system state does not depend on the gain of the observer; the placement of the poles of the regulator is not influenced by the choice of observer. On the other hand, the control error will converge after the convergence of the observation error, which seems natural.

2.5. Linear quadratic (LQ) control

In system control, stabilizing the controllable modes of the system and placing them in a stable region of the complex plane[6] are not the only constraints that can be imposed on the system. Possible state feedbacks stabilizing are different in the choice of gains to achieve a compromise between response speed, accuracy and power or amplitude of the control inputs. The optimal control consists of determining the control that minimizes a criterion as a function of the energy of input and output signals. To minimize the error energy and that of the control signal, we can consider the following measures.

$$\int_0^{\infty} y(t)^T y(t) dt \quad \text{and} \quad \int_0^{\infty} u(t)^T u(t) dt. \qquad [2.115]$$

6 The left half-plane for continuous cases and the unit-radius circle for discrete systems.

Minimizing the energy of the error signal only usually results in an infinite energy control; so to avoid it, the following test is used. A positive scalar ρ enables a compromise to be established between speed and energy spent in the control:

$$J_o(u) = \int_0^{\infty} (y(t)^T y(t) + \rho u(t)^T u(t))dt. \quad [2.116]$$

For continuous systems, the state quadratic criteria is generally considered in the following form:

$$J(u) = \int_0^{\infty} (x(t)^T Qx(t) + u(t)^T Ru(t))dt \quad [2.117]$$

and for discrete systems:

$$J(u) = \sum_{k=0}^{\infty} (x(k)^T Qx(k) + u(k)^T Ru(k)). \quad [2.118]$$

The weighting matrices Q and R are consistent, symmetrical positive and definite positive. They provide a means of differently weighting the states $x_i(t)$ and inputs $u_i(t)$. A usual choice for matrix Q is the one that amounts to weighting the output only, that is $Q = C^T C$.

The controller thus obtained, by minimizing J ($\min_u J(u)$), is linear; this justifies its designation of linear quadratic regulator (LQR).

EXAMPLE 2.13.– *Control of a discrete system defined by* $x_{k+1} = x_k + u_k$ *with* $x_0 = x_o$.

Consider the quadratic criterion $J(\rho) = \sum_{k=1}^{\infty} (x_k^2 + \rho u_k^2)$ *with a weighting* $\rho > 0$.

We want to find a gain K *for the state feedback control* $u_k = -Kx_k$ *minimizing criterion* $J(u)$.

We have $x_{k+1} = (1 - K)x_k = (1 - K)^k x_0$.

Hence,

$$
\begin{aligned}
J(\rho) &= \sum_{k=1}^{\infty}(x_k^2 + \rho K^2 x_k^2) \\
&= (1+\rho K^2)\sum_{k=1}^{\infty} x_k^2 = (1+\rho K^2)x_0^2 \sum_{k=1}^{\infty}(1-K)^{2k}.
\end{aligned}
\qquad [2.119]
$$

If the series converges (i.e. $0 < K < 2$*),* $J(u) = (1+\rho K^2)\frac{x_0^2}{1-(1-K)^2}$ *if* $0 < K < 2$ *and otherwise* $J(u) = \infty$.

$$
J(\rho) = \frac{(1+\rho K^2)x_0^2}{1-(1-K)^2}, \qquad [2.120]
$$

$$
\begin{aligned}
\frac{\partial J(\rho)}{\partial K} &= \frac{2\rho K x_0^2(1-(1-K)^2) - 2(1-K)(1+\rho K^2)x_0^2}{(1-(1-K)^2)^2} \qquad [2.121] \\
&= \frac{2\rho K x_0^2(1-(1-K)^2) - 2(1-K)(1+\rho K^2)x_0^2}{(1-(1-K)^2)^2} = 0, \qquad [2.122]
\end{aligned}
$$

$$
\rho K - \rho K(1-2K+K^2) - 1 - K + \rho K^2 - \rho K^3) = 0, \qquad [2.123]
$$

$$
3\rho K^2 - 2\rho K^3 - 1 - K = 0 \qquad [2.124]
$$

to be verified.

The optimization of the criterion gives

$$
K_{\text{opt}} = \frac{\sqrt{1+4\rho}-1}{2\rho}, \qquad [2.125]
$$

The corresponding cost is then

$$
J(\rho) = \rho\frac{\sqrt{1+4\rho}-1}{2\rho - \sqrt{1+4\rho}+1}. \qquad [2.126]
$$

If ρ is very small or zero, we get a low cost control, which gives us $\lim_{\rho \to 0} K_{opt}(\rho) = 1$*. This corresponds in the CL system to deadbeat response control because* $\det(zI - A + BK_{opt}(0)) = z$.

On the other hand, if the control is very strongly penalized with a very large ρ, we obtain as optimal gain a very small gain $K_{opt} = \frac{1}{\sqrt{\rho}}$*. The CL system thus obtained is then barely stable because its eigenvalue becomes* $\lambda = 1 - \frac{1}{\sqrt{\rho}}$.

2.5.1. *Linear quadratic regulator*

THEOREM 2.18 (Continuous Regulator).– For the controllable system $\dot{x} = Ax + Bu$, with $x(0) = x_o$, the control that minimizes the quadratic criterion

$$J(u) = \int_0^{\infty} (x(t)^T Q x(t) + u(t)^T R u(t)) dt \qquad [2.127]$$

with $Q \geq 0$ and $R > 0$; it is a static state feedback control law defined by: $u_{opt} = -K_{opt} x$ with $K_{opt} = R^{-1} B^T P$, where P is a constant symmetric positive definite matrix, a unique solution of the algebraic Riccati equation (ARE):

$$A^T P + PA - PBR^{-1} B^T P + Q = 0. \qquad [2.128]$$

If in addition $Q > 0$, then $P > 0$ and the CL control system is stable.

PROOF.– Consider the positive definite quadratic function (Lypunov's) $V(x) = x^T P x$, where P satisfies equation [2.128] along the trajectories of the system

$\frac{d}{dt} V(x) = x^T P \dot{x} + \dot{x}^T P x = x^T (A^T P + PA) x + 2x^T PBu,$

$\frac{d}{dt} V(x) = x^T PBR^{-1} B^T P x - x^T Q x + 2x^T PBu = -(u_{opt} - u)^T R(u_{opt} - u) - x^T Q x + u^T R u.$

This gives us for the criterion $J = x_o^T P x_o + \int_0^{\infty} (u_{opt} - u)^T R(u_{opt} - u) dt$ which is really minimum and equals $J = x_o^T P x_o$ for $u = u_{opt}$.

The loop system is stable because it verifies the Lyapunov condition

$(A - BK_{opt})^T P + P(A - BK_{opt}) = -Q - PBR^{-1} B^T P < 0.$

THEOREM 2.19 (Discrete Regulator).– For the controllable system $x_{k+1} = Ax_k + Bu_k$, with $x(0) = x_o$, the control that minimizes the quadratic criterion

$$J(u) = \sum_{k=1}^{\infty}(x(k)^T Qx(k) + u(k)^T Ru(k)) \qquad [2.129]$$

with $Q \geq 0$ and $R > 0$, is a static state feedback control law defined by $u_{opt} = -K_{opt}x_k$ with $K_{opt} = R^{-1}B^T P$, where P is a constant symmetric positive definite matrix, unique solution of the discrete ARE:

$$P = A^T PA - A^T P(R + B^T PB)^{-1}PA + Q. \qquad [2.130]$$

REMARK 2.3.– *The optimal control is a static state feedback control.*

In general, there is a large number of solutions P of the Riccati equation but the one positive definite is unique. The theorem includes the case where Q is positive or zero.

Often, when the pair (A, C) is observable, we choose $Q = C^T C$, whose rank is that of C.

Example: double integrator case.–

2.6. Optimal control (LQ)

In many complex systems, a specification of the poles alone is not enough to design a good control system, as a matter of fact:

– when considering a p-input system, of order n, the gain matrix K such that

$$u(t) = v(t) - Kx(t) \qquad [2.131]$$

has np elements and there are only n possible poles available. There are therefore too many parameters to determine.

– It may be possible that the placement of some poles implies the use of an excessively large control signal, which is not achievable when remaining within the linear domain; for instance, when it is desirable that a mobile robot turn very fast, it is important that the steering wheels can change direction very quickly; however, the actuators have a limited operating speed. Furthermore, the placement of poles may result in some actuators to become saturated, which drives the system out of the linear domain and consequently to make it unstable.

– Very often, the specification imposes a minimal consumption; for example, a system for maintaining the attitude of a satellite will have to not only maintain the satellite in a good position, but in addition it will have to do it by consuming the least possible amount of fuel. Similarly, a temperature regulating system for a building must follow a certain temperature profile with a minimum fuel cost.

For all these reasons, we are brought to consider optimal control by calculating a control system based on the minimization of a criterion (cost function).

Let a process be defined by its state equation:

$$\dot{x}(t) = Ax(t) + Bu(t), \qquad [2.132]$$

we are looking for a control law of the form (regulation):

$$u(t) = -Kx(t) \qquad [2.133]$$

where K is a matrix whose elements should be determined in an optimal way to meet the requirements of the specifications.

In the case of the placement of the poles, we have seen how gains can be found when we know where to place the poles. In this case, we want to find the gains that minimize a cost function, which in turn is the integral of a quadratic function of the state and the control signals. This cost function is defined by

$$J = \int_0^{\infty} \left(x^T(\tau)Qx(\tau) + u^T(\tau)Ru(\tau)\right) d\tau, \qquad [2.134]$$

where Q and R are real symmetric matrices, which greatly simplifies the calculations without losing generalities. Matrix R must be positive definite and Q must be positive semidefinite.

The optimal gain matrix is given by

$$K = R^{-1}B^T P \qquad [2.135]$$

where P is a constant symmetric positive definite matrix, unique solution of the ARE:

$$A^T P + PA - PBR^{-1}B^T P + Q = 0. \qquad [2.136]$$

EXAMPLE 2.14.– *Optimal control of the inverted pendulum. Let the model for a single axis (rotation axis) be:*

$$\begin{aligned}\dot{\theta}(t) &= \omega(t)\\ \dot{\omega}(t) &= \Omega^2\theta(t) + u(t)\end{aligned} \qquad [2.137]$$

with

$$\Omega^2 = \frac{g}{l} \qquad u = \frac{\gamma(t)}{ml^2}. \qquad [2.138]$$

If we choose the state vector $x = \begin{pmatrix} \theta & \omega \end{pmatrix}^T$, we will have the following state–space representation:

$$\begin{aligned}\dot{x} &= \begin{pmatrix} 0 & 1 \\ \Omega^2 & 0 \end{pmatrix} x + \begin{pmatrix} 0 \\ 1 \end{pmatrix} u \\ y &= \begin{pmatrix} 1 & 0 \end{pmatrix} x\end{aligned} \implies \begin{aligned}A &= \begin{pmatrix} 0 & 1 \\ \Omega^2 & 0 \end{pmatrix}; \quad B = \begin{pmatrix} 0 \\ 1 \end{pmatrix} \\ C &= \begin{pmatrix} 1 & 0 \end{pmatrix}.\end{aligned} \qquad [2.139]$$

The chosen cost function is

$$J = \int_0^{\infty} (\theta^2 + \frac{1}{c^2}u^2)d\tau \Longrightarrow Q = \begin{pmatrix} 1 & 0 \\ 0 & 0 \end{pmatrix} \quad \text{and} \quad R = \frac{1}{c^2}. \qquad [2.140]$$

Represented by P is the solution of the Riccati equation (ARE) given by

$$P = \begin{pmatrix} p_1 & p_2 \\ p_2 & p_3 \end{pmatrix}, \qquad [2.141]$$

which makes that the optimal gain is given by

$$K = R^{-1}B^TP = c^2 \begin{pmatrix} 1 & 0 \end{pmatrix} \begin{pmatrix} p_1 & p_2 \\ p_2 & p_3 \end{pmatrix} = \begin{pmatrix} c^2p_2 & c^2p_3 \end{pmatrix}. \qquad [2.142]$$

The Riccati equation is written as:

$$A^TP + PA - PBR^{-1}B^TP + Q = 0. \qquad [2.143]$$

from which

$$PA = \begin{pmatrix} p_1 & p_2 \\ p_2 & p_3 \end{pmatrix} \begin{pmatrix} 0 & 1 \\ \Omega^2 & 0 \end{pmatrix} = \begin{pmatrix} p_2\Omega^2 & p_1 \\ p_3\Omega^2 & p_2 \end{pmatrix}$$
$$A^T P = (PA)^T = \begin{pmatrix} p_2\Omega^2 & p_3\Omega^2 \\ p_1 & p_2 \end{pmatrix} \quad [2.144]$$
$$PBR^{-1}B^T P = \begin{pmatrix} c^2 p_2^2 & c^2 p_2 p_3 \\ c^2 p_2 p_3 & c^2 p_3^2 \end{pmatrix},$$

which yields:

$$A^T P + PA - PBR^{-1}B^T P + Q =$$
$$\begin{pmatrix} 2\Omega^2 p_2 - c^2 p_2^2 + 1 & \Omega^2 p_3 + p_1 - c^2 p_2 p_3 \\ \Omega^2 p_3 + p_1 - c^2 p_2 p_3 & 2p_2 - c^2 p_3^2 \end{pmatrix}. \quad [2.145]$$

It can thus be written that:

$$\begin{pmatrix} 2\Omega^2 p_2 - c^2 p_2^2 + 1 & \Omega^2 p_3 + p_1 - c^2 p_2 p_3 \\ \Omega^2 p_3 + p_1 - c^2 p_2 p_3 & 2p_2 - c^2 p_3^2 \end{pmatrix} =$$
$$\begin{pmatrix} 0 & 0 \\ 0 & 0 \end{pmatrix} \Longrightarrow \left| \begin{array}{l} 2p_2\Omega^2 - c^2 p_2^2 + 1 = 0 \\ p_1 + p_3\Omega^2 - c^2 p_2 p_3 = 0 \\ 2p_2 - c^2 p_3^2 = 0. \end{array} \right. \quad [2.146]$$

which yields

$$p_2 = \frac{\Omega^2 + \sqrt{\Omega^4 + c^2}}{c^2} \quad [2.147]$$
$$p_3 = \frac{\sqrt{2}}{c^2}\left(\Omega^2 + \sqrt{\Omega^4 + c^2}\right)^{1/2}$$

hence the following optimal gain K:

$$K = \left(\Omega^2 + \sqrt{\Omega^4 + c^2} \quad \sqrt{2}\left(\Omega^2 + \sqrt{\Omega^4 + c^2}\right)^{1/2} \right) \quad [2.148]$$

is the evolution matrix of the loop system and it is therefore given by:

$$A - BK = \begin{pmatrix} 0 & 1 \\ -\sqrt{\Omega^4 + c^2} & -\sqrt{2}\left(\Omega^2 + \sqrt{\Omega^4 + c^2}\right)^{1/2} \end{pmatrix} \quad [2.149]$$

whose eigenvalues are

$$\lambda_{1,2} = -\frac{\sqrt{2}}{2}\left(\left(\sqrt{\Omega^4+c^2}+\Omega^2\right)^{1/2} \pm j\left(\sqrt{\Omega^4+c^2}-\Omega^2\right)^{1/2}\right). \quad [2.150]$$

We can see the evolution of these roots according to c:

– If $c \to \infty$ $\lambda_{1,2} \approx -\frac{\sqrt{2c}}{2}(1 \pm j)$, the cost of the control is small, the response time is very short and damping becomes $\frac{\sqrt{2}}{2}$. The fact that the damping tends to this value is not specific to this problem; it is a fairly general result for the second-order systems optimally controlled.

– If $c \to 0$ $\lambda_{1,2} = -\Omega$, the cost of the couple is very significant, the roots tend to $-\Omega$, which corresponds to the stable root of the uncontrolled system (the other root $+\Omega$). This is a general result: when the cost of the control becomes infinite, if the system is initially unstable, the unstable roots tend to their symmetrical with respect to the imaginary axis.

2.7. Comprehension and application exercises

EXERCISE 1.–

Controllability and observability: let the system be described by the schema in Figure 2.5.

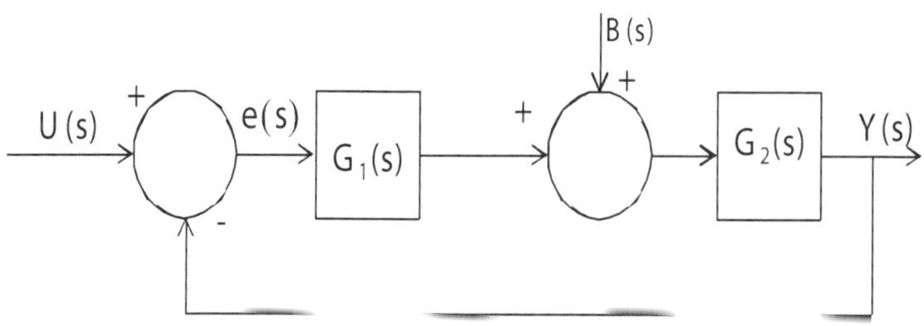

Figure 2.5. *Given noisy system*

The following vector X is taken as state vector

$$X = [x_1, x_2, x_3, x_4]^T. \quad [2.151]$$

1) Write the state–space representation of this system.

2) Study the controllability and the observability.

a) Using the controllability and observability matrices.

b) By changing variables, determine the new state–space representation in modal form. Thereof, derive the controllability and the observability. Give the functional block diagram corresponding to this modal form (parallel decomposition) and study the controllability and the observability.

3) Based on the block diagram of the figure, determine the system transfer function.

EXERCISE 2.–

Controllability and observability: state–space representation of the system. The state vector is given by

$$X = [x_1, x_2, x_3, x_4]^T. \qquad [2.152]$$

The state–space representation:

$$\begin{cases} \dot{X} = AX + BU \\ Y = CX \end{cases} \qquad [2.153]$$

with

$$A = \begin{bmatrix} -1 & 0 & 0 & 0 \\ 0 & -2 & 0 & 0 \\ 1 & 1 & 1 & 0 \\ 1 & 1 & 0 & -3 \end{bmatrix}; \; B = \begin{bmatrix} -2 \\ 3 \\ 0 \\ 0 \end{bmatrix}; \; C = \begin{bmatrix} 0 & 0 & 0.5 & 0.5 \end{bmatrix}. \qquad [2.154]$$

1) Analyze the controllability and the observability of this system.

2) Analyze the controllability and the observability of this system using the diagonal form.

SOLUTION TO EXERCISES 1 AND 2.– Recall the definitions of the controllability and the observability.

DEFINITION 2.12.– *The system is controllable if there is a finite control $u(t)$ that can bring the system from an initial state x_0 to a final state x_f in finite time.*

The system is controllable if the controllability matrix

$$Com = [B, AB, \ldots, A^{n-1}B] \text{ is of rank n.} \qquad [2.155]$$

In the case of a system expressed in the diagonal form, all the elements of vector B just have to be non-zero.

The system is observable if regardless of the initial state $x(0)$ it is possible to determine $x(0)$ from $u(t)$ and $y(t)$ with $0 \leq t \leq \tau$.

The system is observable if the observability matrix

$$Obs = [C^T, A^T C^T, \ldots, A^{T(n-1)} C^T] \text{ is of rank } n. \quad [2.156]$$

In the case of a system expressed in the diagonal form, it suffices that all the elements of vector B be non-zero.

SOLUTION 2.1.– *1) Analyze the controllability and the observability of this system:*

$$\text{Com} = \begin{bmatrix} B & AB & A^2B & A^3B \end{bmatrix} = \begin{bmatrix} -2 & 2 & -2 & 2 \\ 3 & -6 & 12 & -24 \\ 0 & 1 & -3 & 7 \\ 0 & 1 & -7 & 31 \end{bmatrix}, \quad [2.157]$$

Com is of rank=3 $\Rightarrow$ *non-controllable system (a non-controllable state).*

$$\text{Obs} = \begin{bmatrix} C \\ CA \\ CA^2 \\ CA^3 \end{bmatrix} = \begin{bmatrix} 0 & 0 & 0.5 & 0.5 \\ 1 & 1 & 0.5 & -1.5 \\ -2 & -3 & 0.5 & 4.5 \\ 7 & 11 & 0.5 & -13.5 \end{bmatrix}, \quad [2.158]$$

Obs is of rank=3 $\Rightarrow$ *non-observable system (a non-observable state).*

2) Analyze the controllability and the observability of this system using the diagonal form. After computation of the eigenvalues $\lambda_i (i = 1, ..., 4)$ *and the eigenvectors matrix* P*, we obtain the following state–space representation:*

$$\begin{cases} \dot{Z} = P^{-1}APZ + P^{-1}BU \\ Y = CPZ \end{cases} \quad [2.159]$$

with

$$P^{-1}AP = \begin{bmatrix} -1 & 0 & 0 & 0 \\ 0 & -2 & 0 & 0 \\ 0 & 0 & 1 & 0 \\ 0 & 0 & 0 & -3 \end{bmatrix}; \; P^{-1}B = \begin{bmatrix} -2 \\ 3 \\ 0 \\ -2 \end{bmatrix};$$

$$CP = \begin{bmatrix} 0 & 2/6 & 0.5 & 0.5 \end{bmatrix}. \quad [2.160]$$

The third state is non-controllable and the first state is non-observable. The new block diagram is given below.

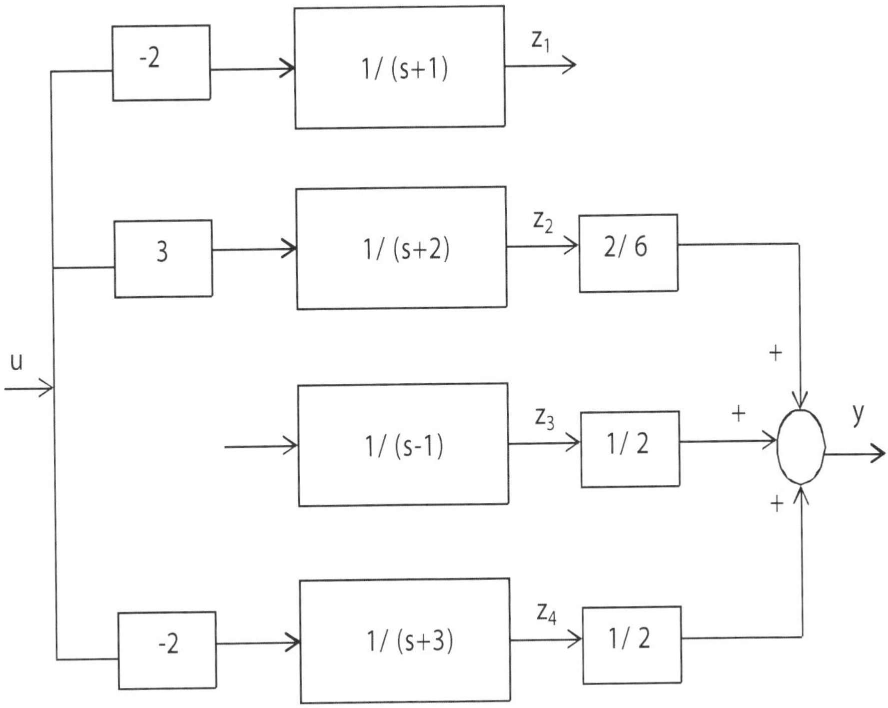

Figure 2.6. *New block diagram*

The transfer function of the system is given by

$$H(p) = \frac{Y(p)}{U(p)} = \frac{1}{(p+3)(p+2)}. \qquad [2.161]$$

In conclusion, the system is of the second order. The minimal representation of the process in the state form only introduces two states.

Caution: The real system is that given by the diagram in Figure 2.6; we do not have the right to simplify the diagram. After simplification, the transfer function $H(p)$ does not actually reflect what is happening in the process.

EXERCISE 3.– *Optimal control.*

–*Preamble*:

To calculate an integral of the form:

$$J = \int_0^{\infty} \varepsilon^2(t)dt, \quad [2.162]$$

we can use Parseval's theorem and replace the integral over time by an integral over the complex frequency s,

$$J = \int_0^{\infty} \varepsilon^2(t)dt = \frac{1}{2\pi j} \int_{-j\infty}^{+j\infty} \varepsilon(s)\varepsilon(-s)ds \quad [2.163]$$

where $\varepsilon(s)$ is the Laplace transform of $\varepsilon(t)$.

If $\varepsilon(s)$ is expressed in the form of a fraction:

$$\varepsilon(s) = \frac{\sum_{k=0}^{n-1} b_k s^k}{\sum_{k=0}^{n} a_k s^k} = \frac{B(s)}{A(s)} \quad [2.164]$$

and if $A(s)$ has all its zeros in the left half-plane, the value of the integral J has been calculated and is given in Table 2.1.

$\varepsilon(s) = \frac{b_0}{a_1 s + a_0}$	$J = \frac{b_0^2}{2a_1 a_0}$
$\varepsilon(s) = \frac{b_1 s + b_0}{a_2 s^2 + a_1 s + a_0}$	$J = \frac{b_1^2 a_0 + b_0^2 a_2}{2a_2 a_1 a_0}$
$\varepsilon(s) = \frac{b_2 s^2 + b_1 s + b_0}{a_3 s^3 + a_2 s^2 + a_1 s + a_0}$	$J = \frac{b_2^2 a_1 a_0 + (b_1^2 - 2b_0 b_2)a_3 a_0 + b_0^2 a_2 a_3}{2a_0 a_3(-a_0 a_3 + a_1 a_2)}$

Table 2.1. *Table of the values of the integral J*

–*First question*:

The process to be controlled has a transfer function of the form:

$$G(s) = \frac{100}{s^2}. \quad [2.165]$$

We insert the transfer function $G(s)$ in a unity feedback loop.

– Give the expressions and appearances of the impulse response and the step response of the CL system.

– Give the values of the permanent errors to the step, ramp and parabola.

– Is the system considered as being stable?

– Give the expression of $\varepsilon(s) = E(s) - Y(s)$ in the case of a step impulse input. From there, derive the value of the integral by making use of the preamble:

$$J = \int_0^\infty \varepsilon^2(t)dt. \tag{2.166}$$

– *Second question*:

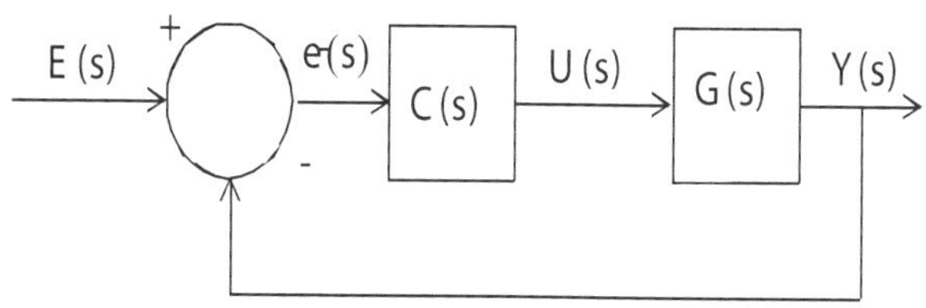

Figure 2.7. *Given system with controller*

We want to insert a corrector $C(s)$ in the direct chain in such a way that the system presents a phase margin of $45°$:

– Give the expression of a controller enabling the targeted condition to be met.

– Plot in the Bode plane the frequency response of the system before and after correction. What is the value of the bandwidth of the system after correction?

– Give the expressions of $\varepsilon(s)$ and of $U(s)$ for an step impulse on input and calculate the values of both integrals J_1 and J_2 defined below using formulas given in the preamble.

$$J_1 = \int_0^\infty \varepsilon^2(t)dt \qquad J_2 = \int_0^\infty u^2(t)dt. \tag{2.167}$$

–*Third question*:

The system is controlled using tachymetric feedback as shown in Figure 2.8.

– Calculate parameters K_1 and K_2 that can be used to obtain the same system characteristics in the second question (phase margin and bandwidth).

– Give the expressions of $\varepsilon(s)$ and of $U(s)$ for a step impulse on input and calculate the values of the integrals J_1 and J_2.

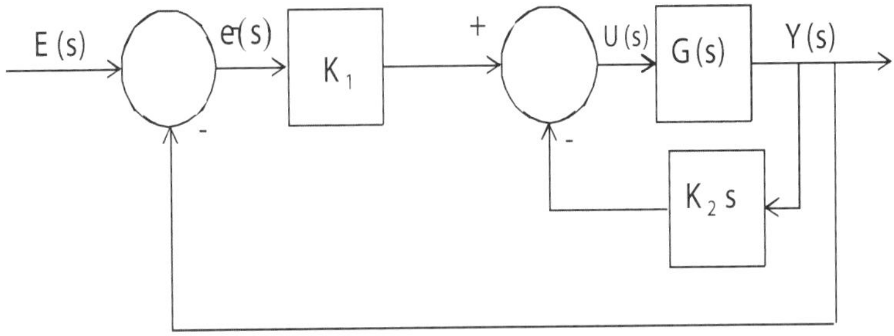

Figure 2.8. *Servo system using tachymetric feedback*

–Fourth question:

We want to calculate the parameters K_1 and K_2 such that to minimize the integrals J_1 and J_2.

– The aim is to minimize the integral of cost $J = J_1 + \lambda J_2$ by taking the value of $\lambda = 0.25$. Determine the values of K_1 and K_2 that minimize J (the derivatives with respect to K_1 and with respect to K_2 will be canceled out).

– Give the expression of the CL transfer function and compare it to that obtained in the previous question (damping value, value of the natural frequency).

– Give the expressions of $\varepsilon(s)$ and of $U(s)$ for a step impulse on input and calculate the values of the integrals J_1 and J_2.

– Compare the values that you have obtained for J_1 and J_2 during the first four questions. Conclusion?

–Fifth question:

We want to control the system by means of a digital controller.

– Draw the block diagram of the system controlled by a digital controller.

– Write the discrete transfer function of the process. Choose the sampling period T_e.

– Determine a controller providing an absolute minimal response time (the output reproduces the input with a delay of a sampling period). Through examination of the controller shape, can you predict the appearance of the control signal $u(nT_e)$ for a Heaviside setpoint y^d?

SOLUTION TO EXERCISE 3.–

–*First question*:

The process is represented by

$$G(s) = \frac{100}{s^2}. \qquad [2.168]$$

The CL transfer function

$$H(s) = \frac{G(s)}{1+G(s)} = \frac{100}{s^2+100}. \qquad [2.169]$$

– The impulse response $h(t)$ is given by the inverse Laplace transform of $H(s)$:

$$h(t) = [10\sin(10t)]u(t), \qquad u(t) = \text{Heaviside}. \qquad [2.170]$$

The step response $y(t)$ is given by the integral of $h(t)$.

$$y(t) = [1 - \cos(10t)]u(t). \qquad [2.171]$$

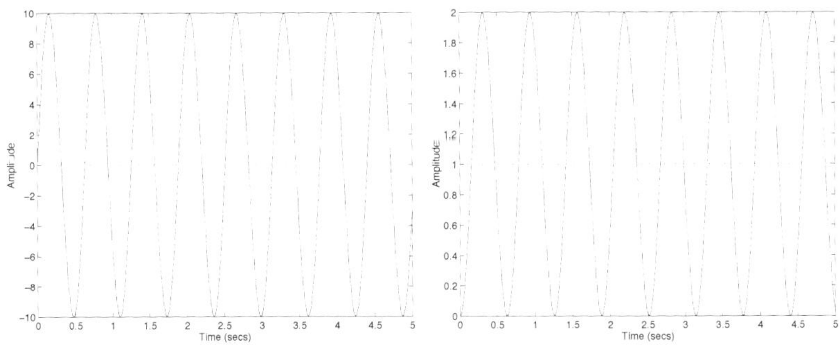

Impulse response Step response

Figure 2.9. *Impulse and step response*

– The errors to the step, ramp and parabola are, respectively, given as:

$$\varepsilon_p(\infty) = \lim_{s\to 0} s\varepsilon(s)\mid_{u(s)=1/s} = \lim_{s\to 0} \frac{1}{1+G(s)} = 0 \quad \text{(integration)}. \qquad [2.172]$$

If the transfer function $\frac{\varepsilon(p)}{u(p)}$ is strictly stable, $s\varepsilon(s)$ has all these poles with negative real parts,

$$\varepsilon_v(\infty) = \lim_{p\to 0} p\varepsilon(p) \mid_{u(p)=1/p^2} = \lim_{p\to 0} \frac{1}{pG(p)} = 0 \quad \text{(double integration).[2.173]}$$

If $s\varepsilon(s)$ has all these poles with negative real parts, the transfer function $\frac{\varepsilon(p)}{u(p)}$ is strictly stable

$$\varepsilon_a(\infty) = \lim_{p\to 0} p\varepsilon(p) \mid_{u(p)=1/p^3} = \lim_{p\to 0} \frac{1}{p^2G(p)} = 0.01. \quad [2.174]$$

If transfer function $\frac{\varepsilon(p)}{u(p)}$ is strictly stable.

REMARK 2.4.– *We note here that $s\varepsilon(s)$ has roots with zero real part, which makes that the theorem of the final Laplace value is no longer applicable. As a result, the limits given above will be valid only after correction of the system so as to become it asymptotically stable.*

– The system is on the boundary of stability (marginally stable = pure oscillator). In effect, the poles of the CL system are $p = \pm j10$ $(p = \pm j\omega_c, \omega_c = 10)$.

– The expression of $\varepsilon(p) = E(p) - Y(p)$ is given by:

$$\varepsilon(p) = \frac{1}{1+G(p)}E(p) = \frac{1}{p(1+G(p))} = \frac{p}{p^2+100}. \quad [2.175]$$

From the preamble, the value of the integral is deduced:

$$J = \int_0^\infty \varepsilon^2(t)dt = \infty, \qquad \text{because } a_1 = 0. \quad [2.176]$$

Second question:

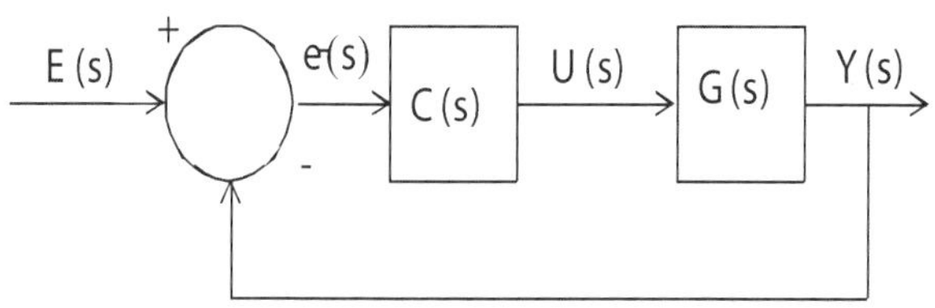

Figure 2.10. *System with phase advance controller*

The phase margin is zero ($\Delta\phi = 180^\circ + \phi(\omega_c) = 180^\circ - 180^\circ = 0$, see plot).

To ensure that the system has a 45°, phase margin, we insert an advance phase controller:

$$C_1(s) = \frac{1 + \alpha Ts}{1 + Ts} \qquad (\alpha > 1) \qquad [2.177]$$

and α given by

$$\alpha = \frac{1 + \sin(\phi_m)}{1 - \sin(\phi_m)} \qquad (\phi_m = 45^\circ) \qquad \alpha = \frac{1 + \sin(45)}{1 - \sin(45)} = 5.82. \qquad [2.178]$$

To determine the value of T, we will use the following relationship $T = \frac{1}{\omega_m \sqrt{\alpha}}$, with ω_m obtained by applying:

$$\mid G(j\omega_m) \mid = 20\log_{10}(\frac{100}{\omega_m^2}) = -10\log_{10}(\alpha) = 20\log_{10}\frac{1}{\sqrt{\alpha}}. \qquad [2.179]$$

This allows us to write:

$$\frac{100}{\omega_m^2} = \frac{1}{\sqrt{\alpha}} \quad \Rightarrow \quad \omega_m = 15.5\,\text{rad/s} \quad \Rightarrow T = \frac{1}{\omega_m \sqrt{\alpha}} = 0.0267. \qquad [2.180]$$

The controller is given by

$$C_1(s) = \frac{1 + 0.1556s}{1 + 0.0267s}. \qquad [2.181]$$

– The transfer function of the corrected system is given by

$$G'(s) = C_1(s)G(s) = \frac{1 + 0.1556s}{1 + 0.0267s}\frac{100}{s^2}. \qquad [2.182]$$

– The bandwidth after correction is given by $\omega_c = \omega_m = 15.5$ rad/s.

– The expressions of $\varepsilon(s)$ and of $U(s)$ are the following:

$$\varepsilon(s) = \left.\frac{Y^d(s)}{1 + C_1(s)G(s)}\right|_{Y^d(s)=1/s} = \frac{0.0267s^2 + s}{0.0267s^3 + s^2 + 15.56s + 100}. \qquad [2.183]$$

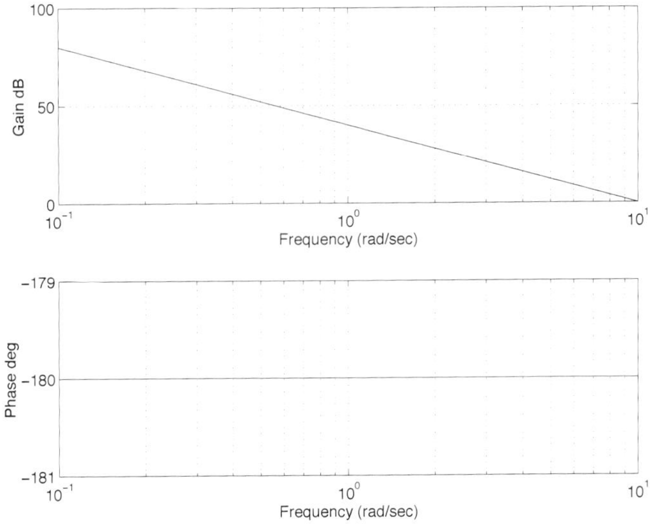

Figure 2.11. *Bode plot of the system*

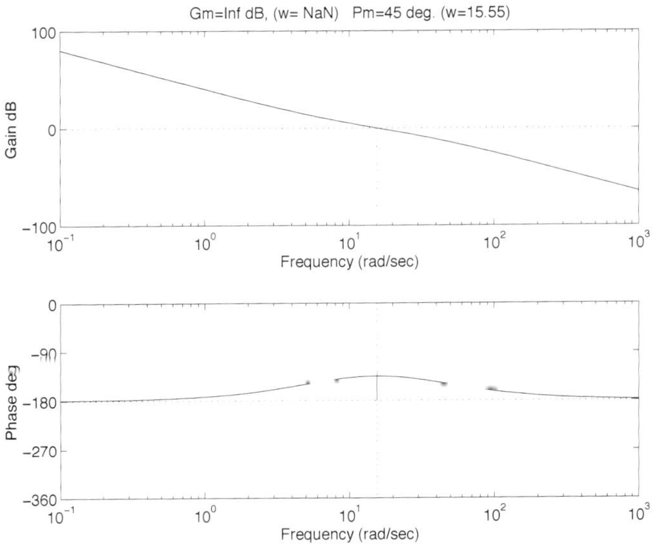

Figure 2.12. *Bode plot of the system after correction*

$$U(s) = C_1(s)\varepsilon(s) = \frac{0.15s^2 + s}{0.0267s^3 + s^2 + 15.56s + 100}. \quad [2.184]$$

Using the formulas given in the preamble, we calculate the values of the two integrals J_1 and J_2:

$$J_1 = \int_0^\infty \varepsilon^2(t)dt = 0.0547, \qquad J_2 = \int_0^\infty u^2(t)dt = 0.589. \quad [2.185]$$

–Third question:

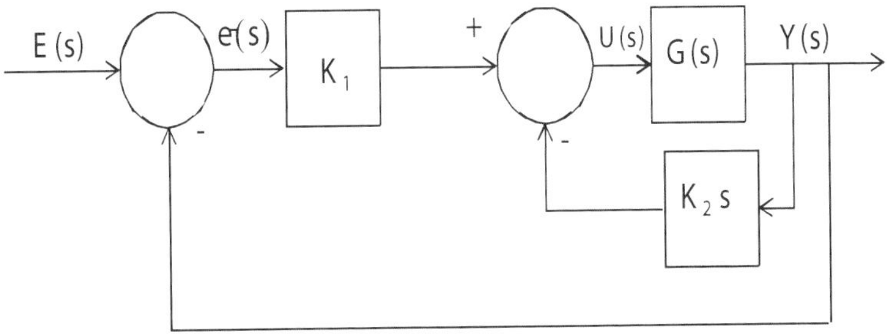

Figure 2.13. *Servo system using tachymetric feedback*

The system is controlled using tachymetric feedback as shown in Figure 2.13.

The new open-loop transfer function is given by

$$G'(s) = \frac{100K_1}{s(s + 100K_2)} = \frac{K}{s(1 + \tau s)} \qquad \begin{cases} K = \frac{K_1}{K_2} \\ \tau = \frac{1}{100K_2} \end{cases} \quad [2.186]$$

from which the modulus and the phase of $G'(s)$ are given by

$$|G'(j\omega)| = \frac{K}{\omega\sqrt{1 + (\tau\omega)^2}}, \qquad \phi(\omega) = -90° - arct(\tau\omega). \quad [2.187]$$

– The parameters K_1 and K_2 used to obtain the same system characteristics as in the second question (phase margin $\Delta\phi = 45°$ and bandwidth $\omega_c = 15.5\text{rad/s}$) are calculated using the fact that $\phi(\omega) = -135°$ for $\omega = \frac{1}{\tau} = 100K_2$. We then chose:

$$\omega_c = 100K_2 = 15.5\text{rd/s} \implies K_2 = 0.155. \quad [2.188]$$

The gain for $\omega = \omega_c$ must be equal to 1 (cross-point at 0 dB):

$$| G'(j\omega_c) | = \frac{100K_1}{\omega_c^2\sqrt{2}} = 1 \quad \Longrightarrow \quad K_1 = 3.415. \qquad [2.189]$$

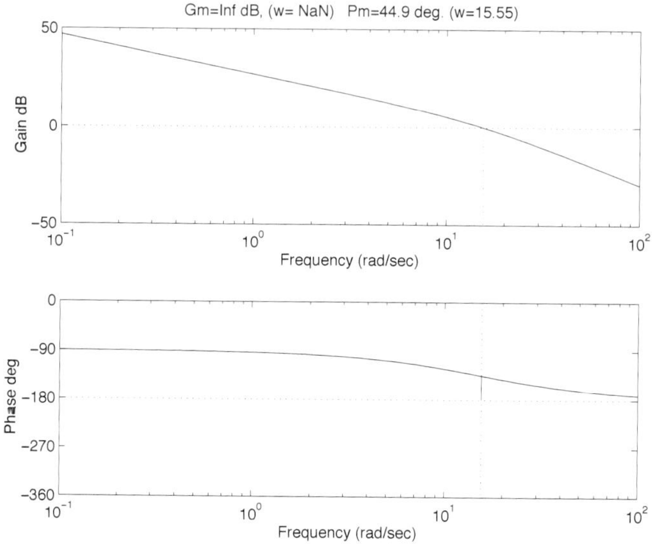

Figure 2.14. *Bode plot of the system after correction*

– The expressions of $\varepsilon(s)$ and of $U(s)$ are given by:

$$\varepsilon(s) = \frac{s + 100K_2}{s^2 + 100K_2 s + 100K_1}, \qquad U(s) = \frac{K_1 s}{s^2 + 100K_2 s + 100K_1}. \qquad [2.190]$$

Using the formulas given in the preamble, we calculate the values of the two integrals J_1 and J_2:

$$J_1 = \int_0^\infty \varepsilon^2(t)dt = \frac{K_1 + 100K_2^2}{200K_1K_2} = 0.0551,$$

$$J_2 = \int_0^\infty u^2(t)dt = \frac{K_1^2}{200K_2} = 0.373. \qquad [2.191]$$

– The CL transfer function is given by

$$H'(s) = \frac{341.5}{s^2 + 15.5s + 341.5} \quad \Rightarrow \quad \omega_0 = 18.5 \text{ rad/s and } \xi = 0.42. \qquad [2.192]$$

This is a second order with the following characteristic parameters: $\omega_0 = 18.5$ rad/s and $\xi = 0.42$.

–*Fourth question*:

We want to calculate the parameters K_1 and K_2 such that to minimize the integrals J_1 and J_2.

– The aim is to minimize the integral of cost $J = J_1 + \lambda J_2$ by taking the value of $\lambda = 0.25$:

$$J = J_1 + \lambda J_2 = \frac{1}{200K_2} + \frac{K_2}{2K1} + \lambda\frac{K_1^2}{200K_2} = \frac{K_1 + 100K_2^2}{200K_1K_2} + \lambda\frac{K_1^2}{200K_2}. \quad [2.193]$$

The values of K_1 and K_2 that minimize J are calculated by cancelling out the derivatives with respect to K_1 and with respect to K_2:

$$\frac{\partial J}{\partial K_1} = 0 \quad \Rightarrow \quad -\frac{K_2}{2K_1^2} + \frac{\lambda}{100}\frac{K_1}{K_2} = 0. \qquad [2.194]$$

$$\frac{\partial J}{\partial K_2} = 0 \quad \Rightarrow \quad -\frac{1}{200K_2^2} + \frac{1}{2K_1} - \frac{\lambda}{200}\frac{K_1^2}{K_2^2} = 0. \qquad [2.195]$$

By solving the equation system, the values of K_1 and K_2 are obtained:

$$K_1 = 2 \quad \text{and} \quad K_2 = 0.2. \qquad [2.196]$$

– The expression of the transfer function of the CL system is written as:

$$H"(s) = \frac{200}{s^2 + 20s + 200} \quad \Rightarrow \quad \omega_0 = 14.14\text{rad/s} \quad \text{and} \quad \xi = 0.707. \quad [2.197]$$

The natural frequency is smaller than in the case of question 3 but the damping is optimal.

– Using the formulas given in the preamble, we calculate the values of the two integrals J_1 and J_2:

$$J_1 = \int_0^\infty \varepsilon^2(t)dt = \frac{K_1 + 100K_2^2}{200K_1K_2} = 0.075,$$
$$J_2 = \int_0^\infty u^2(t)dt = \frac{K_1^2}{200K_2} = 0.1. \qquad [2.198]$$

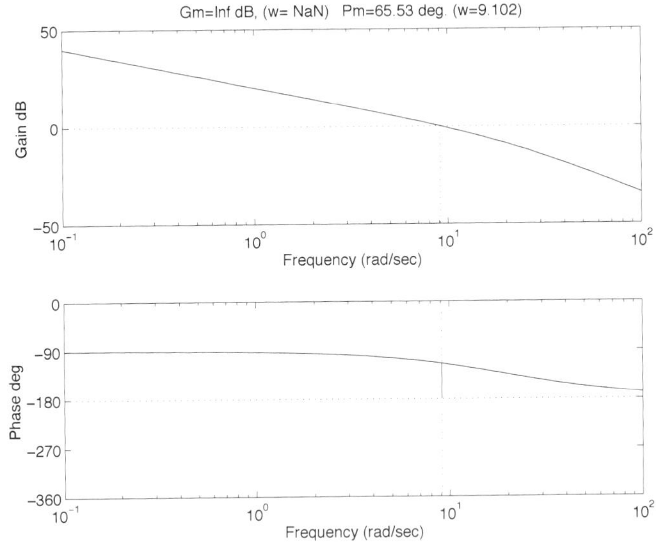

Figure 2.15. *Bode plot of the system after correction*

Comparative table of the different results (criterion minimization) is presented in Table 2.2.

$(J = J_1 + \lambda J_2)$ $(\lambda = 0.25)$	
1) $J = \infty$	
2) $J_1 = 0.054$, $J_2 = 0.59$	$J = 0.2015$
3) $J_1 = 0.055$, $J_2 = 0.375$	$J = 0.14875$
4) $J_1 = 0.075$, $J_2 = 0.1$	$J = 0.1$

Table 2.2. *Criterion minimization comparison*

– Comparison of the step responses and the control signals of the system corrected by three controllers.

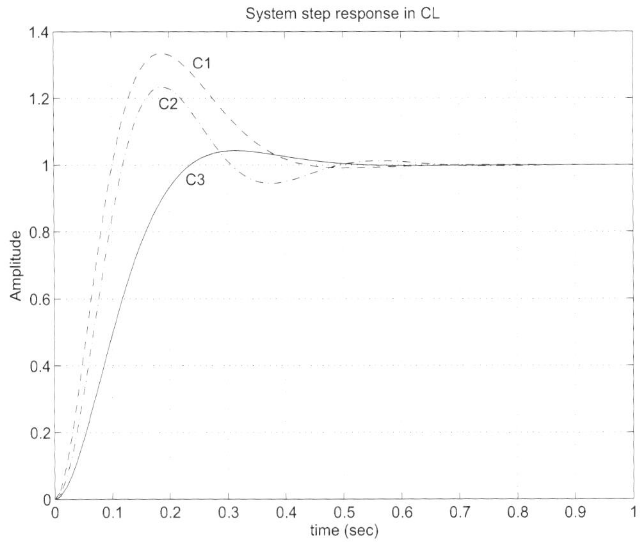

Figure 2.16. *Comparative study (step response)*

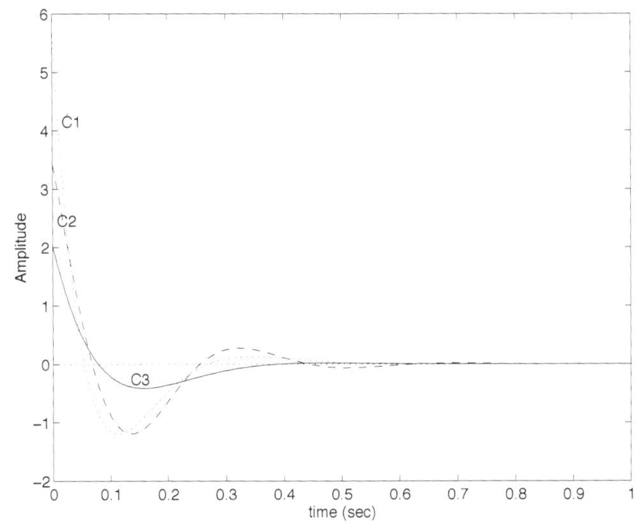

Figure 2.17. *Comparative study (control signal)*

–*Fifth question*

State–space representation:

$$G(s) = \frac{Y(s)}{U(s)} = \frac{100}{s^2} \quad \Rightarrow \quad s^2Y(s) = U(s). \tag{2.199}$$

Assuming

$$\begin{matrix} X_1(s) = Y(s) \\ X_2(s) = sX_1(s) \end{matrix} \Longrightarrow X(s) = \begin{pmatrix} X_1(s) \\ X_2(s) \end{pmatrix} \tag{2.200}$$

then a system state–space representation can be

$$\begin{matrix} \dot{x} = Ax + Bu \\ y = Cx \end{matrix} \quad \text{with}$$

$$A = \begin{pmatrix} 0 & 1 \\ 0 & 0 \end{pmatrix}; \quad B = \begin{pmatrix} 0 \\ 100 \end{pmatrix}; \quad C = (\,1 \; 0\,). \tag{2.201}$$

The characteristic polynomial of the non-looped system

$$P(s) = \det(sI - A) = \begin{vmatrix} s & -1 \\ 0 & s \end{vmatrix} = s^2. \tag{2.202}$$

Commandability

Let the controllability matrix be

$$Com = (B \quad AB) = \begin{pmatrix} 0 & 100 \\ 100 & 0 \end{pmatrix} \quad \Rightarrow \det(Com) = -100^2, \tag{2.203}$$

we correctly have Com of rank 2; the system is thus controllable

Observability

Let the observability matrix be

$$Obs = \begin{pmatrix} C \\ CA \end{pmatrix} = \begin{pmatrix} 1 & 0 \\ 0 & 1 \end{pmatrix} \quad \Rightarrow \det(Obs) = 1. \tag{2.204}$$

Obs is in effect of rank 2; the system is thus controllable.

Poles placement state feedback control

$$u = -Lx + l_1 y^d \quad (L = (l_1, l_2)). \tag{2.205}$$

We thus have the loop system that is written as:

$$\begin{aligned} \dot{x} &= (A - BL)x + Bl_1 y^d \\ y &= Cx \end{aligned} \quad \text{with} \quad A - BL = \begin{pmatrix} 0 & 1 \\ -100l_1 & -100l_2. \end{pmatrix}. \tag{2.206}$$

The polynomial equation is given by:

$$P(s) = \det(A - BL) = s(s + 100l_2) + 100l_1 = s^2 + 100l_2 s + 100l_1. \tag{2.207}$$

The equivalent transfer function of the loop system is given by:

$$\begin{aligned} G'(s) &= C(sI - (A - BL))^{-1}B = \frac{1}{P(s)} \begin{pmatrix} 1 & 0 \end{pmatrix} \\ &\times \begin{pmatrix} s + 100l_2 & 1 \\ -100l_1 & s \end{pmatrix} \begin{pmatrix} 0 \\ 100 \end{pmatrix} \\ &= \frac{100l_1}{P(s)} = \frac{100l_1}{s^2 + 100l_2 s + 100l_1}. \end{aligned} \tag{2.208}$$

In order to get a CL system behaving as in question 3, we have to establish:

$$l_1 = K_1 = 3.415 \quad l_2 = K_2 = 0.155. \tag{2.209}$$

Optimal state feedback control

Let the following criterion be:

$$J = \int_0^\infty (y^2(t) + \lambda u_*^2(t))dt \quad (\lambda = 0.25, \quad u_*(t) = -Lx(t)). \tag{2.210}$$

Matrices Q and R are given by

$$Q = C^T C = \begin{pmatrix} 1 & 0 \\ 0 & 0 \end{pmatrix} \quad R = \lambda = 0.25. \tag{2.211}$$

To obtain the optimal gain matrix, we solve the following Ricatti equation:

$$A^T P + PA - PBR^{-1}B^T P + Q = 0 \tag{2.212}$$

with

$$P = \begin{pmatrix} p_1 & p_2 \\ p_2 & p_3 \end{pmatrix}, \quad \text{symmetric positive definite } (L = R^{-1}B^T P). \tag{2.213}$$

The solution is given by:

$$\begin{aligned} &\begin{pmatrix} -40000p_2^2 + 1 & p_1 - 40000p_2p_3 \\ p_1 - 40000p_2p_3 & 2p_2 - 40000p_3^2 \end{pmatrix} = \begin{pmatrix} 0 & 0 \\ 0 & 0 \end{pmatrix} \\ &\Rightarrow P = \begin{pmatrix} .1 & 0.005 \\ 0.005 & 0.0005 \end{pmatrix} \end{aligned} \tag{2.214}$$

wherefrom the gain L is given by:

$$L = R^{-1}B^T P = \begin{pmatrix} 2 & 0.2 \end{pmatrix} = \begin{pmatrix} K_1 & K_2 \end{pmatrix} \text{ of the fourth question.} \tag{2.215}$$

Observer

The output of the system is always measurable ($y = x_1$); as a result, we must reconstruct the variable (x_2). The proposed observer is then given by:

$$\begin{aligned} \dot{\hat{x}} &= A\hat{x} + Bu + K(y - \hat{y}) \\ \hat{y} &= C\hat{x} \end{aligned} \Leftrightarrow \begin{aligned} \dot{\hat{x}} &= A\hat{x} + Bu + KC(x - \hat{x}) \\ \hat{y} &= C\hat{x}. \end{aligned} \tag{2.216}$$

The observation error equation ($\tilde{x} = x - \hat{x}$) is thus given by

$$\dot{\tilde{x}} = (A - KC)\tilde{x}. \tag{2.217}$$

The characteristic polynomial of the observation error equation is given by:

$$P(s) = \det(sI - (A - KC)) = s^2 + K_1 s + K_2, \tag{2.218}$$

the roots of this polynomial should have negative real parts in order for the observer to be stable.

In state feedback control, we use the reconstructed state:

$$u = -L\hat{x} + l_1 y^d. \qquad [2.219]$$

Therefore for the system, the observer and the control, we get the following equations:

$$\begin{cases} \dot{\hat{x}} = (A - BL - KC)\hat{x} + LCx + Bl_1 y^d \\ \dot{x} = Ax - BL\hat{x} + Bl_1 y^d \\ y = Cx \\ u = -L\hat{x} + l_1 y^d \end{cases}$$

$$\Longrightarrow \begin{cases} \dot{\tilde{x}} = (A - KC)\tilde{x} \\ \dot{x} = (A - BL)x - BL\tilde{x} + Bl_1 y^d \\ y = Cx \\ u = -L\hat{x} + l_1 y^d. \end{cases} \qquad [2.220]$$

The last four equations allow us to define a system having input v, state $\begin{pmatrix} x & \tilde{x} \end{pmatrix}^T$ and the output $\begin{pmatrix} y & u \end{pmatrix}^T$ by:

$$\begin{pmatrix} \dot{x} \\ \dot{\tilde{x}} \end{pmatrix} = \begin{pmatrix} A - BL & -BL \\ 0 & A - KC \end{pmatrix} \begin{pmatrix} x \\ \tilde{x} \end{pmatrix} + \begin{pmatrix} B \\ 0 \end{pmatrix} l_1 y^d \qquad [2.221]$$

$$\begin{pmatrix} y \\ u \end{pmatrix} = \begin{pmatrix} C & 0 \\ L & L \end{pmatrix} \begin{pmatrix} x \\ \tilde{x} \end{pmatrix} + \begin{pmatrix} 0 \\ 1 \end{pmatrix} l_1 y^d. \qquad [2.222]$$

Separation principle: the fact that the matrix of equation [2.221] be triangular indicates that the placement of the observer poles (impose a dynamic to the observation error) is independent of the system control and the placement of the poles for the control.

REMARK 2.5.– *Regarding the system control, it can be observed that the dynamics of the system state does not depend on the gain of the observer, the placement of the poles of the regulator is not influenced by the choice of observer. On the other side, the control error will converge after the convergence of the observation error, which seems natural. Consequently, the dynamics of the observer must be faster than that of the regulator.*

Sixth question:

We want to control the system by means of a digital controller.

– The block diagram of the system controlled by a digital controller.

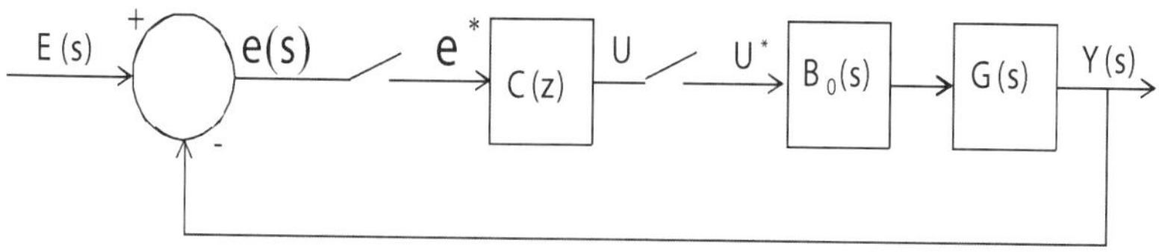

Figure 2.18. *System with digital controller*

– The discrete transfer function

$$G(z) = (1 - z^{-1})TZ[\frac{G(s)}{s}] = 100\frac{T_e^2}{2}\frac{z+1}{(z-1)^2}. \quad [2.223]$$

The sampling period $T_e \simeq T_0/10$, with $T_0 = 2\pi/\omega_0$. ω_0 is the bandwidth of the open-loop system ($\omega_0 = 10$ rad/s $\Rightarrow T_e \simeq 0.06$ s).

The controller providing a response with absolute minimal response time:

$$H(z) = \frac{C(z)G(z)}{1 + C(z)G(z)} = z^{-1} \quad \Rightarrow \quad C(z) = \frac{1}{100}\frac{2}{T_e^2}\frac{z-1}{z+1}. \quad [2.224]$$

The controller is of the derivator type (static gain: $C(1) = 0$).

– Recurrence equation of the controller

$$C(z) = \frac{U(z)}{\varepsilon(z)} = \alpha\frac{z-1}{z+1} \quad \Rightarrow \quad (z+1)U(z) = \alpha(z-1)\varepsilon(z). \quad [2.225]$$

Let the following recurrence equation be:

$$u(k) = -u(k-1) + \alpha(\varepsilon(k) - \varepsilon(k-1)) \quad [2.226]$$

Characteristics of the control signal for a step setpoint:

$$y^d(k) = 1 \text{ (for } k \geq 0\text{), and since } H(z) = z^{-1}\text{, then } y(k) = y^d(k-1) \quad [2.227]$$

$$\text{where } \varepsilon(k) = y^d(k) - y^d(k-1) \quad [2.228]$$

$$\text{and thus } u(k) = -u(k-1) + \alpha(y^d(k) - 2y^d(k-1) + y^d(k-2)). \quad [2.229]$$

$u(0) = \alpha$
$u(1) = -\alpha + \alpha - 2\alpha = -2\alpha$
$u(2) = 2\alpha + \alpha - 2\alpha + \alpha = 2\alpha$
$u(3) = -2\alpha$
$u(4) = 2\alpha$
$\vdots$

Table 2.3. *Characteristics of the control signal for a step setpoint*

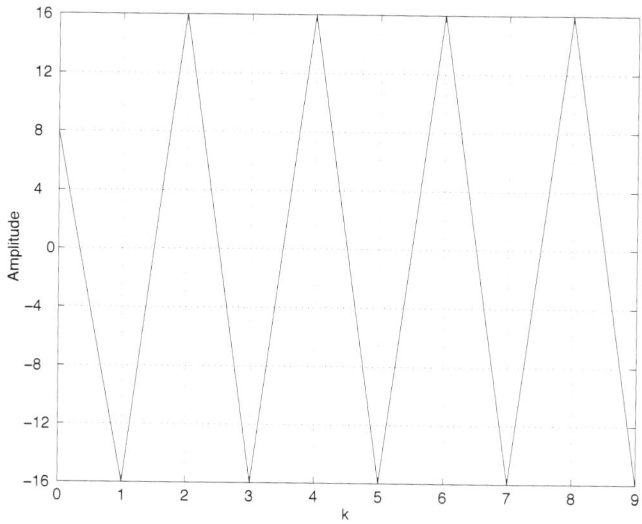

Figure 2.19. $u(t)$ $(\alpha = 8 \; ; T_e = 0.05 \; s)$

3

Actuators: Modeling and Analysis

This chapter explores conversion and energy transfer, friction modeling and the flexible transmission chain.

3.1. Introduction: electric, hydraulic and pneumatic actuators

Machine development has been long associated with the need to extend the physical abilities of human beings in order to perform increasingly complex and difficult tasks.

In robotics, three large families of actuators are mainly used: electric, hydraulic and pneumatic actuators. Recent technological progress in terms of mass power, reliability and flexibility have been made possible due to actuators. Actuators are systems whose main role is to convert energy; it is necessary to analyze the various technological possibilities which exist for each of these functions. The motorization of an actuator includes the engine itself, as well as the converter that allows it to be controlled. The types of engines can be differentiated considering the primary source of energy. In general, three types of energy sources are used: hydraulic, pneumatic or electric.

Hydraulic technology has unmatched couple and mass power; it is the solution for large robots with significant payloads. Pneumatic technology makes it possible to obtain satisfactory performance but implies a complex servo structure. These two technologies have in common the need for a structure external to the robot providing the energy, pneumatic distribution network and oleo-hydraulic plant. There are also particular systems that operate according to different principles. As an example, there are actuators that operate based on the deformation of metals (possibly including form memory or piezoelectric) with ultrasound or electrostatic forces.

The conversion of energy into motion is a key point of actuators (mechanical energy from electric, thermal, chemical, gravitational, hydraulic, pneumatic or magnetic energy, etc.). As actuators or energy converters have been developed, the problem related to the control of these actuators has become of paramount importance. This is due to concerns for accuracy and performance. The development of actuators has thus evolved from a simple problem of energy conversion to an increasingly varied technology including problems concerning the control, the accuracy of sensors, sensor and observer information processing, etc. Integrated actuators are as follows: energy conversion, information acquisition, signal processing, information reconstruction and control. It should be noted that the recent development of materials with new characteristics (SMA, permanent magnets, etc.) and new means of transmitting energy, electronics and computer technologies are also contributing to the development of sensors and actuators.

Finally, we also note the possibility that smart actuators can be achieved with self-calibration, information processing, communications and control systems or servos.

3.1.1. *Representation methods for physical systems*

3.1.1.1. *Analogies between physical systems*

When modeling a complex system, its study is approached by blocks after a decomposition in elemental parts for which differential equations will be established. For each system block, the constitutive equation or the physical law governing it is determined. The association of these equations gives the model of the system. A basic idea (for example from the approach by Bond graphs, whose study is not part of this book) for a method of unified representation is that in most physical phenomena, energy can only be stored in inductive or capacitive form (see analogies) or dissipated. Energy or power (its derivative with respect to time) are quantities common to all areas of physics. A way to describe the dynamic behavior of a physical system is to follow the flow of energy and indicate where it is dissipated, stored through the components of the process (see Blau *et al.*'s works). The connection between building blocks is achieved by nodes (in series or in parallel), transformers (amplifiers and reducers) or gyrators (energy converters). When studying physical systems (mechanical, electrical, hydraulic, thermodynamic, etc.), one encounters several types of energy. For all of these types, two types of generalized variables can be defined: power variables and energy variables. These variables can be classified into two categories: efforts and flows.

3.1.1.2. *Representations and energy transfer*

With regard to systems, we can classify physical quantities into two categories as in thermodynamics: effort quantities and flux quantities (or flow). The study of passive systems shows great flexibility when it is based on the use of these physical quantities, especially for systems composed of interconnected blocks. It has the advantage of representing energy and its evolution in different blocks that build up the system. It can be considered that one of the main motivations for this representation and analysis method is that most physical dynamic systems obey the equation of conservation of energy. The variation of the stored energy is equal to the sum of the power supplied to the system and to the internal power of the latter (produced and/or dissipated).

$$d\,\{\text{stored energy}\}/\text{dt} = \text{provided power} + \text{internal power}$$

Moreover, it should be noted that this energy balance is at the basis of any modeling and representation methodology for dynamic systems: bond graph (BG) representation. This representation differs from signal-flow graphs and in some sense is similar with the graphical representation of chemical formulas. It was introduced around 1950 by H. M. Blau, who established the basic notations. This method of modeling allows the representation of most physical systems using an equivalent passive diagram, and the understanding of its mechanism facilitates the realization of such a diagram.

In BG representation and modeling using energy equations, the main focus is on the transfer of energy in the system. Each line (or link or bond) of a BG represents a transfer of energy by a simple building block of the system. It assumes that there is a pair of signals whose directions are opposed. The product of these signals must define the power supplied to the block or element under consideration. In electrical circuits, these signals can be, for a resistance for example, voltage and current. For a mechanical system, we will take the force and the velocity for a damper. It should be noted that this allows us to define the mechanical impedance of a system, its compliance or its admittance.

Analogies between mechanical, hydraulic, pneumatic and other systems with electrical circuits can be used or illustrated by means of this representation method. This explains the fact that this form of representation can be applied to any kind of physical system and that for each element, two types of signal can be found: namely flux (current, velocity, flow, etc.), and effort (tension, force, pressure, temperature, etc.), whose product gives the power supplied to the element being considered. The interest of this methodology lies in the fact that it allows us to easily connect several types of system (electrical, mechanical, hydraulic, pneumatic, chemical, biochemical, etc.) since it implicitly includes energy conversion.

The analysis of passive systems is often based on the equivalent representation of a system so that the scalar product input–output yTu determines the power supplied to the system. In order to clarify this point, we will present the analogies between systems.

System	Effort: e	Flux: f	Power in W
Mecanic q:quant. mvt	Force in N, Torque in N.m	Velocities in m/s in rad/s	F.v Gw
Hydraulic	Pressure P in Pa	Flow Q in m3/s	PQ
Thermic	Temperature in K	Heart Flux dQ/dt in JK/s	TdQ/dt
Electric	Voltage in V	Current I in A	VI

Table 3.1. *Analogy between systems*

3.1.2. *Modeling of a few constituents of physical systems*

In the following, we recall the definitions of a few constituents and their modeling in order to make possible the modeling, the analysis, the simulation and the study of the dynamic behavior of systems that are of our interest.

3.1.2.1. *Circuits and electric networks*

A network is a set constituting of nodes and branches. It is possible to access a network through a pair of nodes (or pair of terminals). The network is said to be oriented when we choose a traveling direction for each branch. For a dipole, there are two conventions for the orientation: the receiver convention, the current and the voltage are opposites, as well as the generator convention, current and voltage are in the same direction. For the receiver convention (R), the network is oriented as shown in Figure 3.1. A single-access network is a dipole.

3.1.2.2. *Characteristics of direct current machines*

Most machines have permanent magnets or separated inductors. In this case, we consider F as constant. A permanent magnet machine has a network of real characteristics that make it simply utilizable in servos. The velocity is almost proportional to the supply voltage (see figure below). The decrease in characteristics is mainly due to the resistance of the winding of the armature, and to a lesser extent to the magnetic reaction of the armature, which is normally negligible. The limitations are mechanical and thermal. In effect, from a thermal point of view, the direct current machine appears as a source of heat to the rotor and as a result is difficult to dissipate.

As for the collector, it is probably the lowest point of this machine. The collector is a sliding contact that switches on and off the current in the coils that make up the armature. The current is technologically limited to values ranging up to $20 \text{ A} / \text{cm}^2$ for metallographitic brushes, whereas the copper wire is used at densities of the order of $5 \text{ A} / \text{mm}^2$. The size of the collector is relatively significant and increases the length of the machine. Additionally, the inevitable wear of the brushes that establish contact require inconvenient maintenance for a production machine. Mechanically, the collector blades are not strong enough for the excess of speed that may pull them out, although the main risk is the flash of the collector that appears when the potential difference is too significant between two successive blades.

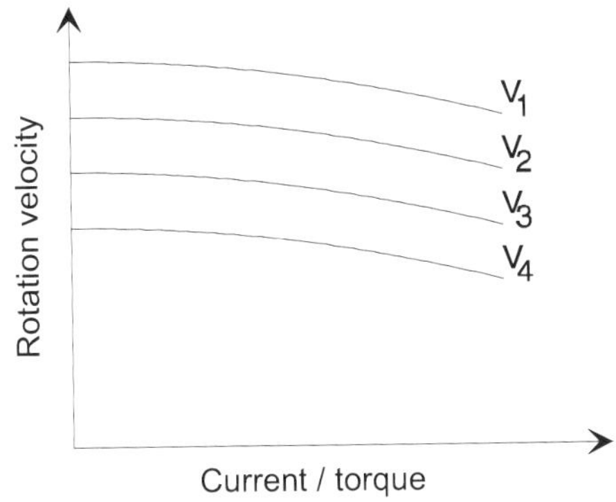

Figure 3.1. *Characteristic velocity = f(torque) at constant voltages*

The DC machine, comprising an iron rotor, withstands temporary torque overloads relatively well. Disc or bell machines, whose rotor is copper, having very low mechanical and thermal inertia are much more sensitive to these overloads, which must be avoided in the electronics of the servo.

3.1.2.3. *Dynamic behavior or transient behavior*

A motor in variable state can be represented schematically by the circuit below. In addition to the previous model, we take into account the inductance of the coil that creates the electric time constant of the motor and a mechanical part.

R: internal resistance of the motor; L: inductance of the winding; $e(t)$: motor-induced EMF; $u(t)$: voltage at the terminals of the motor; J; moment of inertia of the motor shaft; f: viscous friction coefficients on the shaft; $i(t)$: instantaneous current flowing through the engine; $\Delta\omega$: the instantaneous rotation velocity of the motor axle.

The equations of this system are written in the following manner.

Electrical equations:

$$u(t) = R.i(t) + L.di(t)/dt + e(t), \quad [3.1]$$

$$e(t) = K\omega(t). \quad [3.2]$$

Mechanical equations:

$$C = J.d\Delta\omega(t)/dt + \Delta\omega(t) + Cr. \quad [3.3]$$

Cr being the load torque on the output shaft.

$$C = K.i(t). \quad [3.4]$$

Or even substituting $e(t)$ in the first equation and C in the second equation:

$$u(t) = R.i(t) + L.di(t)/dt + K.\Delta\omega(t), \quad [3.5]$$

$$K.i(t) = J.d\Delta\omega(t)/dt + \Delta\omega(t) + Cr. \quad [3.6]$$

By multiplying the first relationship by $i(t).dt$, it is shown that the electric energy borrowed from source $u(t).i(t).dt$ is found in the form of energy dissipated by the Joule effect $R.i(t)2.dt$ of energy stored in electromagnetic form $L.i(t).di$ due to the current and mechanical energy variation $e(t).i(t).dt = K.\Delta\omega(t)i(t).dt$.

Note that for the representation of this system, a state–space representation can be derived (in the form of a differential equation of the first order). To this end, we chose as state vector $X = [i(t); \Delta\omega(t)]T$; in a first stage, we consider the resistive torque as being zero $Cr = 0$. This gives us the following state–space representation:

$$\dot{X} = A.X + B.U, \quad [3.7]$$

$$y = C.X. \quad [3.8]$$

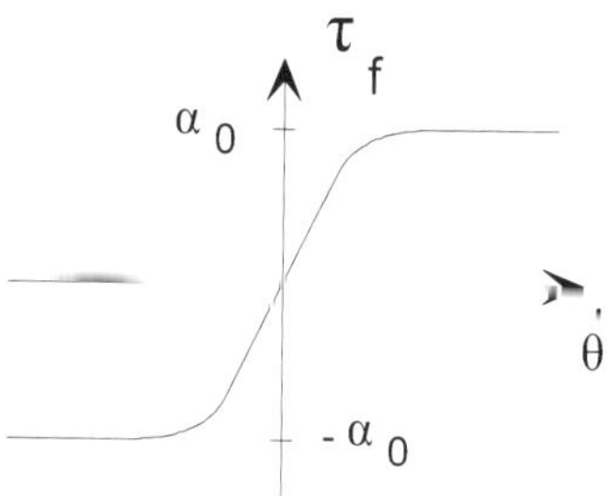

Figure 3.2. *Adherence model*

The indices + and - above the coefficients a_i indicate a positive or negative direction of the motion. It should be noted that the coefficients a_i are represented as α_i in the above figures.

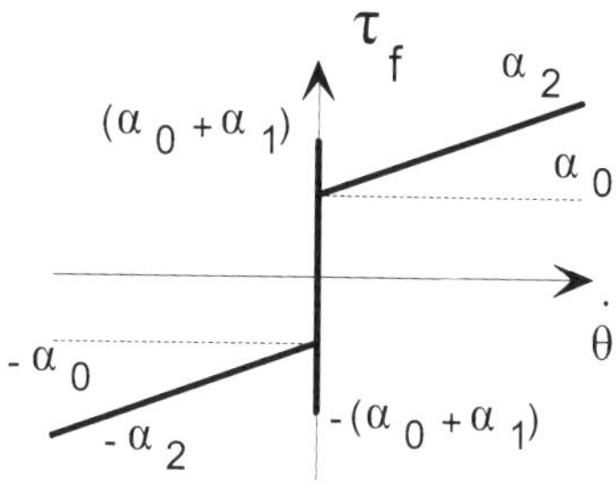

Figure 3.3. *Karnopp model*

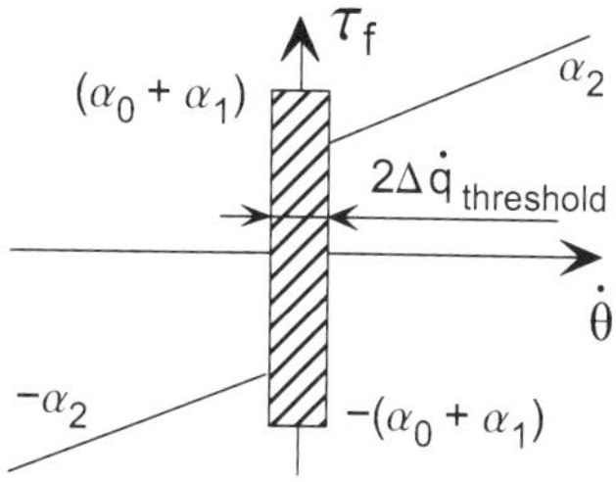

Figure 3.4. *Coulomb model*

The coefficient a_0 corresponds to the amplitude of dry friction. Karnopp has suggested a different modeling presented where a_0 represents the Coulomb friction coefficient, a_2 is the viscous friction coefficient and $a_0 + a_1$ corresponds to the value of the torque generated by the adherence phenomenon. This approach makes it possible to separate the model into two regions. In the first region, the velocity is zero or almost zero. The system moves away from this region and goes into the viscous friction region only if the force applied to the system exceeds the adherence torque. This model is a means to encompass both Coulomb's model and Dahl's model as well as Stribeck's friction effect.

3.1.2.4. *Asymmetries*

The models presented above are all symmetrical. However in practice, this is not always the case. Imperfections in the different elements of the system can result in an asymmetric behavior and performance in different manners according to the rotation direction. In the case of asymmetry, the coefficients a_0 and a_2, of the dry friction and the viscous friction, respectively, will be different depending on the rotation direction; the same occurs for adherence friction.

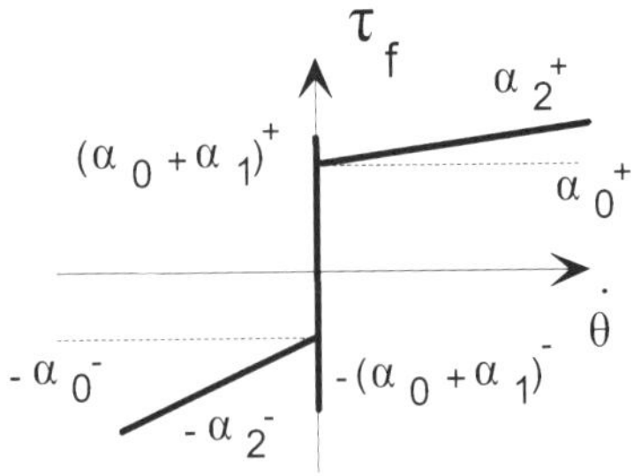

Figure 3.5. *Dahl model*

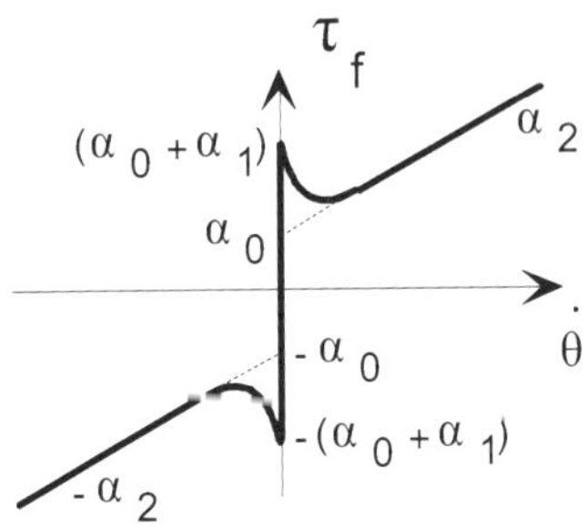

Figure 3.6. *Generalized Stribeck curve*

3.1.2.5. *Position dependence*

The friction torque can also depend on the angular position. For example, this dependence can be a consequence of imperfections in centering elements. These imperfections generate, for instance in a straight gear train reducer, oscillations of period proportional to the reduction ratio.

3.1.2.6. *Stribeck's friction*

This friction obeys the generalized Stribeck curve (see Armstrong's work [ARM 00]) and corresponds to the area of exponential growth (downward bends) of the friction torque, which appears when the velocity increases in the range of very low velocities. The variation phenomenon of the friction coefficient with respect to the velocity has been observed as early as 1902 by Stribeck. In effect, when the stiction torque has been overcome, the friction torque exponentially decreases and reaches approximately 60% of the pull-out torque (or break away torque), then increases in proportion to the velocity. This is explained by the system shifting from a metal lubrication process (direct friction of both surfaces in region 1) to a partially fluid and partially metallic lubrication process (region 2) during the relative motion of the two mechanical parts. Thereafter, the lubrication becomes fully fluid (region 3). These measures confirm the model proposed by Tustin, who includes this area of negative dependence of the friction with respect to the velocity. The presence of this area creates stability problems. It is worth noting that the velocity limiting the regions (2) and (3) corresponds to the threshold velocity of the Karnopp model.

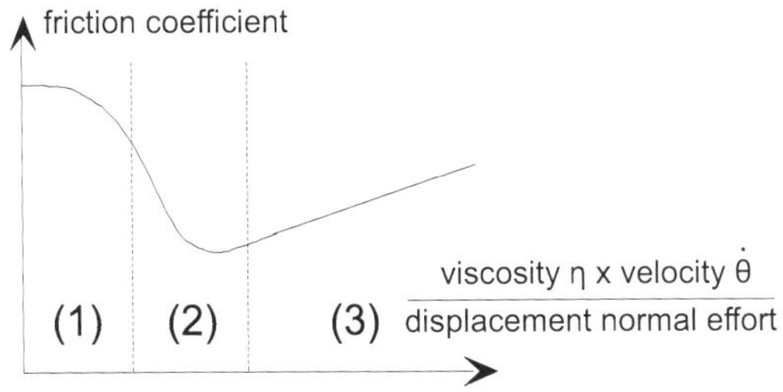

Figure 3.7. *Friction effect in regions 1–3*

3.1.2.7. *Friction modeling*

The model proposed by Tustin in 1947 [TUS 47] recovers the main components of friction previously presented; it has been experimentally validated by Armstrong in the case of a robot for manipulation. Tustin's model can thus be expressed by Canudas de Wit [CAN 98]:

$$\tau(\dot{q}) = (a_0 + a_1.e^{-b|\dot{q}|}. + a_2|\dot{q}|)sgn(\dot{q}) \qquad [3.9]$$

where a_0 represents the Coulomb friction coefficient, a_2 is the viscous friction coefficient and $(a_0 + a_1)$ corresponds to the value of the adherence torque. The value b corresponds to the lubrication coefficient. If the friction presents asymmetries, the ai coefficients can each take two different values following the direction of the motion. Other representations of the same form can be found in Armstrong and Dupont's works [ARM 00, DUP 09] or, for example, in Canudas's work [CAN 98]:

$$\tau(\dot{q}) = (a_0 + a_1.|\dot{q}|^{1/2}. + a_2|\dot{q}|)sgn(\dot{q}). \qquad [3.10]$$

Karnopp's model is expressed in the following way, if we consider the sliding area,

$$|\dot{q}| > \Delta\dot{q} \qquad \tau(\dot{q}) = (a_0.sgn(\dot{q}) + a_2\dot{q}) \qquad [3.11]$$

and if we are in the adherence area, $|\dot{q}| > \Delta\dot{q}$

$$\tau(\dot{q}) = \min(\tau_{\text{calculated}}, \tau_{\text{adherence}}.sgn(\tau_{\text{calculated}})) \qquad [3.12]$$

where a_0 and a_2 represent dry friction and viscous friction coefficients, $\tau_{\text{calculated}}$ is the calculated torque imposed on the system, $\tau_{\text{adherence}}$ is the estimated adherence torque and $\Delta\dot{q}$ corresponds to the estimated adherence area for the system being considered. The function min. selects the nearest lowest value and limits maximum torque to the adhesion torque $(a_0 + a_1)$.

Here, some results are obtained on the test site.

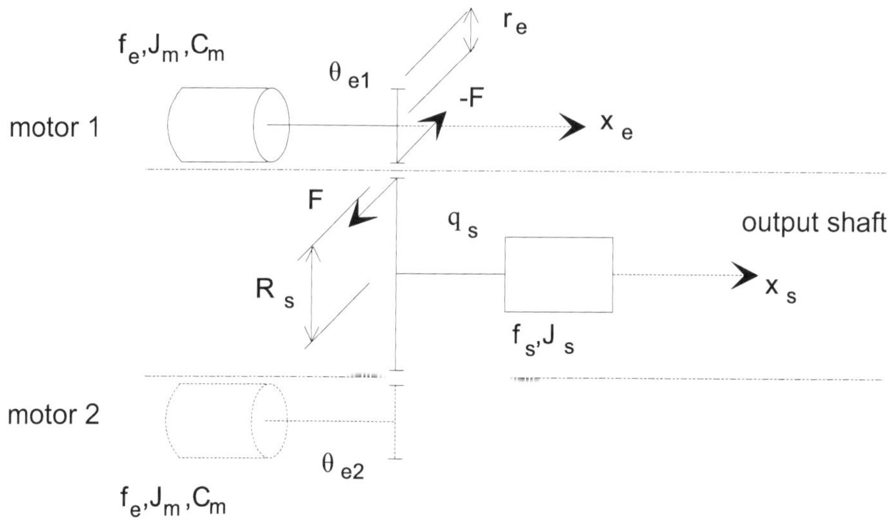

Figure 3.8. *Diagram of the proposed actuator model*

3.2. Transmission chains, actuators and sensors

3.2.1. *Electric actuators in robotics*

A robot or manipulator arm is a mechanical multifunctional system intended to assist people or to replace them in hostile environments or at scales which are either very large or very small (micromanipulation, microrobotics, macrooperations, protheses, etc.). There are several control modes for manipulators: all or nothing (open-loop switch control), programming, remote operation or analog automatic control.

Actuators (muscles) are chosen according to the tasks achieved and required performance. Simple manipulators and small dimensions generally make use of pneumatic actuators (in the open-loop system and nowadays increasingly more in the closed-loop system). When the required effort is not very important, stepper motors are used, which in principle operate in the open-loop system. When great precision, important torque performance, significant velocities and acceleration are desirable, we resort to electric actuators or direct current motors. Rotating and electric machines are by far the most widespread in applications.

An electric motor is a machine that converts electric energy into mechanical energy. It uses the principles of electromechanical conversion (magnetic induction) when an electric current flows through a conductor. This phenomenon uses the following physical laws:

– Laplace's law: a conductor of length L (in meters), through which flows an electric current I (Amperes), is placed in a magnetic field B (in Teslas), and subjected to a force $F = B.I.l$ ((in newtons).

– Faraday law: an electromotive force (EMF) is induced in an electric circuit by the variation of the magnetic induction flow (in time) that flows through this circuit.

– Lenz's law: the current induced in a conductor opposes the cause that creates it.

The use of the above-mentioned physical laws allows that electric energy be converted into mechanical energy (electric engine) or vice versa (tachymetric generator).

For an electric motor, Faraday's law and Lenz's law make it possible to predict that the electric motor will create a counter EMF, which is expressed as $E = B.v.l$ (in volts). v is the velocity of the conductor in meters by seconds.

Rotating machines have two coaxial cylindrical armatures, movable relative to each other. From the electrical point of view, we call them inductors when they create the magnetic field and generators when they generate an EMF induced by the magnetic field.

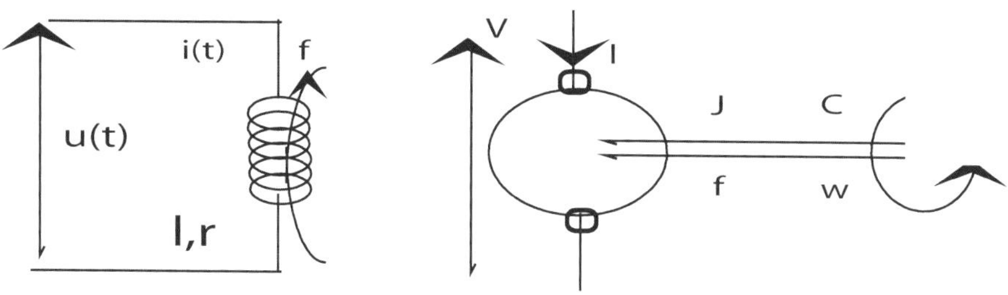

Figure 3.9. *DC motor 1*

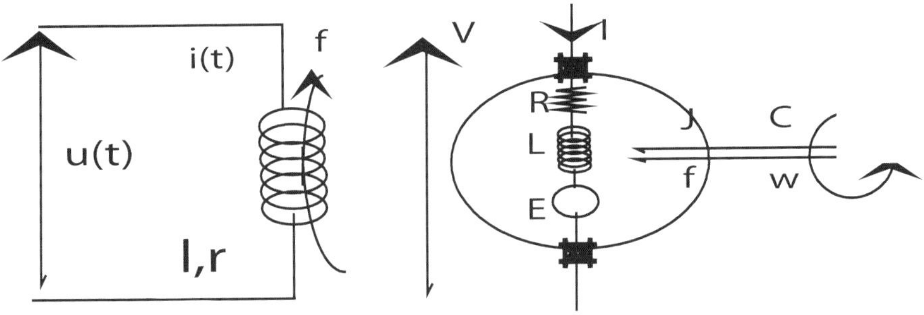

Figure 3.10. *DC motor 2*

3.2.1.1. *Principle*

A direct current motor may consist of two circuits, one inductor and one generator according to the principle below. An inductor current $i(t)$ flows through the inductor of impedance (r, l) that produces the flux. A current I passes through the impedance of the armature (R, L).

We then obtain the following dynamics relationships:

$$V = R.I(t) + L.\frac{dI(t)}{dt} + e(t), \quad [3.13]$$

$$u(t) = r.i(t) + l.\frac{di(t)}{dt}, \quad [3.14]$$

$\phi(t) = f(i)$; nonlinear relation yielding the flux.

$$e(t) = k_1.\phi(t).\omega(t), \quad [3.15]$$

$$C(t) = k_2.\phi(t).I(t), \quad [3.16]$$

$C = J.\frac{d\omega(t)}{dt} + f.\omega(t) + C_r$; C_r being the resistive torque on the output shaft.

Depending on the type of motor, we could impose I = constant (inductor controlled motor; this method is difficult to achieve and is practically not applied) or i = constant (armature controlled motor, flux = constant) because torque C and the EMF E developed are nonlinear functions. For example, if $i = i_o$= constant and $\phi = \phi_o$= constant case of an armature-controlled motor, we then get:

$e(t) = k_1.\phi_o.\omega(t)$ and $C(t) = k_2.\phi_o.I(t)$; armature-controlled motor ($i = i_o$ and $\phi = \phi_o$).

In this case and according to the notations of the diagram below (these notations will be used in the following), we obtain:

$$V = R.I(t) + L.\frac{dI(t)}{dt} + e(t), \quad [3.17]$$

$$e(t) = k_1.\omega(t); \quad C(t) = k_2.I(t), \quad [3.18]$$

$$C(t) = J.\frac{d\omega(t)}{dt} + f.\omega(t) + C_r, \quad [3.19]$$

with $k_1 = k_2$ expressed in the SI system.

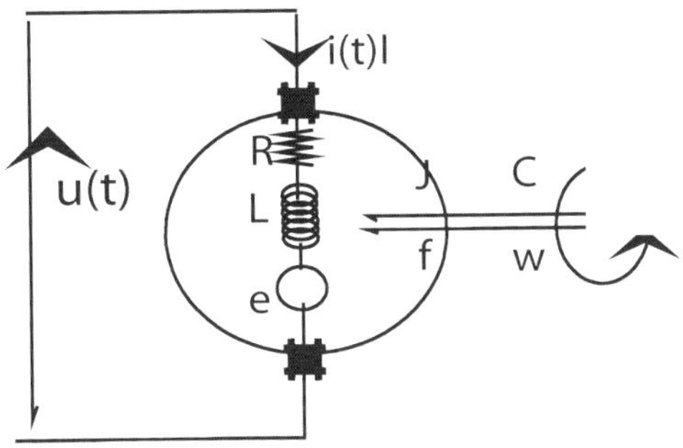

Figure 3.11. *Servo motor*

In the following and before going into detail about the modeling of this dynamic behavior system, we will study the static characteristic of the most-used motors in industrial applications.

A DC motor or separated excitation torque motor is composed of a metal annular piece supporting the permanent magnets intended to produce the magnetic flux. The armature consists of coils constituting several rows. These characteristics are as follows: high torque according to the power and a steady operation at low velocities. The torque is a linear function of the velocity and these time constants are small. These engines are generally compact, robust and practical for mobile machinery or often transported. In the following text, we present the modeling of the dynamic behavior of rotating DC machines.

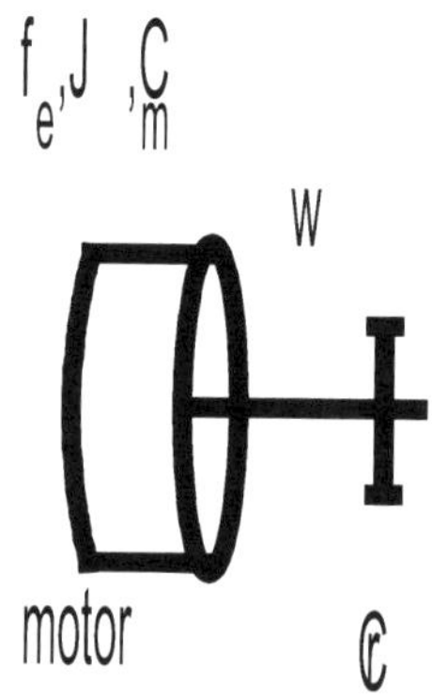

Figure 3.12. *Motor parameters*

3.2.1.2. *Permanent state*

The equivalent electric diagram of a DC motor excited by permanent magnets is that with R, (in ohms) the resistance of the armature and the contact between brushes and slip rings ($R = R_i + R_c$), E (in volts) the counter EMF, U the voltage at the terminals of the motor, C (in $\mathrm{N} \cdot \mathrm{m}$) the electromagnetic torque produced by the armature, I (in amperes) the current flowing through the armature and d/dt the rotation velocity of the armature. C is the produced torque and ω is the speed of rotation of the axis of the engine ($\omega = \frac{d\theta}{dt}$). E is the counter EMF, U is the voltage at the terminals of the motor (volts), I is the current in the armature (amperes), $R = R_c + R_i$ is the resistance (ohms) of the armature R_i in addition to that of the contact between the carbon brushes and the collector R_c viewed from the supply conductors. These quantities are connected by the relationships:

$$
\begin{aligned}
&U = E + R.I \qquad C = I.(p.N.\phi)/(2.\pi.a) = K.I \\
&E = \omega.(p.N.\phi)/(2.\pi.a) = K.\omega
\end{aligned}
\qquad [3.20]
$$

where p is the number of pairs of poles in the inductor, a is the number of winding rows of the armature and N is the number of active conductors; the flow of magnetic induction produced by the inductor is in webers. In many cases where SI units are not used, different values of K_c and K_e are given for the torque constant and counter EMF constant.

3.2.2. *Motor speed torque characteristic*

It is obtained from the relations above using the following expansions:

$$I = \frac{(U-E)}{R} \qquad I = \frac{(U-K.\omega)}{R}$$
$$K.I = C = \frac{(U-K.\omega).K}{R} = K.\frac{U}{R} - \frac{K^2\omega}{R}. \qquad [3.21]$$

It is a family of straight lines in the plane (ω, C) of slope $-K^2/R$ graduated in values of U. The speed gradient is $\frac{\partial\omega}{\partial U}$ and the torque gradient is $\frac{\partial C}{\partial U}$. C_m is the locked rotor maximal torque and ω_m is the maximal speed with zero torque. In practice, the torque is limited to a maximal value, beyond which the current of the armature causes a partial demagnetization of the permanent magnets of the inductor. The characteristic is then limited by a so-called commutation hyperbola, beyond which the lifetime of the carbon brushes is no longer guaranteed. The real characteristic is then reduced to a line segment inside an authorized contour.

The hyperbola is obtained starting from $C = \frac{KU}{R} - \frac{K^2\omega}{R}$ by imposing $P_{max} = C\omega = \frac{RI^2}{4}$ = constant.

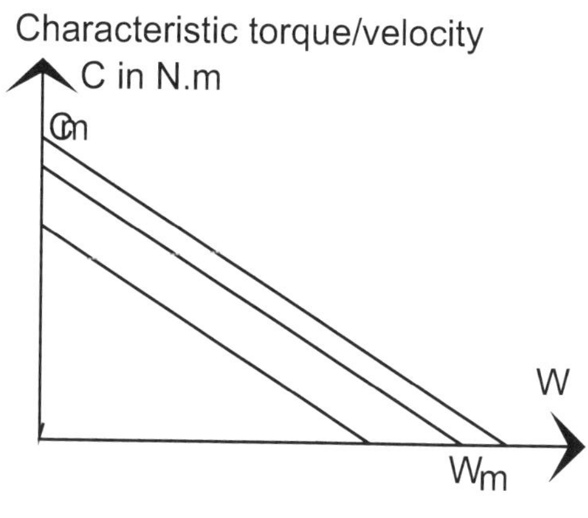

Figure 3.13. *Speed torque characteristic*

3.2.3. *Dynamic behavior or transient behavior*

The motor in variable state can be schematically represented by the circuit below. In addition to the previous model, we take into account the inductance of the coil that creates the electric time constant of the motor and a mechanical part.

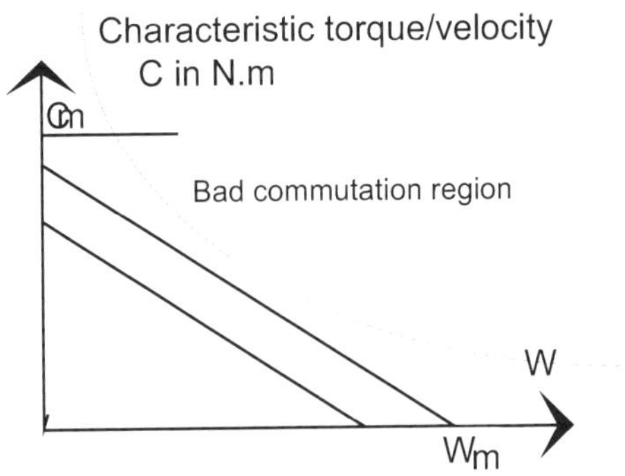

Figure 3.14. *Region of bad commutation*

R: internal resistance of the motor, L: inductance of the windings, $e(t)$: motor-induced counter EMF, $u(t)$: voltage at the terminals of the motor, J: moment of inertia of the motor shaft, f: viscous friction coefficients on the shaft, $i(t)$: instantaneous current flowing through the engine, $\omega(t)$: the instantaneous rotation speed of the motor axle. The following figures illustrate the model and the Bond graph of an engine.

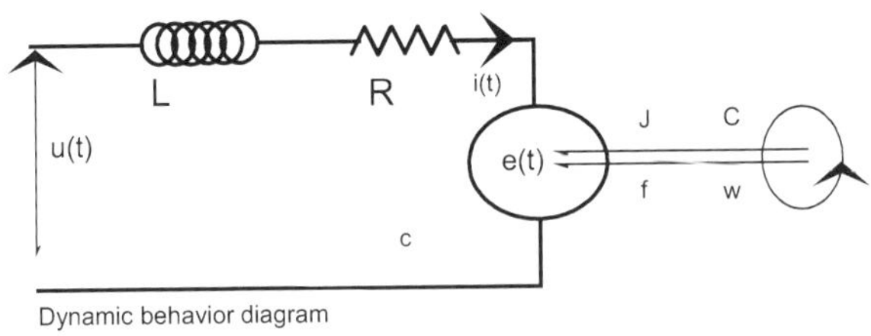

Figure 3.15. *Diagram of a motor in variable state (dynamic)*

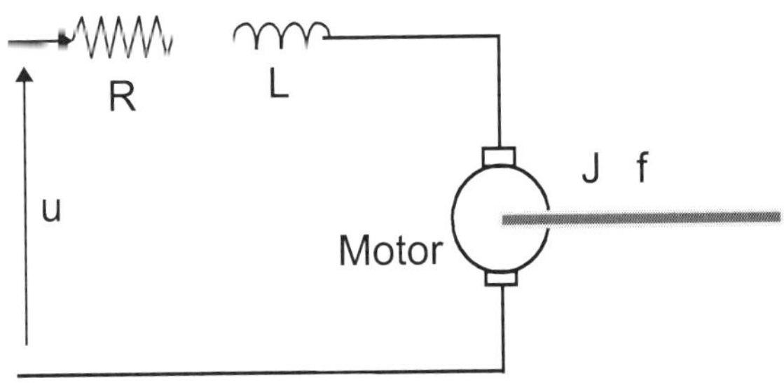

Figure 3.16. *Diagram of a motor in variable state*

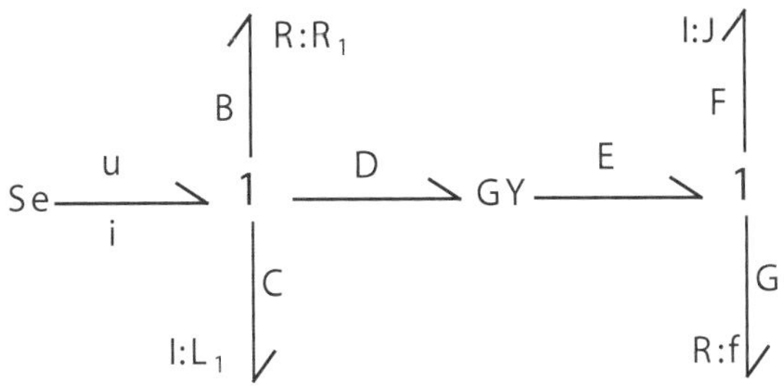

Figure 3.17. *Bond graph for a motor*

The equations of this system are written in the following manner:

– electrical equations:

$$u(t) = R.i(t) + L.\frac{di(t)}{dt} + e(t); \quad e(t) = K.\omega(t) \tag{3.22}$$

– mechanical equations:

$$C = K.i(t) \text{ and } C = J.\frac{d\omega(t)}{dt} + f.\omega + C_r \, . \tag{3.23}$$

where C_r is the resistive torque on the output shaft. Or even by substituting $e(t)$ in the first equation and C in the second equation:

$$u(t) = R.i(t) + L.\frac{di(t)}{dt} + K.\omega(t) \quad \text{and} \quad K.i(t) = J.\frac{d\omega(t)}{dt} + f.\omega(t) + C_r. \tag{3.24}$$

By multiplying the first relationship by $i(t).dt$, it is shown that the electric energy borrowed from source $u(t).i(t).dt$ is found in the form of energy dissipated by the Joule effect $R.i(t)^2$·dt of energy stored in electromagnetic form $L.i(t).di$ due to the current and mechanical energy variation $e(t).i(t).dt = K.\omega(t)i(t).dt$.

For the representation of this system, we can deduct a state–space representation (in the form of a first-order vector differential equation). To this end, we chose as a state vector $X = [i(t); \omega(t)]^T$; first, we consider the resistive torque as being zero $C_r = 0$. This gives us the following state–space representation:

$$\dot{X} = A.X + B.u(t), \tag{3.25}$$

$$y = C.X = \omega(t) = [0; 1].X, \qquad [3.26]$$

$$\frac{d}{dt}\begin{pmatrix} i(t) \\ \omega(t) \end{pmatrix} = \begin{pmatrix} -\frac{R}{L} & -\frac{K}{L} \\ \frac{K}{J} & -\frac{f}{J} \end{pmatrix}\begin{pmatrix} i(t) \\ \omega(t) \end{pmatrix} + \begin{pmatrix} \frac{1}{L} & 0 \\ 0 & -\frac{1}{J} \end{pmatrix}\begin{pmatrix} u(t) \\ C_r(t) \end{pmatrix}. \qquad [3.27]$$

Consider the modeling of a direct current motor as represented by the above figures. Its equivalent Bond graph is shown in Figure 3.17. Link A (Bond) represents the source of energy (u the effort and i the flux), B describes the energy loss in the resistance, C is the energy stored in the inductor in magnetic form and D represents the counter EMF. The gyrator GY converts the transformation of energy in the form of rotation. F translates the kinetic energy due to the inertia of the rotating rotor and G represents the dissipation of energy by friction.

3.2.4. *Electric systems motor load*

Consider a DC motor whose shaft is connected to a speed reducer. This actuator has no flexibility at the level of the transmission of motions. We assume to be negligible any clearance and dry friction (Coulomb). K_m is the current gain. J_m, J_s are the inertia of the motor shaft on input and the output load. f_m, f_s are the viscous friction at the level of the transmission shaft of the motor on input and output. C is the transmission torque and τ materializes a perturbation that would add up to the load inertia and to viscous friction forces (arm inertia, dry frictions, etc.). In the case where the transmission axis is rigid, we have $\dot{q}_e = N\dot{q}_s$ and $q_e = Nq_s$, the position and the input speed are proportional to those on output with the same coefficient multiplying the reducer output torque.

The equations describing the dynamics of the actuator are given by the following:

$$K_m i = J_m\ddot{q}_e + f_m\dot{q}_e + C, \;\; N.C = J_s\ddot{q}_s + f_s\dot{q}_s + \tau, \qquad [3.28]$$

$$\dot{q}_e = N\dot{q}_s \rightarrow NK_m i = J_o\ddot{q}_s + f_o\dot{q}_s + \tau, \qquad [3.29]$$

$$\text{with } J_o = J_s + N^2 J_m \;\text{ and }\; f_o = f_s + N^2 f_m, \qquad [3.30]$$

$$NK_m i = NJ_m\ddot{q}_e + Nf_m\dot{q}_e + NC = NJ_m\ddot{q}_e + Nf_m, \dot{q}_e + J_s\ddot{q}_s + f_s\dot{q}_s + \tau,$$

$$NK_m i = N^2 J_m\ddot{q}_s + J_s\ddot{q}_s + N^2 f_m\dot{q}_s + f_s\dot{q}_s + \tau.$$

Under the hypothesis of rigidity of the transmission and without disruption ($\tau = 0$), the system will be written as follows: $\ddot{q}_s = -a.\dot{q}_s + b.i$ with $a = \frac{f_s + N^2 f_m}{J_S + N^2 J_m}.$ and $b = \frac{K_m N}{J_s + N^2 J_m}.$ ($a_{\min} \leq a \leq a_{\max}, 0 \leq b \leq b_{\max}$).

3.2.4.1. *Flexible transmission chain: electric actuators with flexibility*

The hypothesis of rigid manipulators does not take into account certain physical phenomena that can no longer be overlooked if the robot must perform fast and accurate movements. In effect, at the articulation level there is elasticity between the actuator and the arm, and sometimes in the arm, itself. Hereafter, we only consider the flexibility at the level of the joint and the arm is assumed rigid. Taking into account the transmission elasticities at the level of the joints will result in doubling the dimensions of the system. The order-n system then becomes a $2n$-order system.

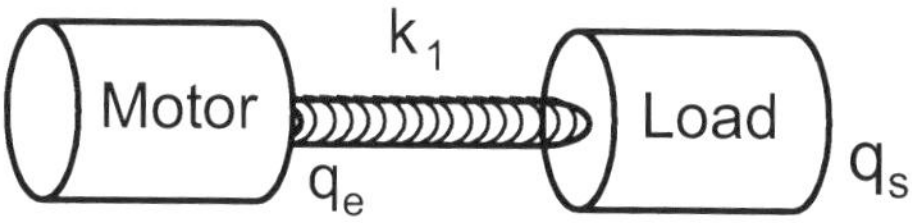

Figure 3.18. *Electric actuators with flexibility*

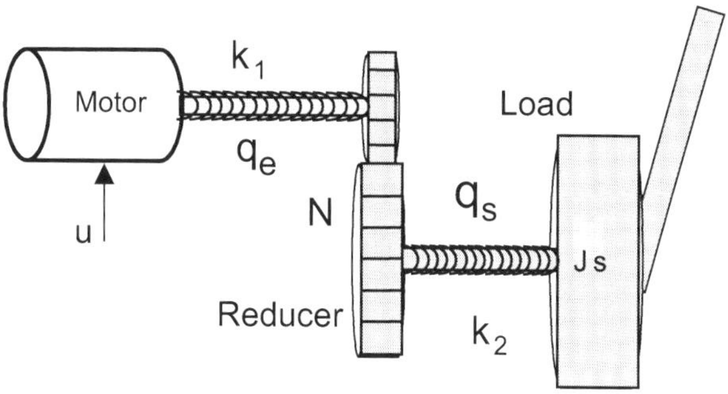

Figure 3.19. *Diagram of the motor with speed reducer and an arm*

Application for an arm having a single degree of freedom:

– r and L, resistance and electric inductance of the motor (Ω and H);

– K_m, motor torque constant ($N.m/A$);

– K_e, counter EMF of the motor ($V/(Rad/s)$);

– J_m, J_s, inertia of the motor shaft on input and of the output load ($kg.m^2$);

– f_m, f_s, viscous friction at the level of the motor transmission shaft the on input and output ($N.m/(Rad/s)$;

– C, transmission torque (Nm);

– τ_m motor torque (Nm);

– τ materializes the load due to the presence of the arm on the reducer output;

– u, power supply voltage of the motor (V);

– i, current flowing through the motor (A);

– $\dot{q}_m$, speed of rotation of the motor shaft (Rad/s).

The dynamic model representing the system above is expressed as follows:

$$u = Z(p)i + K_e\dot{q}_m, \tag{3.31}$$

$$\tau_m = K_m.i, \tag{3.32}$$

$$u = \frac{Z(p)}{K_m}i + K_e\dot{q}_m, \tag{3.33}$$

$$\tau_m = K_m i = J_m\ddot{q}_e + f_m\dot{q}_e + k_1(q_e - Nq_s), \tag{3.34}$$

$$k_2(q_s - q_e/N) = J_s\ddot{q}_s + f_s\dot{q}_s + \tau, \tag{3.35}$$

$$C = k_1(q_e - Nq_s) = -.k_2(q_s - q_e/N)/N = -k_2(Nq_s - q_e)/N^2, \tag{3.36}$$

$$-N^2C/k_2 = (Nq_s - q_e), \tag{3.37}$$

$$q_e = Nq_s + N^2C/k_2, \tag{3.38}$$

$$C/k_1 + Nq_s = q_e, \tag{3.39}$$

$$q_e = (C/2k_1 + N^2C/2k_2)/ + Nq_s, \tag{3.40}$$

$$q_e = C(k_2 + N^2k_1)/2k_1k_2 + Nq_s, \tag{3.41}$$

$$\begin{array}{l} \ddot{\theta}_s = -a.\dot{\theta}_s + b.v - m.g.l.\cos(\theta_s), \\ d = m.g.l. \end{array} \tag{3.42}$$

We then write the vector q (of dimension $2n \times 1$) in the following manner:

$q = [q_e, q_s]$ with $q_e = [q_{e1}, q_{e2}, ..., q_{en}]$ and $q_s = [q_{s1}, q_{s2}, .., q_{sn}]$.

The dynamic equations are then described by:

$$M(q_s)\ddot{q}_s + C(q_s, \dot{q}_s)q_s + k(q_e - q_s) = 0, \quad [3.43]$$

$$J.\ddot{q}_e - k(q_e - q_s) = u, \quad [3.44]$$

where $M(q)$ is the symmetric positive definite matrix:

– k is the links elasticity, and;

– J is the symmetric positive definite matrix representing the inertia of actuators.

3.3. Pneumatic actuators

Because of the complexity of their modeling, pneumatic actuators have little use in robotics. Among their advantages and disadvantages, we can identify their reduced size, better efficiency than electric motors at low speed, lower weight and their compliance. For more details on the development of these components and research in the field of pneumatics, we can highlight Khellal's work [KEL 87] and Scavarda *et al.*'s work. The modeling of the robot is based on the fundamental laws of thermodynamics that reflect the dynamic and the physical behavior of the system. It is generally made aiming at the control of the system. Consequently, the types of models being presented depend on the purposes sought after. During the modeling of such a system, some consider linearization methods to simplify the equations of the model (see Shearer *et al.*'s works [SHE 56]). Generally, the control of such systems is based on a proportional integral derivative approach or an equivalent one (see Scavarda's book [SCA 89] and Pu's work [PU 96]). In this case, the effect of each parameter has been widely discussed in detailed applications in Bobrow *et al.*'s works [BOB 98].

3.3.1. *Pneumatic system modeling*

The structure of the system under study is represented in Figure 3.20. The dynamic model of the system is composed of two stages: a dynamic stage that takes the mechanical aspects into consideration and a second stage corresponding to the pneumatic actuators.

The pneumatic actuator is composed of two conceptually distinct parts: the mechanical part of the straight symmetrical double-acting cylinder type and the control part of the transfer of power. This transfer is carried out by a servovalve that acts as the interface between electrical data and air flow. The servovalve consists of two stages (see Figure 3.21).

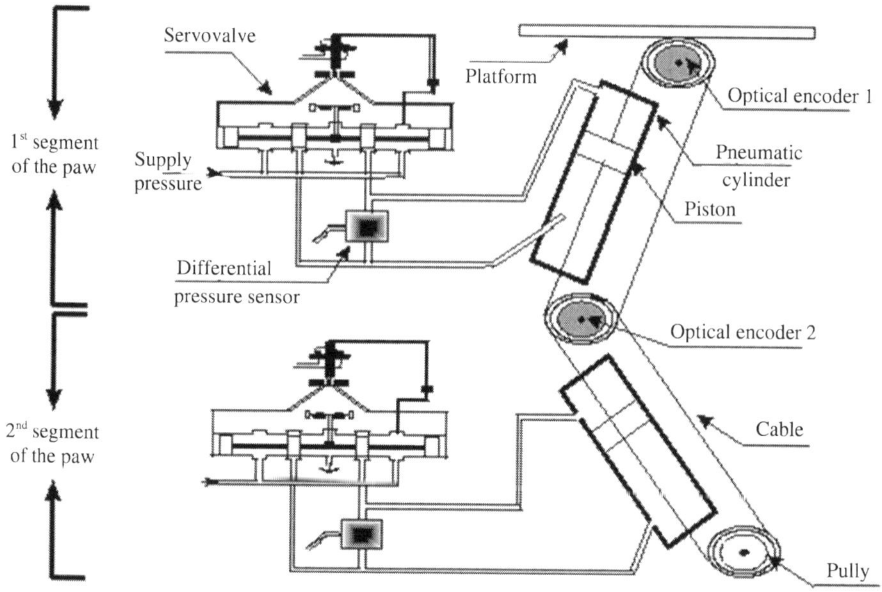

Figure 3.20. *2-DOF arm controlled by electropneumatic servovalves*

The first stage corresponds to the control stage. By means of a steerable nozzle, it makes it possible to obtain the displacement of a drawer proportionally to a control current. The second stage consists of a drawer moving within a casing. The operating principle is as follows: according to the displacement of the drawer, an orifice is uncovered, whereas another one is shut in the same proportions; unbalance is thus created between the two orifices. On the output of the servovalve, we have the actuator or cylinder, the behavior of which depends of the development of pressures in the different chambers that it is made up of.

We define the parameters of the pneumatic system in Figure 3.20 (see Scavarda's book):

– m_{ij} designates the masses of the discharge flows into the holes ij;

– P_p, P_n are the pressures in chambers P and N, respectively;

– P_a, P_r are the supply and output pressures, respectively;

– V_p, V_n are the volumes in chambers P and N, respectively;

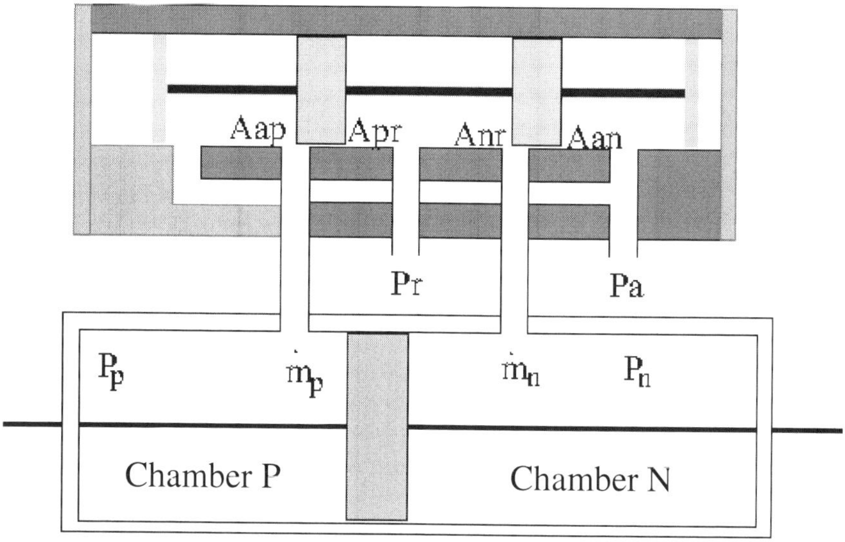

Figure 3.21. *Flow stage of the servovalve*

– $\gamma = 1.405$ is the ratio of specific heat;

– $r = 286\,\mathrm{Jkg^{-1}K^{-1}}$ is the ideal gas constant;

– T is the temperature in kelvin;

– l is the radius of the pulley;

– i is the input current of the motor torque;

– i_0 is the offset of the initial current (servovalve) and $\beta = \text{constant} \geq 0$;

– S is the piston section;

– y is the piston displacement;

– x is the piston displacement;

– $\dot{m}_1$ is the mass flow in kg/s;

– C_{m_i} is the flow parameter;

– x is the passage section in m^2;

– P_{am_i} is the upstream pressure at stop in pascal;

– T_{am_i} is the upstream temperature at stop in kelvin;

– $C_q = 0.7$ is the flow coefficient depending on the upstream and downstream pressure ratio and on the type of orifice.

The servovalve is also composed of a set of drawers and four holes modulated by the position of the drawer. The displacement of the cylinder occurs due to variations in air flow; the diagram of the servovalve is represented in Figure 3.21. The determination of the dynamic model of actuators (Shearer and Richard's work) is based on the study of the flow stage subject to a fixed pressure Pa. The servovalve controls the flow of air that is converted into pressure supplying the two chambers of the cylinder through four holes. The application of the laws of thermodynamics will be made with the following assumptions.

PROPOSITION 3.1.– The fluid is considered to be an ideal gas. Potential and kinetic energies of the fluid are negligible. No leaks appear between the two chambers on each side of the piston.

PROPOSITION 3.2.– The displacement of the piston is due to a slight variation around the central position of the drawer of the servovalve. The pneumatic system is symmetrical. The pressure equations can be linearized around the central position.

The mass of the flow patterns depends on current i in the servovalves, the movement of the cylinder and the pressures at the level of the chambers (details in Kellal's work [KEL 87]).

The application of the first law of thermodynamics gives us:

power supplied to the fluid = required compression power + power dissipated by displacement.

The flow entering and leaving the servovalve depends on the opening of the hole x, the pressure of the supply air, the motion of the cylinder and the geometry of the servovalve. The airflow supersonic condition takes effect when the ratio of the downstream and upstream pressure is less than 0.528 ($\frac{P_n}{P_a}$ where $\frac{P_n}{P_u} \leq 0.528$).

We consider the temperature to be constant, adiabatic exchange: the exchange of heat within the actuator is negligible. The expression of energy conversion can be written as in the following manner (Shearer *et al.* [SHE 56]):

$$\dot{m}rT = \frac{V}{\gamma}\frac{dP}{dt} + P\frac{dV}{dt}; \qquad [3.45]$$

$\dot{m}_1$, flow rate entering chamber 1,

$\dot{m}_2$, flow rate entering chamber 2,

$$\dot{m}_1 r T_1 = \frac{V_1}{\gamma}\frac{dP_1}{dt} + P_1\frac{dV_1}{dt}, \qquad [3.46]$$

$$\dot{m}_2 r T_2 = \frac{V2}{\gamma}\frac{dP_2}{dt} + P_2\frac{dV_2}{dt}, \qquad [3.47]$$

x is the opening of the hole. P_{am_1} is the upstream pressure. The mass flow through each section is:

$$\dot{m}_1 = C_q C_{m_1} x \frac{P_{am_1}}{\sqrt{T_{am_1}}}, \qquad [3.48]$$

$$\dot{m}_2 = -C_q C_{m_2} x \frac{P_{am_2}}{\sqrt{T_{am2}}}, \qquad [3.49]$$

therefore with the help of the above equations, it yields:

$$\frac{V_1}{\gamma}\frac{dP_1}{dt} = \dot{m}_1 r T_1 - P_1\frac{dV_1}{dt}, \qquad [3.50]$$

$$\frac{V2}{\gamma}\frac{dP_2}{dt} = \dot{m}_2 r T_2 - P_2\frac{dV_2}{dt}, \qquad [3.51]$$

which gives:

$$\dot{P}_1 = \frac{\gamma r T_1}{V_1}\dot{m}_1 - \frac{\gamma}{V_1}P_1\frac{dV_1}{dt} = \frac{\gamma r T_1}{V_1}\left[\dot{m}_1 - \frac{P_1}{rT_1}\frac{dV_1}{dt}\right], \qquad [3.52]$$

$$\dot{P}_2 = \frac{\gamma r T_2}{V_2}\dot{m}_2 - \frac{\gamma}{V_2}P_2\frac{dV_2}{dt} = \frac{\gamma r T_2}{V_2}\left[\dot{m}_2 - \frac{P_2}{rT_2}\frac{dV_2}{dt}\right]. \qquad [3.53]$$

Dynamic equations of the piston and cylinder:

$$M\ddot{y} = (P_1 - P_2)\,S - ky - f\dot{y} - F_e, \qquad [3.54]$$

we establish $l = l_a - l_b = l_1 + l_2; V_1(t_o) = l_1 S$ and $V_2(t_o) = l_2 S$,

$$V_1 + V_2 = (l_a - l_b) S = lS, \qquad [3.55]$$

$$V_1 = (l_1 + y) S = l_1 S + yS, \qquad [3.56]$$

$$V_2 = (l_2 - y) S = l_2 S - yS, \qquad [3.57]$$

$$\frac{dV_1}{dt} = S\frac{dy}{dt} = S\dot{y}, \qquad [3.58]$$

$$\frac{dV_2}{dt} = -S\frac{dy}{dt} = -S\dot{y}, \qquad [3.59]$$

$$\dot{P}_1 = \frac{\gamma r T_1}{V_1}\left[C_q C_{m_1} x \frac{P_{am_1}}{\sqrt{T_{am_1}}} - \frac{P_1}{rT_1} S\dot{y}\right], \qquad [3.60]$$

$$\dot{P}_2 = \frac{\gamma r T_2}{V_2}\left[-C_q C_{m_2} x \frac{P_{am_2}}{\sqrt{T_{am2}}} + \frac{P_2}{rT_2} S\dot{y}\right], \qquad [3.61]$$

$$\ddot{y} = \frac{S}{M}P_1 - \frac{S}{M}P_2 - \frac{f}{M}\dot{y} - \frac{k}{M}y - F_e, \qquad [3.62]$$

$$\dot{m}_1 = C_q C_{m_1} x \frac{P_{am_1}}{\sqrt{T_{am_1}}}, \qquad [3.63]$$

$$\dot{m}_2 = -C_q C_{m_2} x \frac{P_{am_2}}{\sqrt{T_{am2}}} \qquad [3.64]$$

The air flow rate in chambers P and N can be expressed according to those in the various orifices A_{ij} as being:

$$\Delta\dot{m}_p = (\dot{m}_{ap} - \dot{m}_{pr}) \text{and } \Delta\dot{m}_n = (\dot{m}_{an} - \dot{m}_{nr}). \qquad [3.65]$$

This leads to the following four equations for the servovalve:

$$\begin{cases} \dfrac{dP_p}{dt} = -\dfrac{\gamma P_p}{V_p}\dfrac{dV_p}{dt} + \dfrac{rT\gamma}{V_p} \cdot \Delta\dot{m}_p \\ \dfrac{dP_n}{dt} = -\dfrac{\gamma P_n}{V_n}\dfrac{dV_n}{dt} + \dfrac{rT\gamma}{V_n} \cdot \Delta\dot{m}_n. \end{cases} \quad [3.66]$$

The assumptions taken into account (in Shearer and Richard's work [SHE 56, RIC 96]) allow for the linearization of the discharge flow mass, which can be considered proportional to the position of the drawer x (opening of the valve) and also to the pressures in the chambers. The airflow mass can be connected to the opening x of the valve, the current and the pressures by the following relationships:

$$\frac{dm_{ij}}{dt} = f_{ij}(x, P_{ij}) \text{ and } i = f(x) + i_0. \quad [3.67]$$

In the case where the dynamics of the drawer is negligible (high bandwidth of the servovalve compared to that of the system), this leads to the following equations:

$$\begin{cases} \Delta\dot{m}_p = f_p\left(x, \Delta P\right) = C_x x - C_p \Delta P_p \\ \Delta\dot{m}_n = f_n\left(x, \Delta P\right) = C_x x - C_n \Delta P_n \\ \quad C_x = \dfrac{\partial \dot{m}_{ap}}{\partial x}\,|_{x_0} - \dfrac{\partial \dot{m}_{ap}}{\partial x}\,|_{x_0} \\ \quad C_x = \dfrac{\partial \dot{m}_{an}}{\partial P}\,|_{P_0} - \dfrac{\partial \dot{m}_{nr}}{\partial P}\,|_{P_0}. \end{cases} \quad [3.68]$$

The volume in the chamber can be written in the following manner:

$$V_p = V - V_n = S.y = S.l.q. \quad [3.69]$$

In the following equations, index i indicates the initial conditions. The relations between volumes, the discharge flow and the piston velocity are:

$$\frac{dV_p}{dt} = -\frac{dV_n}{dt} = S\dot{y} \quad [3.70]$$

$$\begin{cases} \Delta\dot{P}_p = \dfrac{\gamma rT}{V_{p_i}}(C_x x - C_p \Delta P_p) - \dfrac{P_{p_i}}{V_{p_i}}\gamma Sl\dot{q} \\ \Delta\dot{P}_n = \dfrac{\gamma rT}{V_{n_i}}(-C_x x - C_n \Delta P_n) - \dfrac{P_{n_i}}{V_{n_i}}\gamma Sl\dot{q} \end{cases} \quad [3.71]$$

with

$$\Delta P_p = P_p - P_{p_i}; \Delta P_n = P_n - P_{n_i} \quad [3.72]$$

$$\Delta V_p = V_p - V_{p_i}; \Delta V_n = V_n - V_{n_i}. \quad [3.73]$$

The dynamic appearance of the piston and external loads can be described as follows:

$$m\frac{d\dot{y}}{dt} = S(P_p - P_n) - V(y, \dot{y}) - F_r \quad [3.74]$$

where:

– $V(y, \dot{y})$ represents dry and viscous frictions and some nonlinear functions that could have been negligible during the identification of the model;

– F_r is the resistive force produced by the load (proportional to the difference of pressures in both chambers);

– m is the mass of the cable and piston, which is negligible compared to the inertia of the segment;

The relationship between the opening x of the valve and the current can be simplified by setting $i = \beta.x + i_0$, but it should be noted that in practice there is a dead region and saturation. This simplified model will then be used for identification and control. It should also be highlighted that the parameters of the actuators are varying in time and depend on initial conditions (ambient temperature, pressures, etc.).

The model of the actuator is then defined by equations [6.159]–[3.74]. The equations of the electropneumatic actuators are nonlinear and have time-variant parameters. Therefore, the full model of the system justified by bibliographic and experimental results (see Shearer *et al.* [SHE 56]) is presented Table 3.2. These equations will be simplified for ease of learning and implementation considering the hypotheses A and B mentioned earlier.

The dynamics of the pneumatic actuator can thus be expressed by means of some simplifications in the form:

$$\dot{\tau} = Ji - B\tau - E\dot{q},$$

$$\text{with } \tau = K(\Delta P_p - \Delta P_n)$$

where:

$$\begin{cases} J = \dfrac{\gamma r T C_i}{Sl}(\dfrac{1}{V_{p_i}} + \dfrac{1}{V_{n_i}}) \\ B = \dfrac{\gamma r T C_p}{V_{p_i}} \\ E = \gamma(\dfrac{P_{p_i}}{V_{p_i}} + \dfrac{P_{n_i}}{V_{n_i}}). \end{cases} \qquad [3.75]$$

Electric control of the drawer motion	
x = opening of the holes	
Effect of the current on the cylinder:	$i = f(x) + i_0 = \beta.x + i_0$
Mass flow in the openings (equation [3.67])	
Flow rate of the fluid in the chambers:	$\frac{dm_{ij}}{dt} = f_{ij}(x, P_{ij})$
Pressure variation in chambers due to mass flow	
Flow rate of the fluid for the actuator	$\begin{cases} \dfrac{dP_p}{dt} = -\dfrac{\gamma.P_p}{V_p}\dfrac{dV_p}{dt} + \dfrac{rT\gamma}{V_p}.\Delta\dot{m}_p \\ \dfrac{dP_n}{dt} = -\dfrac{\gamma.P_n}{V_n}\dfrac{dV_n}{dt} + \dfrac{rT\gamma}{V_n}.\Delta\dot{m}_n \end{cases}$
Variation in the volumes of chambers according to the motion of the actuator	
Relationship of the piston:	$\begin{cases} V_p = V - V_n = S.y \ \tau = l.F_r \\ \dfrac{dV_p}{dt} = -\dfrac{dV_n}{dt} = S\dot{y} \ y = l.q \end{cases}$
Piston dynamics	
Piston dynamics:	$\begin{cases} V(y,\dot{y}) = b_v(y,\dot{y})\dot{y} + b_c(y,\dot{y})sign(\dot{y}) \\ m\dfrac{d\dot{y}}{dt} = S(P_p - P_n) - V(y,\dot{y}) - F_r \end{cases}$

Table 3.2. *Thermodynamic and mechanical equations of the actuator*

J, B, E are the diagonal matrices of dimensions (2×2) called thermodynamic parameters depending on the temperature of fluid T, respective initial and final volumes and pressures V_{p_i}, V_{n_i}, P_{p_i}, and P_{n_i} in the chambers, the thermodynamic constants γ and r of the characteristics and initial conditions.

3.3.2. *Frictions model*

Several models have been proposed to describe viscous and dry frictions. The curves below represent a visualization of the friction torques according to the velocity. The frictions have been modeled by Karnopp as the sum of a window, a constant and a straight line passing through the origin, designating the adherence, the dry friction and the viscous friction. Often the model stipulated does not prove to be accurate in reality. For this purpose, a system must first be simulated with compensation of dry friction in the presence of viscous friction according to the various possible models. The results of this simulation will enable us to choose the compensation algorithm to be implemented in practice after experimentation.

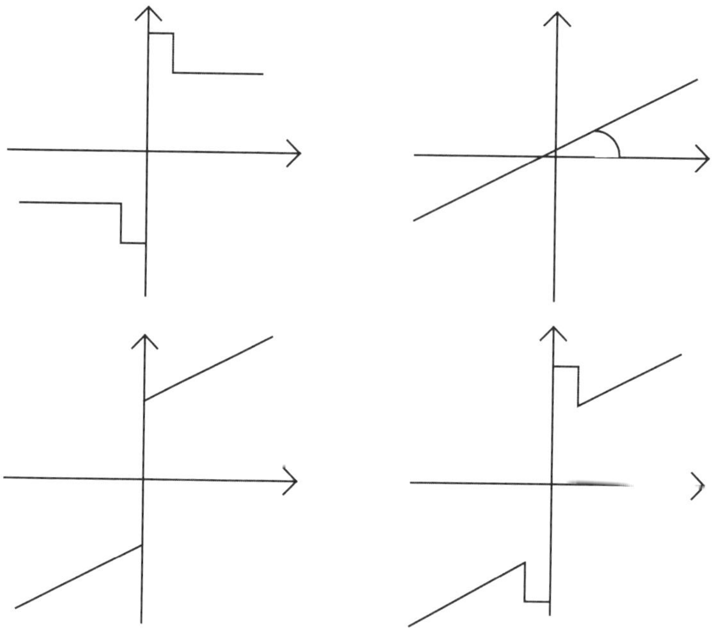

Figure 3.22. *Friction model curves*

In practice, it is possible to have a stable behavior and an error that remains bounded or the occurrence of a slightly oscillating behavior. These overruns above the setpoint are sometimes due to transient overcompensation of dry friction. In effect, overcompensation occurs after the estimate strongly increases during phases at high speed. Undercompensation of the dry friction introduces a static error. On the other hand, the compensation of friction using well-known algorithms always results in an improvement in tracking but sometimes proves insufficient at very low speeds. The accuracy of the estimations and the friction model is crucial in these cases, due to the sensitivity to disturbances of the estimation algorithms. In order to overcome these problems, it is necessary to estimate the perturbation caused by frictions and at the same time adapt the control. The friction problem cannot be addressed independently of the rest of control.

In most applications, the Coulomb friction model is the one retained due to its simplicity. In 1984 Walrath, when trying to define the transient behavior of friction produced by the change in direction of the motion, proposed a dynamic model [WAL 84] which shares similar aspects with Dahl's model but is simpler overall.

$$a.\frac{d}{dt}\tau(t) + \tau(t) = a_0.sgn(\dot{q}) \qquad [3.76]$$

where the parameter a is dependent on velocity $\dot{q}$. This model introduced an additional dynamic but no static characterization.

3.3.2.1. *Adherence or stiction*

The phenomenon known as "stiction" appears at zero velocity; it thus opposes any slip as long as the torque applied on the system remains lower than the adherence torque. This phenomenon is due to gluing effects that occur when there is contact between two materials. The two surfaces thus form a junction that must be broken. This effect gives rise to the stick–slip phenomenon at very low velocities. As a matter of fact, for low velocities (close to zero, quasi-static regime) the frictional force applied on the system oscillates around the dry friction value.

In 1985, Karnopp suggested another model where a_0 represents the Coulomb friction coefficient, a_2 the viscous friction coefficient and $a_0 + a_1$ corresponds to the value of the torque generated by the adherence phenomenon. This approach separates the model into two regions. In the first region, the velocity is zero or almost equal to zero. The system moves away from this region and transits into the viscous friction region (second region) only if the force applied to the system exceeds the adherence torque. This model is a means of encompassing both Coulomb's model and Dahl's model, as well as Stribeck's friction effect.

Asymmetries: the models presented above are all symmetrical. However in practice, this is not always the case. The imperfections in the different elements of the system can lead to asymmetric behavior and act differently according to the direction of the rotation. In the asymmetric case, the coefficients a_0 and a_2, respectively, of the dry friction and the viscous friction will differ depending on the rotation direction; the same occurs for adherence friction.

The dependence of position: friction torque can also depend on the angular position. This dependence can be, for example, a consequence of imperfections in centering the elements. Furthermore, in a straight gear train reducer these imperfections generate oscillations of period proportional to the reduction ratio.

Stribeck's friction: this friction obeys the generalized Stribeck curve (see Armstrong's works [ARM 00]) and corresponds to the area of exponential decay (downward bends) of the friction torque that appears when the speed increases in the range of very low velocities. The variation phenomenon of the friction coefficient with respect to the velocity has been observed as early as 1902 by Stribeck. In effect, when the stiction torque has been overcome, the friction torque exponentially decreases and reaches approximately 60% of the breakaway torque (of rupture of adherence), then increases proportionally to the velocity. It is explained by the system transiting from a metal lubrication state (direct friction of both surfaces in region 1) into a state of partially metal partially fluid lubrication (region 2) during the relative movement of the two mechanical parts. Subsequently, the lubrication becomes completely fluid (region 3). These measures confirm the model proposed by Tustin [TUS 47], which includes this negative dependence region of friction according to velocity. The presence of this region creates stability problems. It should be noted that the velocity limiting regions (2) and (3) corresponds to the threshold velocity of the Karnopp model [KAR 06].

Friction modeling: the model proposed by Tustin resumes the main components of friction previously presented; it has been validated experimentally by Armstrong for a project of a manipulator robot. The Tustin model can thus be expressed as by Canudas de Wit:

$$\tau(\dot{q}) = (a_0 + a_1.e^{\ b|\dot{q}|}. + a_2|\dot{q}|)sgn(\dot{q}) \qquad [3.77]$$

where a_0 represents the Coulomb friction coefficient, a_2 the viscous friction coefficient and where $(a_0 + a_1)$ corresponds to the value of the adherence torque. The value b corresponds to the lubrication coefficient. If the friction presents asymmetries, the coefficients a_i can each take two different values following the direction of the motion. Other representations of the same form can be found in Armstrong's [ARM 00] and Dupont's [DUP 09] or, for example, in Canudas de Wits' [CAN 98] and Fukuda's [FUK 98] works:

$$\tau(\dot{q}) = (a_0 + a_1.|\dot{q}|^{1/2}. + a_2|\dot{q}|)sgn(\dot{q}). \qquad [3.78]$$

Karnopp's model is expressed in the following manner, if we are considering the sliding region: $|\dot{q}| > \Delta\dot{q}$

$$\tau(\dot{q}) = (a_0.sgn(\dot{q}) + a_2\dot{q}) \qquad [3.79]$$

and if we consider the adherence area: $|\dot{q}| > \Delta\dot{q}$

$$\tau(\dot{q}) = min(\tau_{\text{calculated}}, \tau_{\text{adherence}}.sgn(\tau_{\text{calculated}})) \qquad [3.80]$$

where a_0 and a_2 represent the dry friction and viscous friction coefficients, $\tau_{\text{calculated}}$ is the calculated torque imposed on the system, $\tau_{\text{adherence}}$ is the estimated adherence torque and $\Delta\dot{q}$ corresponds to the estimated adherence area for the system being considered. The function min. selects the nearest lowest value and limits the maximum torque to the adhesion torque $(a_0 + a_1)$.

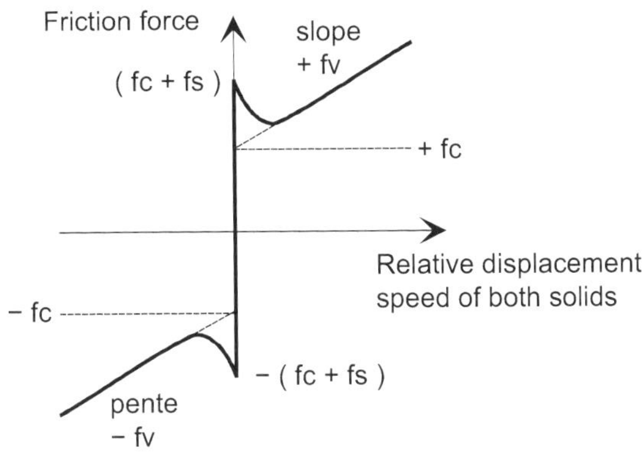

Figure 3.23. *Friction force with respect to the relative velocity*

We will consider the mass m of the arm as a point located a distance l from the axis of rotation. Quantity t is the torque applied on the axis of rotation and q is the angle of the arm with the horizontal axis. The moment of inertia of the arm with respect to the axis of rotation is $J = ml2$ (see Figure 3.25). For dry friction, we consider the simplified model: $fs = bcsign(q)$.

The dynamic equation of the system can be written taking into account that b is the coefficient of viscous friction (non-linear) and k, k define the stiffness of the axis of the arm.

$$ml2\ddot{q} + bv|q|\dot{q} + bcsigne(q) + koq + k1q3 + mglsin(q) = t. \qquad [3.81]$$

3.4. Hydraulic actuators

3.4.1. *System description*

The robot considered here is the Slingsby TA9; a detailed description and a modeling of this manipulator can be found in Clegg's work. Each joint is powered by a servovalve and the double effects of linear jacks. The supply pressure is controlled by an electrohydraulic servovalve (MOOG E777-006) controlled by a current input.

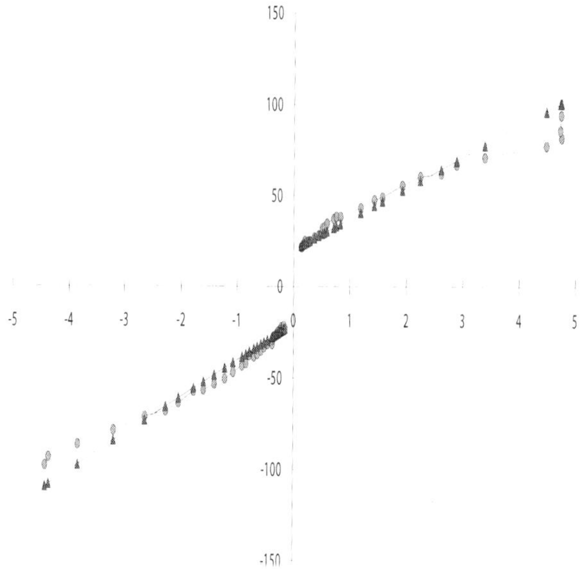

Figure 3.24. *Example of an arm having a single degree of freedom*

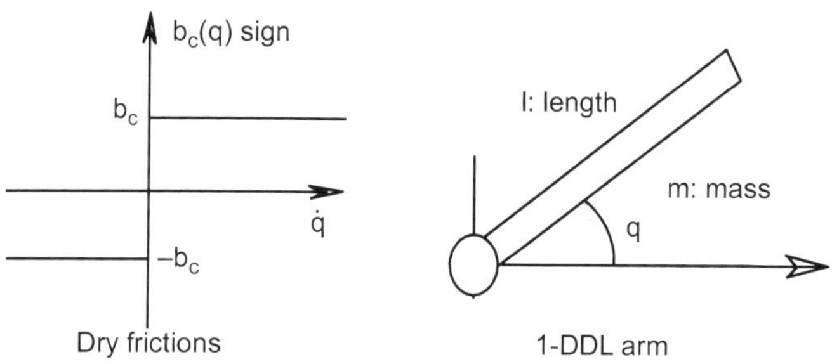

Figure 3.25. *1-DOF arm with dry frictions*

The current is converted into movement of the coil valve through a DC motor. A solenoid converts the actual current input into a displacement of the drawer that provides the input pressure to a chamber and opens another to the tank. The valve controls the oil flow that is converted into supply pressure for both cylinder chambers by means of restrictions. The model of the robot is composed of two stages: a dynamic one for the mechanical aspect, and a second one that will be called the hydrodynamic stage according to the hydraulic part.

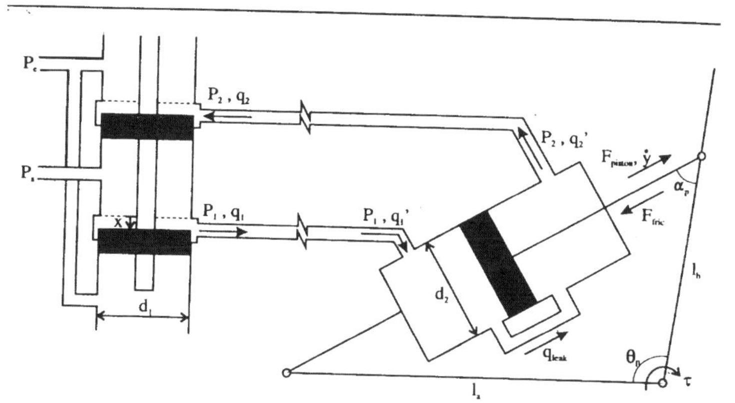

Figure 3.26. *Robot model: hydrodynamic stage*

3.4.2. *Mechanical model*

The dynamic model of a rigid link manipulator can be described by using Lagrange's method (Khalil *et al.* [KHA 86]). We remind that q, $\dot{q}$, $\ddot{q}$, τ denotes, respectively, the $(n \times 1)$ vectors of the common position, velocity, acceleration and torsion moment.

$$M(q)\ddot{q} + C(q,\dot{q})\dot{q} + G(q) + F_v(\dot{q}) = \tau. \qquad [3.82]$$

$M(q)$ is the $(n \times n)$ generalized inertia matrix, $C(q,\dot{q})\dot{q}$ is the centripetal and Coriolis forces vector and $G(q)$ is the $(n \times 1)$ vector of gravitational forces. The matrix $C(q,\dot{q})$ is usually achieved through the use of Christoffel symbols and the matrix $\frac{1}{2}\dot{M} - C$ is symmetric and diagonal. The following physical properties of rigid robots (with joints) can be used for control (Takegak and Ortega's works [ORT 08]):

1) $\exists \alpha_0, \alpha_1 \in R$ such that $\alpha_0 I_n < M(q) < \alpha_1 I_n, \forall q$.

2) $\exists \alpha_2 \in R$ such that $\|C(q,z)\| < \alpha_2 \|z\|, \forall q, \forall z$.

3) $\exists \alpha_3 \in R$ such that $\|G(q)\| < \alpha_3, \forall q$.

4) The friction and load noises of torsion moments are limited by Craig: $\exists \alpha_4, \alpha_5, \alpha_6 \in R$, such that $\|F_v(\dot{q})\| < \alpha_4 + \alpha_5 \|\dot{q}\|, \forall \dot{q}$.

5) The actual systems are limited in velocity and accelerations $\|\dot{q}\| < \dot{q}_{\max}$ and $\|\ddot{q}\| < \ddot{q}_{\max}$.

τ is the torsion moment required to make the linking movement. This makes it possible to connect the linkage equations of the manipulator to the actuators. The torsion moment of input τ can be evaluated based on the measure of the differential pressure P, which is the pressure applied to the pistons. For each robot link, the pressure is converted into a force (F_p is the force applied by the piston and F_{v_i} is the frictions and perturbation moments of torsion) and then in a moment of torsion through the use of geometric transmission actuators and its Jacobian $J_{p_i}(q_i)$, where q_i is the angular displacement of linkage i.

$$J_{p_i}(q_i) = \frac{l_a l_b \sin(q_i)}{\sqrt{l_a^2 + l_b^2 - 2 l_a l_b \cos(q_i)}}. \quad [3.83]$$

The lengths l_a, l_b are the geometric characteristics of the attachments of the cylinder actuator and the links. For each articulation, we can write that y is the displacement of the piston and A_2 is the surface traversed by the piston.

$$J_{p_i}(q_i)F_p = J_{p_i}(q_i)A_2 P = \tau + m\frac{d\dot{y}}{dt} + J_{p_i}(q_i)F_{v_i}. \quad [3.84]$$

Pressure P can be obtained from the analysis of the behavior of the actuator. According to the piston mass m, the friction effects will be neglected for ease of presentation. Common moments of torsion and their derivatives can be expressed as the force function as follows:

$$\tau = J_{p_i}(q_i)F_p = J_{p_i}(q_i)A_2 P, \quad [3.85a]$$

$$\dot{\tau} = J_{p_i}(q_i)\dot{F}_p + \dot{J}_{p_i}(q_i)F_p. \quad [3.85b]$$

3.4.3. *Hydraulic actuator model*

The modeling is based on the study of the flux stages provided by a fixed pressure P_s. The flow rate mass depends upon the input current of the motor with torsion motion, i, the jack movement and the pressure of the chamber (Cellier's works [CEL 96]). The action of the current on the position of the valve jack x is defined by $x = f(i - i_o) = K_i.(i - i_o)$, with K_i the displacement gain of the valve and i_o the initial compensation current of the servovalve. This compensation will be neglected now assuming that it is experimentally compensated ($x = K_i i$).

The following parameters are defined:

$P_s = 175 \times 10^5 \mathrm{N.m}^{-2}$: pressure supply;

$\rho = 870\,\mathrm{kg.m}^{-3}$: oil density coefficient;

$\beta = 7 \times 10^8 \mathrm{N.m}^{-2}$: bulk modulus;

d_1: coil diameter;

k_f : leakage coefficient;

V_t: total oil volume inside chambers and connection tubes;

C_d : constant factor taking into account the turbulent flux through the hole.

Let P denote the differential pressure in the chambers and x the displacement of the valve coil. We take the valve flow as being positive on outputs and by means of the use of the square root law for the turbulent flow through the orifice, assuming there is conservation of energy and that the heat exchange be negligible, we can write the following expression where Q_s is the bidirectional flow through the holes:

$$Q_s = K_v x (P_s - sign(x)P)^{\frac{1}{2}}$$

with

$$K_v = \frac{C_d \pi d_1}{\sqrt{\rho}}$$

and

$$Q_s = A_2 \dot{y} + k_f P + \frac{V_t}{4\beta} \dot{P}.$$

This gives the differential equation for the evolution of the pressure:

$$\dot{P} + \frac{4\beta k_f}{V_t} P + \frac{4\beta A_2}{V_t} \dot{y} = \frac{4\beta K_v \sqrt{P_s - sign(x)P}}{V_t} x. \qquad [3.86]$$

The force applied by the hydraulic actuator is the function of the surface section of the piston A_2, ($F_p = A_2 P$) and obeys the following differential equation:

$$\dot{F_p} + \frac{4\beta k_f}{V_t} . F_p + \frac{4\beta A_2^2}{V_t} \dot{y} = \frac{4\beta K_v A_2 \sqrt{P_s - sign(x)P}}{V_t} K_i i.$$

The pressure differential equation can be written as:

$$\dot{F}_p + B_1.F_p + E_1\dot{q} = J_1.k_o.i. \qquad [3.87]$$

The expressions of the parameters B_1 , E_1 , J_1 and k_o can be obtained from the two previous equations. For the modeling and the control, we are rather interested in the expression of torsion moments.

Using relations [3.85a] and [3.85b], we can obtain from the above equation a differential equation for the dynamic of the moment of torsion:

$$\dot{\tau} + (\tfrac{4\beta k_f}{V_t} - \dot{J}_{p_i}(q_i)J_{p_i}(q_i)^{-1})\tau + \tfrac{4\beta A_2^2}{V_t}J_{p_i}(q_i)\dot{q}$$

$$= \tfrac{4\beta K_v A_2}{V_t}J_{p_i}(q_i)\sqrt{P_s - sign(x)P}K_i i$$

$$\dot{\tau} + B.\tau + E\dot{q} = J.k_o.i \qquad [3.88]$$

with $B = B(q) = \frac{4\beta k_f}{V_t} - \dot{J}_{p_i}(q_i)J_{p_i}(q_i)^{-1}$,

$$E = \frac{4\beta A_2^2}{V_t}J_{p_i}(q_i), k_o = \frac{4\beta K_v A_2}{V_t}K_i$$

and $J = J(x, P) = J_{p_i}(q_i)\sqrt{P_s - sign(x)P}$.

By defining $J_p(q) = diag(J_{p_1}(q_1), .., J_{p_n}(q_n))$ (the positive diagonal matrix) for the gain transfer between the force and the torsion moment, we can now generalize this equation for an n linkage manipulator. Then J, k_o, B, E become $(n \times n)$ diagonal matrices called the parameters of the actuator according to the temperature, the oil characteristics, positions and initial conditions. Note that J and E are not linearly dependent on variables of the actuator. From the dynamic equations of the robot [3.82] and [3.85a] and introducing a term F_v, the inclusion of all friction effects (of angular, linear and nonlinear terms and perturbations), we can obtain the full dynamic model of the hydraulic operated manipulator.

$$\tau = M(q)\ddot{q} + C(q, \dot{q})\dot{q} + G(q) + F_v \qquad [3.89]$$

$$\dot{\tau} + B.\tau + E\dot{q} = J.k_o.i. \qquad [3.90]$$

For the control law proposed here, we consider as state variables, positions, velocities and accelerations, then the robot model is rewritten in a single stage to eliminate the moment of torsion of the equations of the model. The derivative of the moment of torsion is analytically obtained by the derivative of equation [3.89] (some arguments are removed for ease of notation):

$$\dot{\tau} = M\dddot{q} + \dot{M}\ddot{q} + C\ddot{q} + \dot{C}\dot{q} + \dot{G} + \dot{F}_v. \qquad [3.91]$$

For the overall system, we obtain an independent equation of the applied torsion moment or measures of differential pressure as follows:

$$\begin{aligned} J.k_o.i &= M\dddot{q} + (\dot{M} + C + BM)\ddot{q} + \dot{G} + \dot{F}_v + \\ &\quad +(\dot{C} + BC + E)\dot{q} + BG + BF_v. \end{aligned} \qquad [3.92]$$

This equation is rewritten as follows for simplicity:

$$J.k_o.i = M\dddot{q} + C\ddot{q} + \beta\ddot{q} + \gamma\dot{q} + \delta \qquad [3.93]$$

with $\beta = \dot{M} + BM$; $\gamma = \dot{C} + E + BC$ and $\delta = \dot{G} + \dot{F}_v + BG + BF_v$. Equation [3.93] will be used for the design of the control and stability analysis.

3.5. Application exercises

EXERCISE 1.– Accuracy of a sensor.

Explain what are the limitations of a sensor. Base your answer on examples that you know.

EXERCISE 2.– Transfer function of a DC motor.

The DC motor with permanent magnets has the following equations:

$$\begin{aligned} E(t) &= K\Omega(t) \\ C_m(t) &= Ki(t) \\ u(t) &= E(t) + Ri(t) + L\frac{di}{dt} \\ J\frac{d\Omega}{dt} &= C_m(t) - C_r(t) \end{aligned} \qquad [3.94]$$

where u and i are the voltage and armature current.

EXERCISE 3.– Measurements on an asynchronous motor.

On the power lines feeding a three-phase asynchronous motor, the following measurements are carried out:

– voltage RMS value between two phases, 400 V;

– current in one phase, 10 A;

– total power absorbed by the engine, 3,460 W.

Determine:

1) the power factor;

2) the phase shift between voltage and current. Specify which is ahead of the other.

EXERCISE 4.– Current control of a DC motor.

We want to control the current of the armature in a DC motor with permanent magnets powered by a chopper. We assume that current measures are available and that we can directly impose the voltage at the terminals of the armature. We propose the following strategy to be followed: the EMF E is considered as a constant perturbation (case of significant inertia where velocity variations are slow compared to the current fluctuations). The controller is of the form PI ($C(p) = Kp + Ki/p$); the zero of the controller is selected such that to compensate for the pole of the system; the bandwidth of the control is chosen to be equal to one-tenth of the chopper frequency f_h.

Give the armature equation of a DC motor with permanent magnets using the notation u and i, respectively, the voltage and the armature current.

The following numeric values are provided: $R = 1\Omega$, $L = 2\,\text{mH}$, $K = 1\,\text{N.m/A}$, $f_h = 10\,\text{kHz}$:

1) give the structure of the current control;

2) give the transfer function between the voltage and current;

3) determine the gains Kp and Ki;

4) give the Bode diagrams of the following transfers:

- system to be controlled;

- controller;

- open-loop transfer function (process + controller);

- closed-loop system (transfer between reference and measure).

EXERCISE 5.– Tensile test machine with electrohydraulic system.

The operation principle of this testing machine is based on the exact control of the "effort–displacement" relationship. It is also possible to operate the displacement, the effort or the speed of the machine. It is therefore important to obtain an output quantity exactly identical to the one desired in order to use the results.

1) First case: displacement control. We have:

– a setpoint: voltage $e(t)$, system input;

– a comparator;

– a current generator amplifier;

– a servovalve: it delivers on output a flow rate $Q(t)$, proportional to the input current $I(t)$;

– a cylinder: it gives on output a displacement $x(t)$, proportional to the input flow rate $Q(t)$. Its displacement corresponds to the elongation of the specimen;

– a position sensor: inserted in the feedback chain, it measures this elongation.

Question: Establish carefully the block diagram of this control (number of steps setpoint; output displacement of the cylinder compared to the piston $\lfloor(t)$). Between each block, specify the name and unit of the quantity that flows. What is the nature of this control (regulation – follower system)?

2) Second case: effort control.

For this type of test, it is desirable to obtain a constant effort throughout the traction. An effort sensor is then used. It is placed at the point of attachment of the test piece. Displacement $x(t)$ thus occurs under the control of this sensor.

In this case, establish the block diagram. What is the particularity of such a system?

Observation: the full principle of the tensile test. Will be seen later in materials resistance. Nevertheless, we can succinctly give the principle:

A specimen (test piece) is subjected to a traction effort, which causes deformation. If in a first phase the deformation law is linear, then an acceleration of the elongation can be observed. To obtain a deformation following a precise law (setpoint), the system under study is then indispensable.

EXERCISE 6.– Self-balanced vehicle.

This vehicle has two wheels with the same axis of rotation, with its gravity center located above the common axis of the wheels, so that the question is how to stay in equilibrium once mounted on the platform. Just like the brain allows the individual to stand up without falling due to the inner ear, the system comprises an inclination servo device maintaining the platform of the vehicle horizontally, or even the support bar, assumed to be orthogonal to this platform, vertically. For this purpose, the vehicle has sensors and microprocessors controlling the two electric engines fitted to both wheels.

The structural description of the vehicle is as follows.

The system includes:

– a cart (chassis + two wheels), carrying the driver;

– two geared motors (electric motor + reducer) driving the wheels (one per wheel);

– a set of sensors consisting of a gyrometer and a pendulum, delivering information about the inclination angle of the chassis with respect to a vertical axis and about its derivative;

– an electronic computer that computes, based on the information originating from the sensors and from user instructions, control instructions for geared-motor groups;

– incremental encoders (angular sensors) providing the computer with a picture of the rotation velocity of the engines;

– batteries storing energy and providing it to the various components;

– a control box located on the handlebars.

1) Among the seven elements outlined in the structural description, what are those that show that the vehicle is an automated system?

2) Regulation of the inclination of the chassis.

Useful quantities are as follows:

$\psi(t)$, inclination of the chassis with respect to the vertical;

$\psi_C(t)$, inclination setpoint of the chassis with respect to the vertical;

$\alpha(t)$, inclination of the driver with respect to the support bar, perturbation;

$u(t)$, control voltage of the gear motor.

In addition, we have: $\beta(t) = \psi(t) - \alpha(t)$.

The action chain that enables the inclination to be regulated is concretized by:

– an amplifier and gear motor that delivers a torque $C_m(t)$ (mechanical action that tends to drive a solid into rotation). The resulting torque is proportional to the control voltage $u(t)$;

– a composite cart + driver that is governed from the dynamic point of view by the following differential equation:

$$\beta\frac{da^2(t)}{dt^2} = bC_m(t) + \beta c(t). \qquad [3.95]$$

The control aspect comprises:

– a comparator that by unity loop achieves the difference $\varepsilon(t) = \psi_C(t) - \psi(t)$;

– a controller that adapts the gap and provides a voltage $\omega(t)$ that controls the system;

– there are then two loops: a first loop (internal), a drift control that is performed by the gyrometer, a second loop (around the first one), position control achieved by the pendulum.

$$Ku(t) = \frac{d\psi(t)}{d(t)} + Ku(t)\psi(t). \qquad [3.96]$$

Question: find the control voltage of the gear motor and establish the functional block diagram of this regulation of the inclination of the chassis of the vehicle. Specify the names and units of the quantities that flow.

4

Digital Control and Polynomial Approach

This chapter explores controllers and digital regulation, and the solution of polynomial equations.

4.1. Introduction to digital control

The principle of digital systems control is very similar to analog control. The block diagram of an analog process controlled by means of a digital controller is presented in Figure 4.1.

The controller is implemented on a digital computer as a microcontroller, a microprocessor or a complete computer equipped with input–output cards. The problem therefore consists of determining the structure of a digital corrector and the values of the parameters of this structure in order to satisfy constraints regarding the behavior of the looped system. These constraints, given by specifications, are provided in terms of stability of the system, performance (response time, overshoot, accuracy, rejection of disturbances), and in terms of robustness to errors in modeling (stability margins). They can be expressed in the temporal domain (response time, overshooting) or in the frequency domain (phase margin and gain margin).

There are two main methods for the synthesis of digital controllers. The most common is using transposition from a continuous representation to a digital one: its first stage consists of determining an analog controller and then transforming it into a digital controller with equivalent performance. This type of method widely used in the industry fully justifies the study of analog system control even if the great majority of controllers used are now digital.

The second type of synthesis method is called direct digital synthesis. Less used in the industry than transposition synthesis, it allows, however, for finer adjustments when the models of processes to control are well known.

There are a large number of theoretical tools available to us to synthesize a controller while satisfying given specifications, which are now well known. In order to determine the stability of the loop system, we can notably use algebraic criteria (Jury criterion, Routh criterion) and the Nyquist criterion. Furthermore, to comply with performance constraints, Evans's locus is a very powerful tool. Frequency diagrams such as Bode diagrams and methods for error calculation in steady state can also be included. With regard to the robustness of the loop systems, we will mainly make use of harmonic diagrams (Bode). It is in this area that simulation becomes of paramount importance in order to test the behavior of the loop system with regard to different sources of errors in models (uncertainties, nonlinearities, unmodeled dynamics, etc.). In effect, the analytical tools available to us are not powerful enough to be able to simply address these errors.

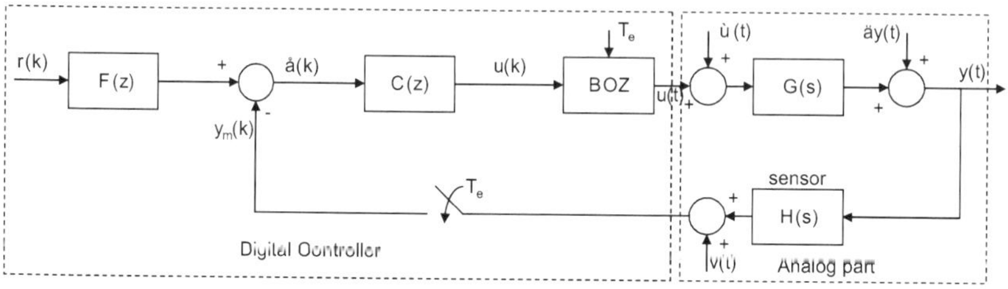

Figure 4.1. *Digital control principle for an analog system (where $G(s)$ is the process, $H(s)$ is the sensor, $C(z)$ is the controller, $F(z)$ is the prefilter, r is the signal setpoint, y is the output signal (to be regulated), y_m is the output measurement, ε is the deviation or control error, u is the command, $w(t)$ is the input disturbance, $v(t)$ is the measurement noise and $\delta y(t)$ is the output disturbance)*

4.1.1. *Digital controller synthesis by transposition*

This method of synthesis is initially based on the determination of an analog controller that meets the specifications being imposed. For this purpose, one may use frequency synthesis methods based on Bode diagrams, syntheses through the placement of the poles of the closed-loop system by making use of the root locus. One may also choose a standard controller structure (proportional (P), proportional integral (PI), proportional derivative (PD), proportional integral derivative (PID)), and determine the parameters of the structure by means of empirical methods.

The problematic aspects of controller synthesis using a transposition method lie in the choice of a transposition method. The objective of the transposition is to find a digital controller for which the behavior of the digital control is as close as possible to the behavior of the analog control. Due to sampling and quantization effects, one may already suspect that it is not possible to obtain an exactly identical behavior.

REMARK 4.1.– *If the analog controller has been correctly synthesized, the analog servo has optimal behavior vis-à-vis the specifications provided. It is thus possible to note that the digital control resulting from the transposition will be at best equivalent to the analog control, but certainly not better. We can therefore very quickly identify the main weakness of transposition synthesis methods for digital controllers.*

Several transposition techniques are commonly used. Each of them have their advantages and disadvantages and are more or less adapted to the problems you will have to address. There is no better technique than any other in the general case and that can be applied with eyes shut. The methods presented here are:

– sampling, zero-order holder (zoh);

– Euler transformation;

– bilinear transformation (without and with prewarping);

– poles and zeroes conservation transposition.

The first-order sampling method can also be found.

4.1.1.1. *Sampling hold transposition*

This transposition method consists simply of replacing a continuous system with a digital system consisting of the continuous system preceded by a sampler and a zeroth-order holder.

The transfer function of the equivalent digital system is thus simply obtained with calculating the Z-transform of the continuous system preceded by a zeroth-order blocker. For example, when it is desirable to transpose a controller C(s), the result of the transposition will be:

$$C(z) = Z\{B_0(s)C(s)\} = (1 - z^{-1})Z\left\{\frac{C(s)}{s}\right\}. \qquad [4.1]$$

REMARK 4.2.– *When this transposition is applied to an analog process preceded by a zeroth-order holder, the digital transfer function of the process is obtained.*

By comparing diagrams in Figure 4.2 presenting a servo with a continuous controller and the servo obtained by transposing the controller through sampling hold, it can be denoted that the effect of the analog digital converter (zeroth-order holder) that must be introduced between the digital controller and the process to be controlled is not taken into account. This will be one of the limitations related to transposition methods (regardless of the conversion technique being used). We recall that the zeroth-order holder reduces the frequencies close to the Nyquist frequency and that it introduces a delay of half the sampling period.

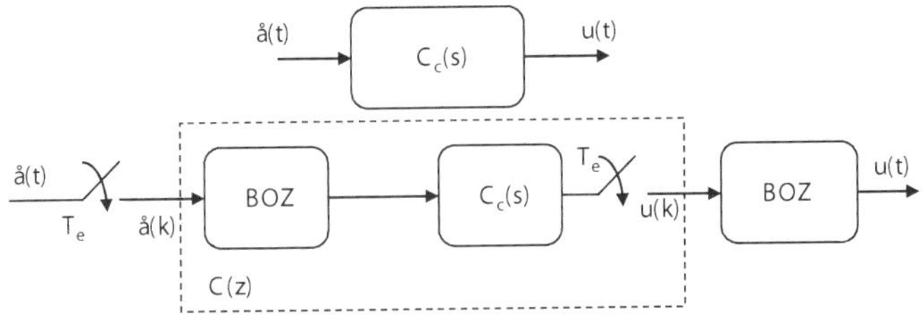

Figure 4.2. *Principle of sampling hold transposition*

Below are some properties of the sampling hold transposition technique:

– the poles of the controller $C(z)$ are in $z = e^{T_e p_i}$ where p_i are the poles of the analog controller $C(s)$. It is said that the poles are preserved;

– if the controller $C(s)$ is stable, then the obtained controller $C(z)$ is equally stable. This follows directly from the conservation of the poles of the controller. On the other hand, there is no guarantee about the stability of the closed loop;

– the zeros of the controller are not preserved. If $C(s)$ has no zeros, then $C(z)$ may very well have some or vice versa;

– the static gain of the controller (and therefore of the closed loop) is preserved.

4.1.1.2. *Effect of the sampling period*

The digital approximation by sampling hold will be better when the sampling period is small. The output of the digital controller is equal to the output of the analog controller at sampling times and is then maintained during a sampling period, as it is shown in Figures 4.3–4.5. The digital output is therefore always lagging behind the analog output.

In practice, this transposition technique gives correct results only if the sampling period is very small compared to the response time of servo systems. In this case, from a temporal point of view, continuous signals are correctly approximated and in addition we are working with signals whose frequencies remain far below the Nyquist frequency, so that the effect of the zeroth-order holder not being considered in the transposition is negligible. If it is not possible to sample very quickly, it is highly recommended to use another transposition method less sensitive to the increase in the sampling period.

4.1.2. *Euler's transposition*

The sampling hold transposition previously presented is often too brutal. Other more complex methods are therefore necessary.

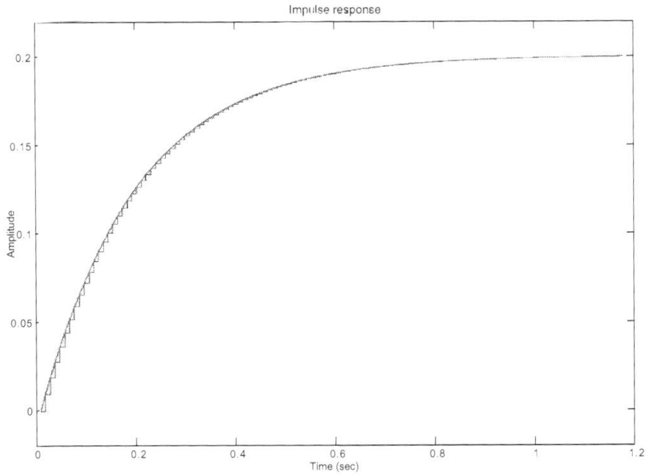

Figure 4.3. *Effect of the sampling period. Holding* $Te = 0.01s$ *(continuous and zoh)*

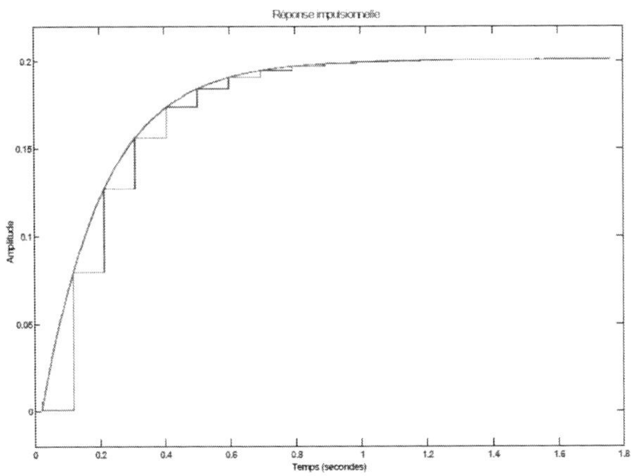

Figure 4.4. *Effect of the sampling period. Holding* $Te = 0.1s$ *(continuous and zoh)*

For analog systems, input and output signals are connected by differential equations. For digital systems, there are difference equations that connect these signals.

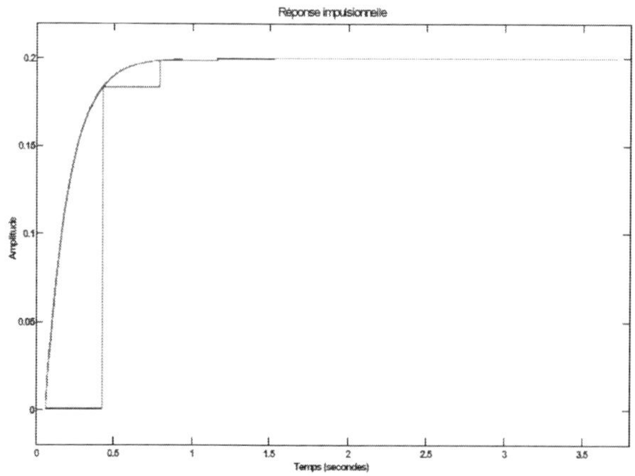

Figure 4.5. *Effect of the sampling period. Holding $Te = 0.5s$ (continuous and zoh)*

Euler's and bilinear transformation methods, which will be presented thereafter, are based on approximations of differential equations using difference equations.

Euler's transposition methods are based on the numerical approximation of the derivation. Backward approximation consists of approximating the derivation by:

$$D = \frac{dx}{dt}(t) \sim \frac{x(kT_e) - x((k-1)T_e)}{T_e}. \qquad [4.2]$$

From a frequency point of view, the continuous derivation is expressed by a multiplication by the Laplace variable s : $\frac{D(s)}{X(s)} = s$. Using Z-transformation, equation [4.2] becomes:

$\frac{D(s)}{X(s)} = \frac{1-z^{-1}}{Te}$.

By analogy, we can propose the change in variable

$$s \longrightarrow \frac{1 - z^{-1}}{T_e} = \frac{z-1}{T_e z}, \qquad [4.3]$$

that is $z = \frac{1}{1-T_e s}$.

In practice, we thus replace s by $z - 1/Tez$ in the expression of $C(s)$ to obtain $C(z)$.

This transposition method has the following properties:

– If the analog controller is stable, then the digital controller obtained is also stable. However, the reverse is not true. It is also possible to transform an unstable analog controller into a stable digital controller. In effect, the poles p_i of the continuous controller are transformed into poles $\frac{1}{1-p_i T_e}$ of the digital controller.

If p_i has a real negative value, then the corresponding numerical poles are inside a circle of radius 0.5 centered in $z = 0.5$ and are therefore stable (see Figure 4.6). However, not all of the unit circle is covered by the transposition.

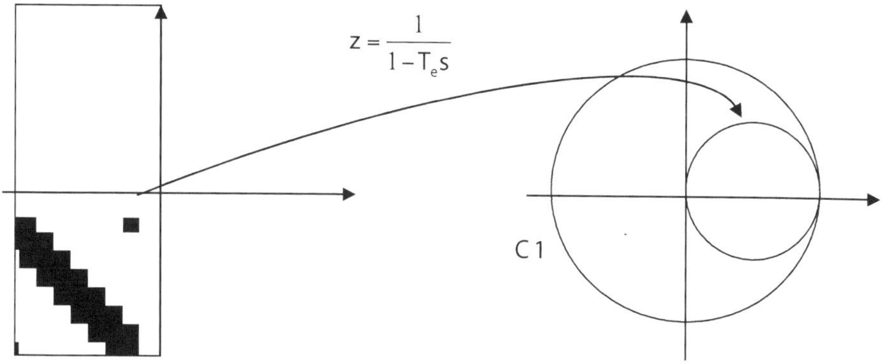

Figure 4.6. *An image of the stability region by Euler's transformation toward the rear*

REMARK 4.3.– *It can be observed that the poles in* 0 *(integrators) become poles in* 1 *regardless of the sampling period. Therefore, there is preservation of the integrative effects contained in the controller.*

– The static gain of the controller and therefore that of the loop system are preserved. As a matter of fact, when $s = 0$ and by change in variable, we obtain $z = 1$. Consequently, $lim_{z \longrightarrow 1} C(z) = lim_{s \longrightarrow 0} C(s)$.

As for the sampling hold transposition, the zeroth-order holder is not taken into account in the transposition.

4.1.2.1. *Effect of the Euler transposition on the integration*

By changing variable, the continuous integration $\frac{1}{s}$ becomes $\frac{T_e z}{z-1}$. By inverse Z-transform, $I(z) = \frac{T_e z}{z-1} X(z)$ gives $I(kT_e) = I((k-1)T_e) + T_e X(kT_e)$. Graphically, integration is therefore performed by summing the area of rectangles of width T_e and with value $X(t)$ at sampling times (see Figure 4.7).

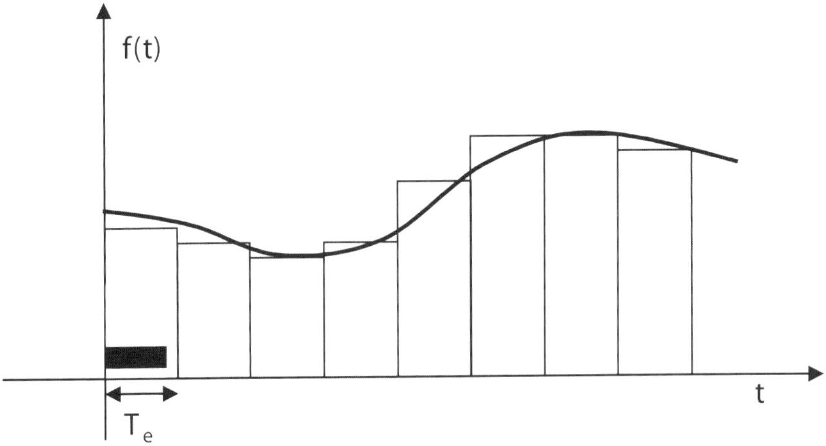

Figure 4.7. *Representation of the integration obtained Euler's transformation*

4.1.2.2. *Frequency effect of Euler's transposition*

We are now going to compare the frequency behavior of the digital controller obtained by Euler's transposition to the frequency behavior of the original controller.

REMARK 4.4.– *The comparison is achieved here with very simple transfer functions. A similar analysis can be carried out with any transfer functions.*

To obtain the harmonic behavior of a digital system, $z = \exp(jwT_e)$ is imposed in the transfer function (see Bode diagrams).

We thus obtain:

$$s \approx \frac{z - 1}{T_e z} = \frac{e^{j\omega Te} - 1}{T_e e^{j\omega Te}} = j\omega e^{-j\omega \frac{Te}{2}} \frac{sin(\omega \frac{Te}{2})}{\omega \frac{Te}{2}} \qquad [4.4]$$

to compare with $s = j\omega$ for the continuous harmonic behavior.

As a result, the Euler transposition introduces a pure delay of half a sampling period and a filtering by $F(j\omega) = \frac{sin(\omega \frac{Te}{2})}{\omega \frac{Te}{2}}$. Euler's transposition therefore introduces a significant distortion for frequencies close to the Nyquist frequency.

REMARK 4.5.– *It can be noted that there is similarity between the distortion introduced by Euler's transposition and the distortion introduced by the zeroth-order holder in the reconstruction of a continuous signal.*

– Effect on an integrator: Significant phase advance in the proximity of the Nyquist frequency (see Bode diagrams in Figures 4.8 and 4.9 that follow).

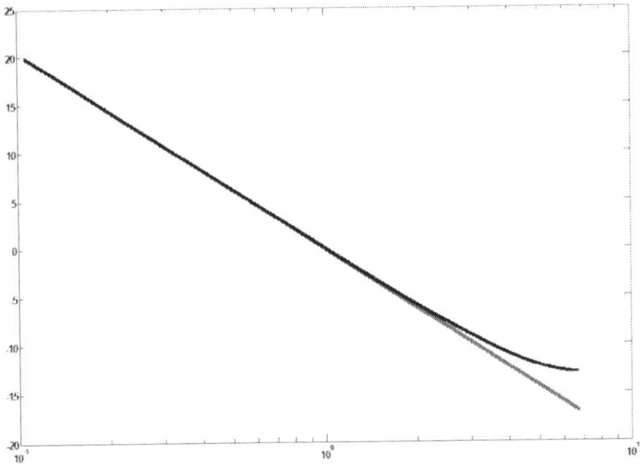

Figure 4.8. *Effect on an integrator (continuous and Euler's)*

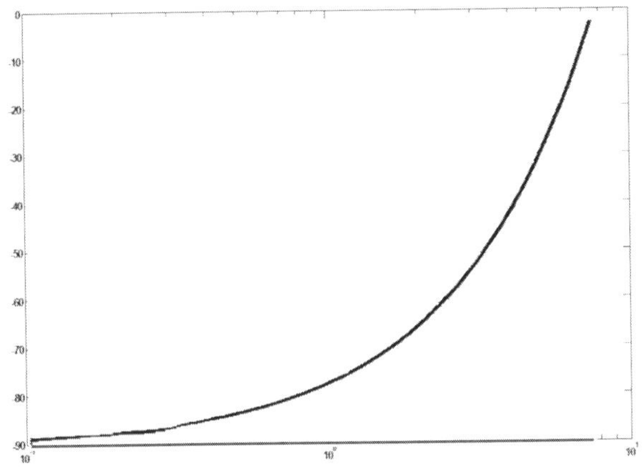

Figure 4.9. *Effect on an phase advance integrator (continuous and Euler's)*

– Effect on a selective filter: We consider a filter $C(s) = \frac{1}{1+s^2}$. After Euler's transformation, we get $C(z) = \frac{T_e z^2}{(Te+1)z^2-2z+1}$. Figure 4.10 hereafter represents the amplitude Bode diagram of $C(s)$ and $C(z)$. It can be noted that the selectivity of the filter has been completely erased by the transposition.

4.1.2.3. *Conclusion*

Euler's transposition is well suited for controllers of the high-pass type, because the behavior in derivation is good (since at the origin it is continuous derivation that was sought to numerically approach). Therefore, one will be able to use it for PD- or PDD-based controllers. On the other hand, it is very poorly suited for selective controllers and when phase behavior must be preserved as, for instance, for phase advance or phase delay controllers.

Furthermore, there is a transform known as bilinear transform. This transform allows us to effectively transpose controllers of the low-pass type. Associated with prewarping, it is also effective for selective filters. Its very good phase behavior makes it the first choice to approximate phase advance or phase delay-based controllers.

4.1.3. *Choice of the sampling period (Shannon's theorem)*

As we have just mentioned, the overall stability and behavior of the closed loop are dependent on the sampling period. It is therefore crucial to properly choose it.

Two limitations are necessary for this choice:

– the sampling period should be long enough to allow that the calculation of the control to send to the process can take place. It thus depends on the hardware (microprocessor, microcontroller) used for this computation. To be more precise, if the computation time T_c is just smaller than the sampling period T_e, we introduce a delay of a sampling period in the direct chain compared to the analog controller. This delay was not taken into account during the transposition of the controller and it can have a destabilizing effect on the closed-loop system.

In order to reduce this delay, we can proceed with the desynchronization of reading of Analog to Digital Converter (ADCs) and of writing in the Digital to Analog Converter (DACs). We write the control for the DACs as soon as the computation of the control is over. If the computation time becomes very small compared to the sampling period ($T_c \ll T_e\tilde{}T_c < \frac{T_e}{10}$), it can be considered from a theoretical point of view that sampling and holding remain synchronous. In this case, we no longer introduce delay in the direct chain.

– The sampling period should be chosen as short enough to allow a good approximation of the behavior of the analog system. Ideally, the Nyquist frequency $f_N = \frac{f_e}{2}$ should be larger than all the useful frequencies of the signals to be sampled. This condition is difficult to evaluate analytically, all the more that the signals to sample are dependent on the controller $C(z)$ obtained by transposition, this latter being dependent on the selected sampling period.

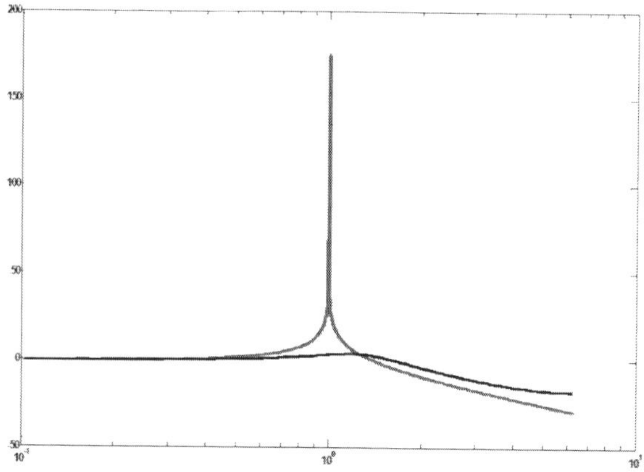

Figure 4.10. *Effect on a selective filter (continuous and Euler's)*

In practice, a general rule of thumb is used for the choice of T_e. One considers that the approximation of the analog system is good if $T_e \leq \frac{\tau}{10}$ where τ is the time constant of the analog closed-loop system. This rule is based on the general observation of the behavior of systems and it should not prevent the system designer from verifying the validity of the choice of T_e through adequate simulations.

4.2. PID controller synthesis and its equivalent digital RST

4.2.1. *Standard controllers*

The most commonly form of encountered standard controllers in the industry, such as PI, PD and PID, can easily be obtained using one or another transposition technique. Table 4.1 gives the analog form and the digital form of these controllers obtained with Euler's transposition. We recall that PDs and PIDs are "real" when the pure derivation is replaced by an approximated derivation consisting of a zero at zero and of a pole with the highest frequency. This "modification" of pure derivation makes it possible to limit the controller gain in high frequency and avoids too strong an amplification of high-frequency noises always present in electronic systems.

REMARK 4.6.– *Concerning digital system control during the computation of the control law and implementation of digital controllers. Transposition synthesis methods (but also direct synthesis methods) provide controllers in the form of their Z-transmittance. The unaware user who would apply this control to the system* $G(s) = \frac{1}{(s+4)^2}$ *would not obtain the expected result as shown in Figure 4.11. Detrimental effect from rounding for digital controllers. The solid line is the step response of the closed-loop system with* $C(z) = \frac{8.67z^2-16.66z+8.003}{z^2-1.667z+0.6667}$*, the dotted line is the response of the system with* $C(z) = \frac{8.67z^2-16.66z+8.003}{z^2-1.6667z+0.6667}$.

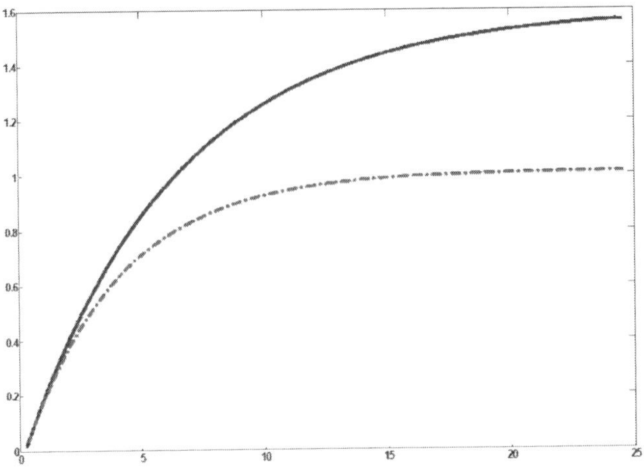

Figure 4.11. *Detrimental effect of rounding in digital controllers (with and without integrator)*

4.2.2. *Study of digital PIDs*

The study that we propose here for digital PIDs is also valid for digital PD and PI controllers. It can also be applied to other forms of PIDs obtained by other transposition methods. It makes it possible to highlight the refinements that are often applied to standard controllers in order to improve their behavior.

A conventional form of digital PIDs has been given in Table 4.1. The control can be decomposed into three terms corresponding to the proportional, integral and derivative parts of the control as shown in Figure 4.12.

Type	Continuous	Digital
P	$C(s) = Kp$	$C(z) = Kp$
PI ideal	$C(s) = Kp(1 + \frac{1}{T_i s})$	$C(z) = Kp(1 + \frac{1}{T_i}\frac{T_e z}{z-1})$
Real PD	$C(s) = Kp\left(1 + \frac{T_d s}{\frac{T_d}{N} s}\right)$	$C(z) = Kp\left(1 + \frac{N(z-1)}{\left(1+\frac{N T_e}{T_d}\right)z-1}\right)$
Real PID	$C(s) = Kp\left(1 + \frac{1}{T_i s} + \frac{T_d s}{1+\frac{T_d}{N} s}\right)$	$C(z) = Kp\left(1 + \frac{1}{T_i}\frac{T_e z}{z-1} + \frac{N(z-1)}{\left(1+\frac{N T_e}{T_d}\right)z-1}\right)$

Table 4.1. *Conventional forms of standard digital controllers obtained using Euler's transposition*

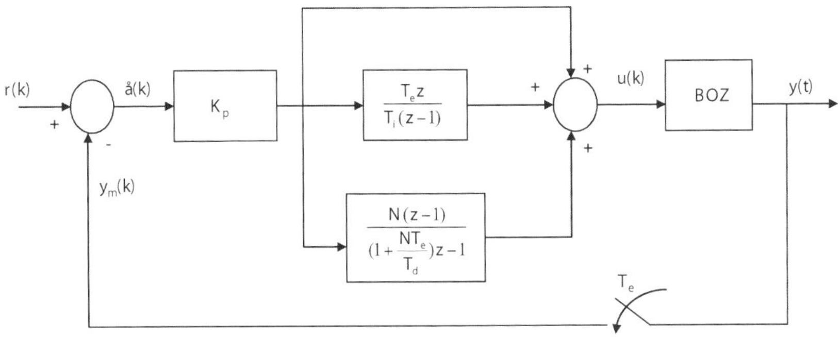

Figure 4.12. *Decomposed representation of a digital PID*

Similarly, the value of the control at every step can be decomposed into three terms obtained using the inverse Z-transform of each term of the transfer function:

$$u(k) = u_p(k) + u_i(k) + u_d(k) \tag{4.5}$$

with

$$u_p(k) = K_p \varepsilon(k), \tag{4.6}$$

$$u_i(k) = u_i(k-1) + K_p \frac{T_e}{T_i} \varepsilon(k), \tag{4.7}$$

$$u_d(k) = \frac{1}{1 + \frac{NT_e}{T_d}} (u_d(k-1) + K_p N(\varepsilon(k) - \varepsilon(k-1))). \tag{4.8}$$

It is in this form that the PID controller will be generally implemented, because:

– it limits the number of calculations compared to a brutal expression of the control law such as presented in the previous section;

– it enables the derived term and the integral term be separately addressed, which makes it possible to apply thereto the techniques presented in the following sections.

4.2.2.1. *Setpoint form without derivation*

The derivation of the difference signal between the setpoint and the measurement is equivalent to the derivation of the setpoint minus the derivation of the measurement. In some cases, the derivation of the setpoint is detrimental to the proper functioning of the system. To understand this, consider an engine that is position controlled. We would like to bring the motor to different positions successively, waiting at each time for the motor to be stable. The setpoint is therefore a succession of steps. Each time the motor is stabilized, the application of a step on the setpoint creates by derivation a peak in the derivation term that results in the saturation of the actuators (variable speed drives). In this particular case, it is preferable to derive the measurement only. Apart from sudden changes in position, the setpoint is constant and it is therefore equivalent to deriving the difference and the opposite of the measure.

In this case, the PID controller takes the form given in Figure 4.13 and the time derivative term becomes:

$$u_d(k) = \frac{1}{1+\frac{NT_e}{T_d}}(u_d(k-1) - K_pN(y_m(k) - y_m(k-1))). \quad [4.9]$$

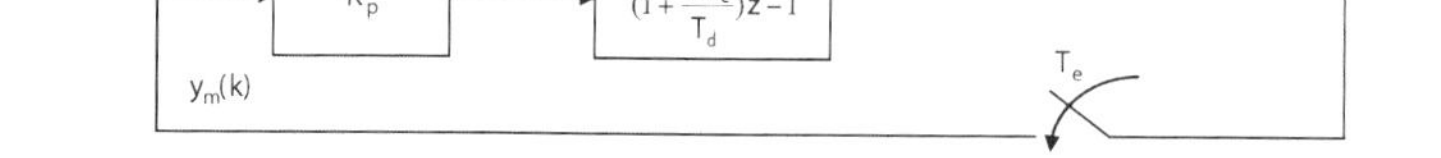

Figure 4.13. *Digital PID without setpoint derivation*

REMARK 4.7.– *The application of the derived effect to the measurement and not to the difference should be reserved for cases where the setpoint is of the step type. In fact, for other types of setpoints (ramp, sinusoid, etc.) it is important to preserve the derivation term of the setpoint that provides information about the future value of the setpoint. In addition, in the case where the setpoint signal is continuous (sudden jump step), there is no occurrence of a detrimental peak on the control.*

4.2.2.2. *Effect of saturations on the integral term*

During the implementation of a digital control law, it is very important to take the physical limitations of actuators into account and to possibly even include software limitations in order to avoid bringing the actuators outside of their linear operating area.

The saturation of the control (hardware or software) may be mathematically represented by a continuous piecewise differentiable function:

– if $u(k) > U_M$, $u_a(k) = U_M$;

– if $u(k) < U_m$, $u_a(k) = U_m$;

– otherwise $u_a(k) = u(k)$;

in which U_M is the upper saturation, Um is the lower saturation, $u(k)$ is the control calculated at step k and $u_a(k)$ is the control actually sent to the actuator (applied control) (see Figure 4.14).

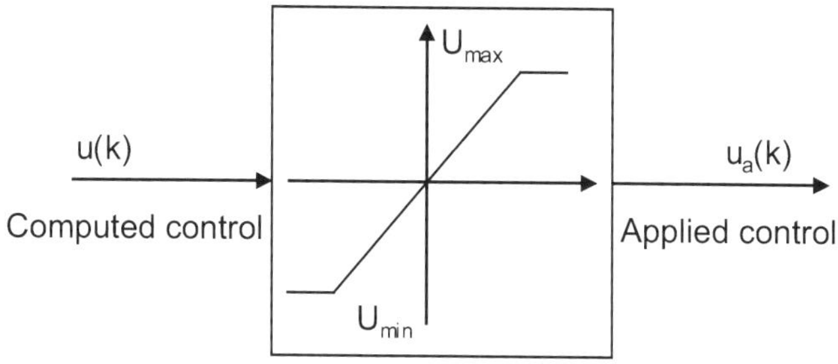

Figure 4.14. *Representation of saturation on control*

When the controller contains an integral action, the saturation of the control has a detrimental effect resulting in significant overshooting and lengthened settling down times. To understand this phenomenon, we are going to focus on the integral action of the controller.

The integral part of the control becomes zero when the integral of the difference between setpoint and measurement since the system has been powered up is zero. In geometric terms, when the integral action stops, the area between the setpoint and the measure is equal to the area between the measure and the setpoint, as it is represented in Figure 4.15. When the control is saturated, the slope of the output signal is limited. As a result, the surface $S1$ between the setpoint and the measurement increases. Therefore, the surface $S2$ must also increase, which is reflected by an increase in the overshoot and settling time (the control becomes zero and the response stabilizes when the area $S2$ is equal to the area $S1$). Figures 4.16 and 4.17 allow us to compare the responses obtained with and without control saturation for the same system. It can be observed that there is a longer settling down time and an increase in overshooting.

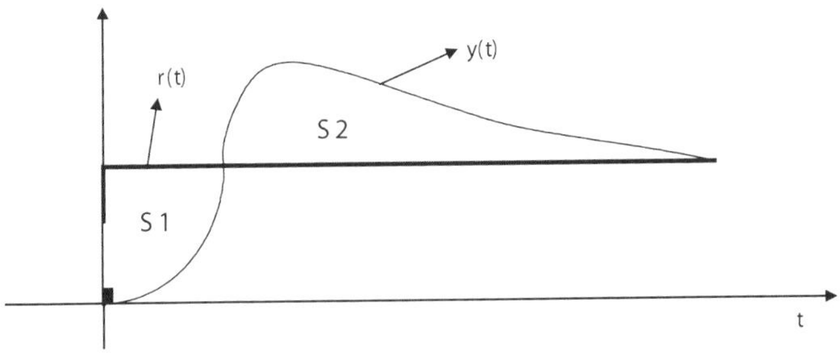

Figure 4.15. *Response of a loop system with a controller containing an integrator*

4.2.2.3. *Integral term antiwindup device*

In order to avoid these problems, some solutions have been proposed to prevent the windup of the integral term, that is to say that it increases while the activation of the system is blocked by saturation. The simplest solution consists of stopping the integration as soon as the control saturates. The complete computation of the control is then achieved in the following manner: the value $u_0(k)$ is determined that corresponds to the conventional value of the control without saturation. If $u_0(k)$ is in the area outside saturations, we apply u_0. On the other hand, if it is detected that u_0 is beyond the limits of saturation, the integral term $u_i(k)$ is recalculated.

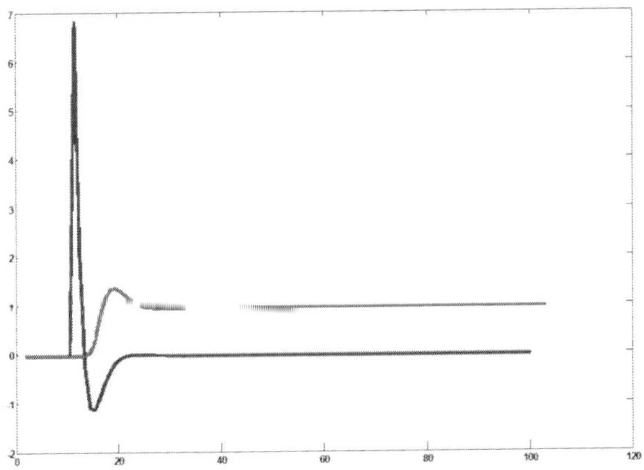

Figure 4.16. *Step response and control for a system without saturation*

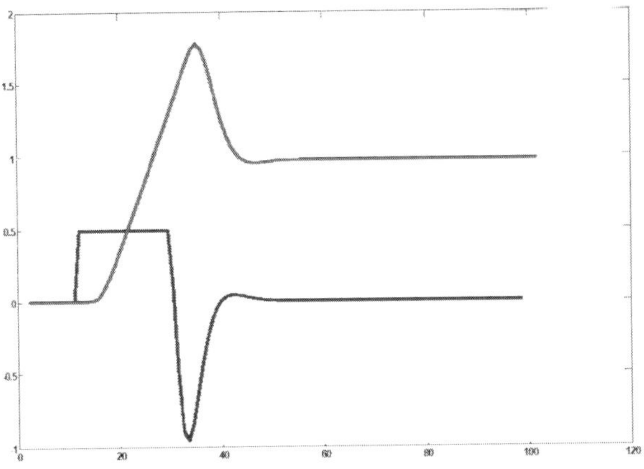

Figure 4.17. *Step response of a loop system with a controller for a system with control saturation*

In other words, the previous value of the integral term is preserved, that is the integration is stopped.

$$u_0(k) = u_p(k) + u_i(k-1) + K_p \frac{T_e}{T_i} \varepsilon(k) + u_d(k), \qquad [4.10]$$

$$u_i(k) = \begin{cases} (u_i(k-1) + K_p \frac{T_e}{T_i} \varepsilon(k) \text{ if } U_m < u_0(k) < U_M \\ u_i(k-1) \text{ if } u_0(k) > U_M \text{ or } u_0(k) < U_m \end{cases} \qquad [4.11]$$

$$u(k) = u_p(k) + u_i(k) + u_d(k). \qquad [4.12]$$

This antiwinding method is very simple, but it can lead to a suboptimal control. In effect, if the term $u_i(k)$ is predominant compared to u_p and u_d, blocking the integration strongly decreases the control being applied and the latter may then be quite distant from the saturation value. In order to obtain the maximal control at the saturation limit, the following device can be used that consists of recalculating the integral term that brings the applied control right to the saturation value u_s.

$$u_0(k) = u_p(k) + u_i(k-1) + K_p \frac{T_e}{T_i} \varepsilon(k) + u_d(k), \qquad [4.13]$$

$$u_i(k) = \begin{cases} U_M - (u_p(k) + u_d(k)) \text{ if } u_0(k) > U_M \\ u_i(k-1) + K_p \frac{T_e}{T_i} \varepsilon(k) \text{ if } U_m < u_0(k) < U_M \\ U_m - (u_p(k) + u_d(k)) \text{ if } u_0(k) < U_m, \end{cases} \qquad [4.14]$$

$$u(k) = u_p(k) + u_i(k) + u_d(k) = u_s. \qquad [4.15]$$

4.2.3. *Digital RST controller synthesis*

An RST controller is synthesized by an algebraic method of pole placement. The goal is to give to the closed loop the behavior of a system described by a model transfer function $Fm = \frac{Nm(z^{-1})}{Am(z^{-1})}$ expressed as a rational fraction in z^{-1}. Usually, this function is of the second order and we will choose Am as a monic polynomial, that is to say such that $Am(z^{-1} = 0) = 1$ (this is always possible). The structure of an RST controller is shown in Figure 4.18 where *R*, *S* and *T* are polynomials in z^{-1}.

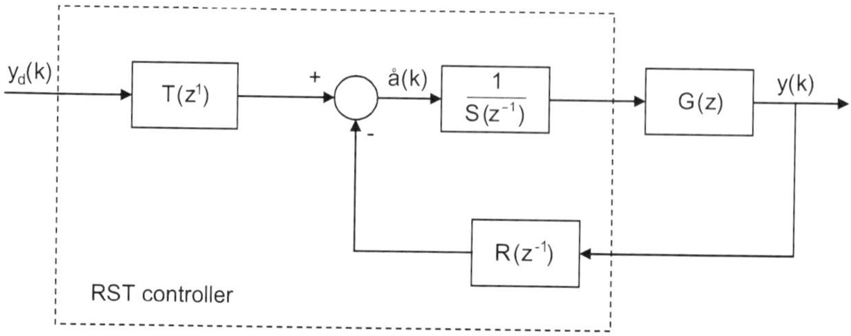

Figure 4.18. *RST controller structure*

In the following, we consider that the process to be controlled is given by the transfer function $G(z)$, which is written in power of z^{-1}:

$$G(z^{-1}) = \frac{z^{-d}B(z^{-1})}{A(z^{-1})} \qquad [4.16]$$

where z^{-d} is the pure delay of the process.

4.2.3.1. *Definition of the model transfer function*

The model transfer function is defined based on criteria such as:

- damping;
- static gain;
- natural frequency;
- overshooting;
- time of first maximum.

These criteria are generally used to define the denominator A_m of the model transfer function. On the other hand, the choice of N_m cannot be fully imposed. Since the influence of the zeros of the transfer function is less than that of the poles, these constraints will not have a significant effect on the behavior of the system.

The transfer function of the system regulated by an RST controller whose structure, shown in Figure 4.18, is equal to:

$$FTBF = \frac{z^{-d}BT}{AS + z^{-d}BR}. \qquad [4.17]$$

A simple analysis shows that the numerator of the transfer function must necessarily contain:

– on the one hand, the pure delay $z - d$ of the process. As a matter of fact, it is not possible to compensate a pure delay because it would be necessary to design a non-causal system. Therefore, N_m contains $z - d$ or possibly $z^{-(d+t)}$ with $t > 0$;

– on the other hand, all the zeros of the process that will not be compensated by S. These zeros are denoted B^-.

Finally, we thus get $N_m = z^{-d}B^- B_m^+$ where only B_m^+ can be freely chosen. In the absence of specific constraints, B_m^+ will often be defined of minimal degree (zero).

4.2.4. Choice of poles and zeros to compensate

It is possible to choose to compensate or not some of the poles and zeros of the process. The process will be divided in the following manner:

$$G(z^{-1}) = \frac{z^{-d}B(z^{-1})}{A(z^{-1})} = \frac{z^{-d}B^+(z^{-1})B^-(z^{-1})}{A+(z^{-1})A^-(z^{-1})} \qquad [4.18]$$

where the polynomial B^+ contains the zeros that we want to compensate and A^+ the poles that we want to compensate.

To compensate a pole, we have to put it in the polynomial R. To compensate a pole, we have to place it in the polynomial S. This is easy to understand by looking at the structure of the controller.

It will thus yield

$$R = A^+R', \qquad [4.19]$$

$$S = B^+ S^{'} . \qquad [4.20]$$

How can we choose the decomposition of B and A?

According to the principle of internal stability of the system, it is already forbidden to compensate poles or zeros outside the unit circle. As a result,

– B^- contains all the unstable zeros of the process;

– A^- contains all the unstable poles of the process.

The other poles and zeros can be compensated if the designer of the controller decides so. However, it can be noted that the compensated zeroes appear as poles in the transfer function between the setpoint Y_c and the control U. The real negative stable zeros therefore lead to real negative stable poles for setpoint control transfer.

It is known that such poles create alternate oscillating behaviors: the control will then alternate between positive and negative values. This phenomenon is to be avoided because it creates premature wear in mechanical systems and in power electric systems (choppers switching). Consequently, we will avoid compensating real negative stable zeros.

4.2.5. *Computation of polynomials R, S and T*

The computation of polynomials *R*, *S* and *T* is carried out by identifying the model transfer function with the transfer function of the controlled loop system:

$$\frac{z^{-d}B^+(z^{-1})B^-(z^{-1})}{A^+(z^{-1})A^-(z^{-1})} = \frac{z^{-d}B_m^+B^-}{A_m}. \qquad [4.21]$$

After simplification and by equalizing (up to a polynomial A_0) numerator and denominator, we obtain:

$$\begin{aligned} A^-S^{'} + z^{-d}B^-R^{'} &= A_0 A_m \\ T &= A_0 A^+ B_m^+. \end{aligned} \qquad [4.22]$$

The first equation is a Diophantine equation where $S^{'}$ and $R^{'}$ are the unknowns. The polynomial A_0 is usually chosen to equal 1, unless the specifications in the requirements impose a filtering (filter not reject) of the disruptions beyond a certain frequency.

4.2.5.1. *Solving of Diophantine equations*

Diophantine equations are of the form:

$$AX + BY = C$$

where A, B and C are known polynomials and X and Y are the polynomials sought after. These equations have an infinite number of solutions but the main interest will be on minimal degree solutions for X and Y.

Two cases are possible:

– $deg(A) + deg(B) > deg(C)$, the equation is regular and there is a solution that is minimal for both X and Y. We then get $deg(X) = deg(B) - 1$ and $deg(Y) = deg(A) - 1$.

– $deg(A)+deg(B)deg(C)$, there is a minimal solution for X for which $deg(X) = deg(B) - 1$ and $deg(Y) = deg(C) - deg(B)$ and a minimal solution for Y for which $deg(Y) = deg(A) - 1$ and $deg(X) = deg(C) - deg(A)$.

Once the degrees of X and Y are known, the Diophantine equation can be expressed in the form of a matrix equation that can be simply solved.

When $S^{'}$ and $R^{'}$ have been determined, *R*, *S* and *T* are obtained by:

$$\begin{aligned} R &= R^{'}A^{+}, \\ S &= S^{'}B^{+}, \\ T &= A_0A^{+}B_m^{+}. \end{aligned} \qquad [4.23]$$

4.2.6. *Additional objectives for synthesis*

One of the objectives of synthesis will be that there is often a zero position static error, in other words a static unity gain. For such a controller in series, this type of constraint is satisfied if the process contains an integrator, or by adding one in the controller. Here, because of the prefilter T, the presence of an integrator does not make it possible to guarantee that the static error be cancelled.

For this purpose, we just actually impose that the model transfer function have a unity static gain, that is to say, it is desirable that:

$$A_m(1) = B_m^{+}(1)B^{-}(1). \qquad [4.24]$$

Another objective may be to reject disturbances. To this end, we proceed as for a conventional series controller: the right integrator number has to be inserted in the system through the controller. *R*, *S* and *T* being polynomials, only the block $\frac{1}{S}$ can contain integrators. To impose an integrator in S, the Diophantine equation is simply rewritten in the following way:

$$A^{-}(1 - z^{-1})S_1 + z^{-d}B^{-}R^{'} = A_0A_m \quad [4.25]$$

with $S^{'} = (1 - z^{-1})S_1$.

4.3. Digital control by poles placement

The purpose of this method is to place the poles of the closed loop in the map of poles and zeros by acting on the gain of the controller and by placing zeros and poles of the controller. The techniques are very similar to pole placement methods for analog systems. It is also possible to decompose the adjustment procedure into five phases:

1) We first calculate the digital transfer function of the process with the chosen sampling period.

2) It is convenient to express the specifications in terms of the position of the poles of the closed loop, by giving a model (often of the second-order kind) of this closed loop:

- the concepts of speed can be expressed in the form of response time;

- the concepts of stability can be expressed in the form of damping (or overshooting);

- one can add thereto reject disturbances constraints;

- accuracy constraints (static gain, cancellation of n-order errors with respect to the setpoint, etc.).

Nonetheless, software tools such as rltool allow us to visualize in time real the effect of poles displacement on the step response, Bode diagram, etc., such that it becomes possible to make the adjustment for other types of constraints.

3) Based on the specifications, we can partially decide on the structure of the controller. For example, constraints of unity static gain may be reflected by the presence of an integrator in the controller. The controller can be expressed in the form $C(z) = C_{imp}(z)C_{free}(z)$ with one part dictated by a first simple reading of the specifications and one part to be determined.

4) The main tool of pole placement syntheses is Evans's locus. Therefore, the root locus of the digital process or, most often, of the process set $+C_{imp}$ will be plotted here.

5) Based on the Evans's locus, one will be able to determine the complete structure of the controller, which should be kept as simple as possible, and the controller parameters will be determined from the specifications and from the nomographs of the root locus.

4.3.1. *Choice of the sampling period*

The sample period must be compatible with the constraints of the specifications. We will notably be able to rely on the expected response times, whereas the specifications imposes, for example, a settling time of 2% of a second, it is desirable to have a sampling period lower than $100\,\mathrm{ms}$. However, unlike transposition syntheses, we can synthesize a controller with a response time of the order of only two or three sampling periods. These controllers, called "finite settling time controllers", however require strong control that can be hampered with control saturation. It should also be remembered that the poles of $G(z)$ are in $e^{T_e p_i}$ where pi are the poles of $G(s)$. By setting T_e too small, we will concentrate the poles of $G(z)$ very close to the unit circle on the map of digital poles and zeros. The result thereof is a difficult adjustment of the controller on the Evans locus. The choice of T_e is also a constraint when the process contains a pure delay. Although in the analog domain it was not possible to carry out root locus synthesis for delay systems, it becomes achievable with digital controllers provided that the delay of the process is a multiple of the sampling period. The Z-transform of the delay then creates system poles in 0. By setting T_e too small compared to the delay τ, a large number of poles are created that make the adjustment using Evans's locus impossible.

4.4. Diophantine, Bézout, greatest common divisor, least common multiple and division

4.4.1. *Polynomial arithmetic*

In Algebra, the arithmetic of polynomials describes the properties of polynomials that can be inferred from arithmetic and are somewhat similar to those of integer numbers. For instance, the polynomial ring $K(X)$ with an indeterminate X and with coefficients in a commutative field K has an Euclidean division. If the reader is not familiar with field and ring structures, K can be considered as a letter symbolizing the set of real or complex numbers. The Euclidean division is at the origin of key elementary arithmetic theorems. The same holds for polynomial arithmetic. Furthermore, Bézout's identity, Euclid's lemma or an equivalent of the fundamental arithmetic theorem are demonstrated in a similar manner; unity and irreducible polynomials are then taking the place of prime numbers.

These results are no longer applied in the same way if the coefficients are chosen in a set A such as that of integer numbers, where elements are not always reversible for multiplication. The study of this configuration requires the use of a paraphernalia of more powerful mathematical tools. They provide a means of showing that if Bézout's identity is no longer verified, an equivalent of the fundamental theorem of arithmetic remains still valid. This property holds if the ring comprises several indeterminates. Moreover, if A is a factorial ring, the polynomial ring with coefficients in A is also factorial, regardless of the number of indeterminates. In some cases, the ring A is not factorial but just Noetherian; given that the polynomial ring does contain only a finite number of indeterminates, it is also Noetherian.

These different outcomes are at the origin of founding theorems of various branches of algebra. Galois theory relies on the Euclidean structure of $K(X)$; the algebraic theory of numbers makes use of the factorial and of the Noetherian character of a polynomial ring with one or more indeterminates over a factorial ring. Finally, theorems such as that of Hilbert's basis or Nullstellensatz (Hilbert's theorem of zeros), essential in algebraic geometry, are the direct consequences of polynomial arithmetic.

4.4.2. *Diophantine equation* $ax + by = c$ *and Bachet–Bézout theorem*

Claude-Gaspard Bachet de Méziriac has proposed a method for solving Bézout's identity.

A few basic techniques make it possible to solve a first family of diophantine equations. An example is given by the first-degree linear equation with two indeterminates:

$$(a, b, c) \in \mathbb{Z}^3 \text{ and } a.x + b.y = c. \qquad [4.26]$$

This equation [4.26] is called Bézout's identity, from the name of the mathematician who generalized this result to polynomials. Its solution uses only the Euclidean division and Euclid's algorithm. This identity has a dual status. It corresponds to a Diophantine equation and represents one of the pillars supporting the building of elementary arithmetic. Euclid's lemma is demonstrated using this identity and the fundamental theorem of arithmetic by means of Euclid's lemma. The fundamental theorem provides a means to determine the properties of the greatest common divisor and the smallest common multiple operators as well as those of coprime numbers.

In elementary arithmetic, if Diophantine questions quickly become difficult, there are still some exceptions that can be solved with a minimum of theoretical tools and a short and simple demonstration.

4.4.3. *Bézout's identity*

In arithmetic, Bézout's identity designates:

– Bachet–Bézout's theorem, which proves an existing solution to the equation $ax + by = GCD(a, b)$;

– the equality $ax + by = GCD(a, b)$;

– Bézout's identity, in the case of polynomials is addressed in the polynomial arithmetic article.

4.4.4. *Greatest common divisor*

In elementary arithmetic, the greatest common divisor (GCD) of two natural integer numbers is the largest natural integer that simultaneously divides these two integers.

For example, the GCD of 42 and 56 is 14. In effect, $\frac{42}{14} = 3$, $\frac{56}{14} = 4$ and 3 and 4 are coprimes (there is no natural number apart from 1 that both divides 3 and 4).

4.4.5. *Least common multiple*

In mathematics and more specifically in arithmetic, the least common multiple (LCM) of two natural integer numbers a and b is the smallest integer that is a multiple of both these numbers. It is denoted $a \vee b$, $lcm(a, b)$, or sometimes just (a, b).

The LCM of a and b can also be defined as a common multiple of a and b that divides all the common multiples of a and b.

The definition extends to relative integers. In the second form, it should be then added that it must be positive. The second form of the definition is in fact generalized to any commutative ring, but the existence and uniqueness is usually lost, it is then referred to as an LCM of two elements. The existence is ensured in factorial rings.

The LCM can be more generally defined for any number of elements: for example, within the natural numbers, the LCM of n integers is the smallest integer simultaneously multiple of these n integers.

4.5. A few comprehension and application exercises

EXERCISE 1.–

The objective is to achieve the synthesis of a digital PID controller. The first part of the topic concerns the study of the system without controller. The second part of the topic addresses the calculation of the digital controller. The third part discusses the simulation of the step response of the system using a controller.

A) *Study of the system without controller*

We want to control a system in a digital loop in Figure 4.18 where $C(z)$ is a digital controller.

$$G(p) = \frac{1}{(1+\tau_1 p)(1+\tau_2 p)}. \quad [4.27]$$

For digital applications, we will assume

– τ_1 = 0.5s,

– τ_2= 1s,

– T_e = 1s, sampling period.

1) Show that $GB_o(z^{-1})$ is written as

$$GB_o(z^{-1}) = Kz^{-1}\frac{1+b_1 z^{-1}}{(1-a_1.z^{-1})(1-a_2.z^{-1})}. \quad [4.28]$$

2) Calculate the static error of the position and the velocity of the controlled system.

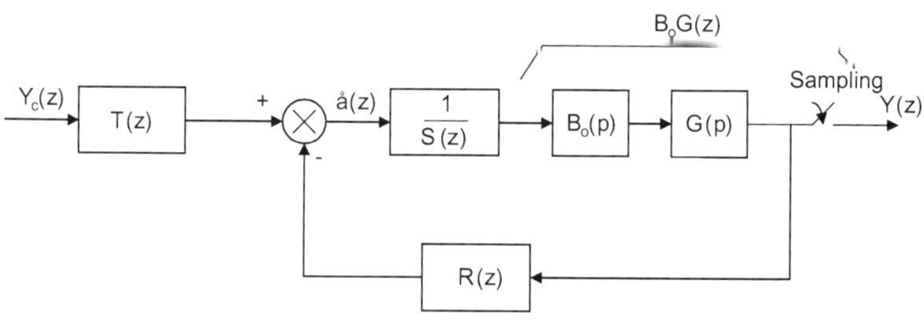

Figure 4.19. *System with digital loop without controller*

B) *Digital PID controller synthesis*

It is desirable to control this system with a digital PID controller modeled in an RST structure. The targeted performances of the servo system after being controlled are:

– a rise time $tm = 3\,\text{s}$,

– a zero first-order static error,

– an overshooting $D\% = 25$.

1) Calculate the two conjugate complex poles that impose the dynamic established by the specification requirements.

2) Therefrom derive the numeric expression of the denominator of the transfer function of the controlled system.

3) Give the structure of the controller and comment on each of the terms comprised in it.

4) Calculate the transmittance $T(z)$, $S(z)$ and $R(z)$ in the case of a structure 1 PID and structure 2 PID.

C) *Calculation of the first samples of a step response*

1) Calculate the value of the first 5 samples of the output for a step response, in the case of PID1 and in the case of PID2. Conclude on the functioning of both controllers.

D) *Study of bias in the system model*

We consider now that there is a bias β_K between the gain of the actual system and its estimated value K ($K_{reel} = K + \beta_K$). Calculate the first-order static error in the case of PID1 and PID2 and conclude.

EXERCISE 2.–

We consider the process given by its transfer function $G(s) = \frac{10s}{(s+10)}$. A controller is synthesized that makes it possible to obtain an overshoot smaller than 5% and response time smaller than $1\,\text{s}$. A simple gain $K = 5.2$ satisfies these requirements. The controller is then transposed into a digital one. We obtain $C(z) = 5.2$ regardless of the chosen transposition method. The response time tr being of the order of a second, the characteristic time constant of this system is of approximately $\tau = \frac{tr}{3}\tilde{}0.3\,\text{s}$. The rule of thumb thus recommends that (if possible) a sampling period $T_e < 0.03\,\text{s}$ be chosen. The step responses for various choices of T_e (0.01, 0.03, 0.1 and 0.3 s in addition to the continuous response) are shown in Figure 4.20. A good choice for the sampling period with regard to the useful frequencies of the signals is not enough to guarantee the validity of Shannon's theorem. As a matter of fact, high-frequency noise signals can overlap with the useful signals that, due to spectral aliasing, can be located in the bandwidth of the process. It is thus always necessary to add an antialiasing filter before the sampler (if this one is not directly contained in the analog-digital converter being used).

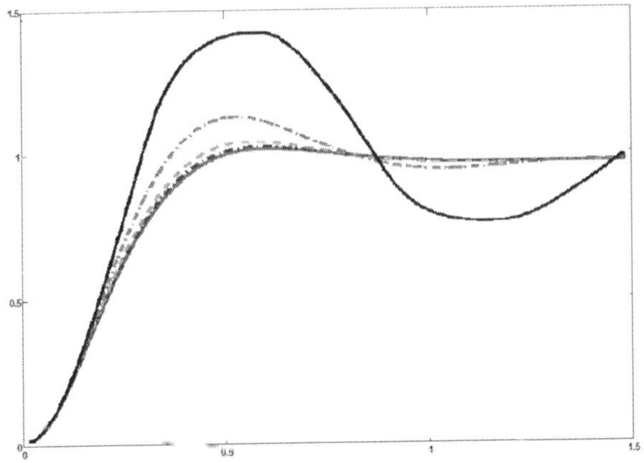

Figure 4.20. *Effect of the sampling period on the step response of a looped system*

EXERCISE 3.–

We aim at controlling process $G(s)$ given by the transfer function $G(s) = \frac{10}{(s+10)(s+3)}$ according to Figure 4.21 as follows:

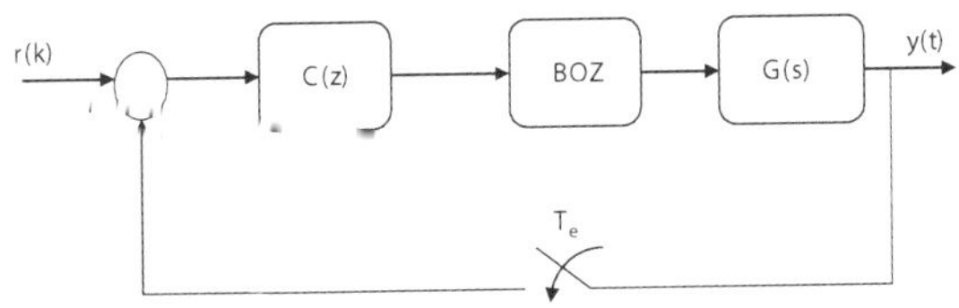

Figure 4.21. *System to be controlled*

The sampling period is set at $T_e = 0.1s$ and requirements are as follows:

1) zero static error;

2) zero error with regard to constant output disruptions;

3) second-order type behavior with $\zeta = 0.7$;

4) 5% response time smaller than 2 s.

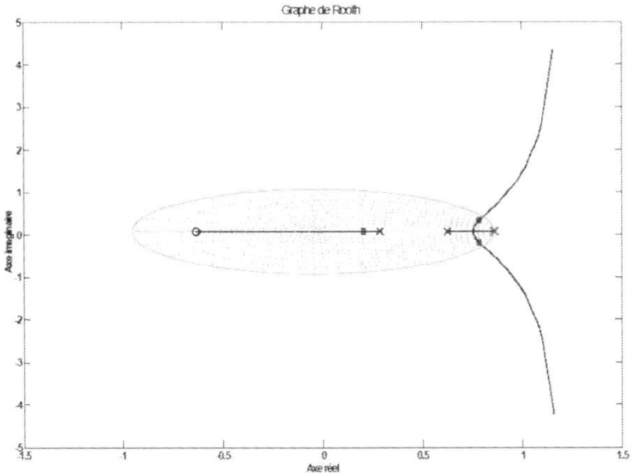

Figure 4.22. *Roots locus*

SOLUTION TO EXERCISE 3.–

1) The sampling period seems compatible with the response time imposed by the requirements. We calculate

$$G(z) = \frac{0.033(z + 0.649)}{(z - 0.3679)(z - 0.7408)}. \qquad [4.29]$$

2) The specifications have a form compatible with pole placement synthesis.

3) The constraint of zero static error imposes an integrator in the controller. This integrator is then sufficient to reject constant output disruptions. The controller will thus be of the form $C(z) = \frac{C_{free}}{z-1}$.

4) The Evans locus of $G(z)$ [4.22] is plotted by imposing an integrator inside the controller. By placing the dominant poles on the iso-damping curve $\zeta = 0.7$, we obtain the step response shown in Figure 4.23.

The settling time is of the order of 5 s and therefore insufficient to meet the specifications: we need to accelerate the system. On the other hand, the two branches of the Evans locus going toward infinity tend to quickly leave the unit circle. A possible method to bring them back into the circle consists of placing a zero of the controller near their origin, for example by compensating the dominant pole of the process. The new Evans's locus obtained is shown in Figure 4.24.

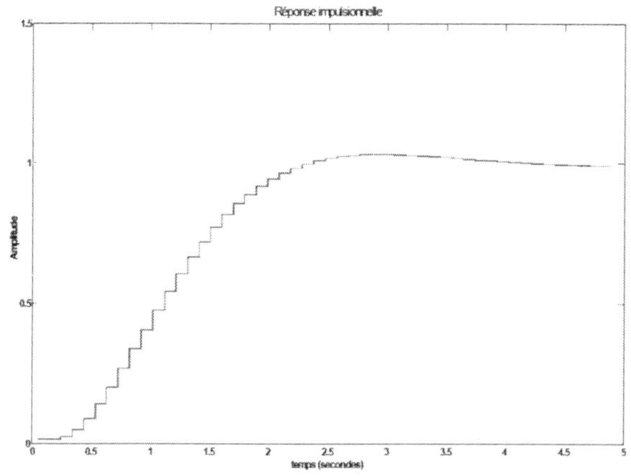

Figure 4.23. *Step response after adding an integrator*

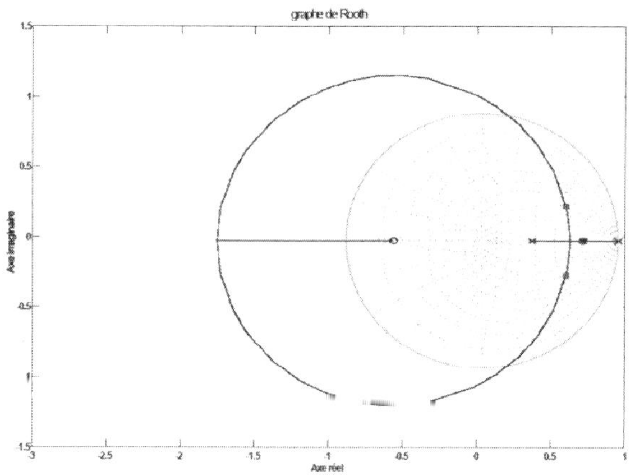

Figure 4.24. *Evans's locus after compensation of the dominant pole of the process*

The locus now has only two branches. By placing the poles of the closed-loop system on the iso-damping curves $\zeta = 0.7$, we obtain the step response shown in Figure 4.25. The response time is now less than 2 s and the controller is thus sufficient to satisfy these requirements.

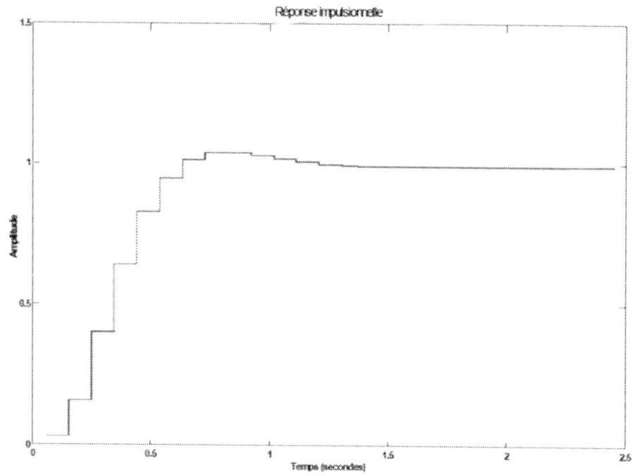

Figure 4.25. *Impulse response of the loop system after controller adjustment*

5

NAO Robot

This chapter explores the NAO robot and its application to home care.

5.1. Introduction

Advances in robotics are such that today we can have humanoid robots, integrating a multitude of interactive functionalities (recognition and speech synthesis, tactile sensors, luminous eyes, etc.). We can thus imagine and program a helping device for individuals or for home care assistance. Under this context, it would be interesting if it was possible to ask this "companion" robot to entertain people and to provide support throughout the day.

Nowadays, since life expectancy is constantly increasing, the well-being and comfort of the elderly has become an important preoccupation, for which we do not necessarily have the right solutions. Retirement homes, however adapted they may be, are only suited to some people. On the one hand, they represent a significant cost to elderly persons and their family. On the other hand, a large number of older people do not appreciate having to leave their house or apartment in exchange for a room or a retirement residence and sometimes they have problems accepting it. There is, therefore, a significant resulting development of so-called home care or home assistance. Home care offers individuals a non-negligible comfort because they no longer have to move out or leave their neighbors and friends. This allows them to end their days in the environment that suits them best. However, this solution has nonetheless a number of limiting factors. Certain elderly people are in need of daily care. This represents a considerable and very burdensome workload for nurses. For example, they must call the person several times or pay visits during the day to verify that drugs have been taken. In addition, a third person is needed for all matters related to daily tasks, which have to be carried out in a home and that are difficult to perform by the elderly person. That is the reason why companies such as Aldebaran

Robotics have turned to a whole new solution for home care: robotics. In effect, the creation of a robot to assist an elderly person on a daily basis is of interest, because in the end it would help to relieve nurses and home support workers. This robot will obviously not be able to perform the job of a nurse such as administering injections or bandages but it may, for example, take care of all the drugs reminders that make up a significant part of their working time. It will also be able to keep company to these people who sometimes may feel alone. This may include reading an e-book, playing several memory games or dancing performances. Naturally, it still remains a robot, which is why one of the paramount tasks will be to also make it friendly because accepting a robot at home is probably not simple to do. It is not here to replace human companionship. Finally, it should be able to detect if the person is having fits and in this case to call a relative, a neighbor or the doctor.

For all these reasons, Aldebaran Robotics has created Nao, the robot discussed in this chapter. The company is finalizing Romeo, a robot with a more significant size and which will be able to carry out this mission with better results. In this chapter, we present a study called "home care for an elderly person", and working with the Nao robot, the goal is to develop several programs that will be put into practice during a day in the company of an elderly person. This will thus range from drugs reminders to reading an e-book or even verifying that the person is awake and in good health.

In this chapter, we will develop software interfaces for humanoid robots. Furthermore, we will explain how to use the software Choregraphe and other interactive software useful to the operation of the Nao robot before developing the programs that we have created with their illustrated features. Finally, we clarify the regulation and control relationship of the Nao robot using these software programs.

5.2. Home care project

5.2.1. *Choregraphe software*

5.2.1.1. *Graphic interface*

Choregraphe is a software program developed by Aldebaran Robotics for the users of the Nao robot. It has been designed in order to make it possible to create and edit movements in order for the robot to become alive by making it move and react interactively. We can create animations and behaviors for the Nao robot, test them, monitor and control them. Choregraphe is a software program compatible with Windows, Linux and Mac OS. There are several versions but the one that was available for this study was version 1.10.37 (Figure 5.1 shows the standard logo of Choregraphe).

5.2.1.2. *Wi-Fi connectivity*

We can use Choregraphe in two different ways in order to work:

– either by connecting it to a local emulator (a virtual 3D Nao): Nao Qi. We can normally command the robot and the three-dimensional (3D) simulation responds like a real physical Nao;

– or by connecting it to the physical Nao robot.

Figure 5.1. *Choregraphe logo*

It is possible to connect to Nao via a Wi-Fi or Ethernet network. First of all, we have used this second connection. This wired solution posed a limitation on distance and constraints of movements because Nao could be blocked by the cable or incur the risk of falling. However, we did not have big problems in proceeding in this manner. Nevertheless in the end, we have mainly used the version employing the Wi-Fi connection.

To connect to Nao

When Choregraphe connects to Nao, we have to select "connection" in the menu, and then select the name of the robot (seen through cabled Ethernet or through Wi-Fi). The user should just wait that the connection is activated.

The different connection states:

	This logo indicates that Nao is stimulated by the Nao Qi program
	The robot is disconnected. It is then not possible to establish a connection
	Choregraphe is connected to the physical robot (connected to Nao)

5.2.1.3. *Choregraphe usage and Nao control*

The software allows us to create behaviors as a sequence of actions or gestures from the main platform. There is a library of preset movements (boxes) and it is possible to create customized boxes to save them and to store the actions that we have programmed. The boxes can be modified and are coded by means of a script in Python (but we can also code in C++, Java, Urbi, etc.).

To create a behavior, the user must select a movement box, make it slide in the main area, modify it if needed and connect it to define the desired action sequence.

For instance: “Say Hey!”:

– Selection of the box “Say” (audio/voice).

– Insertion of the box and modification of the text by double clicking on “Say”.

– Click on the input of the box and connect it with the action input on the left at the top of the main area.

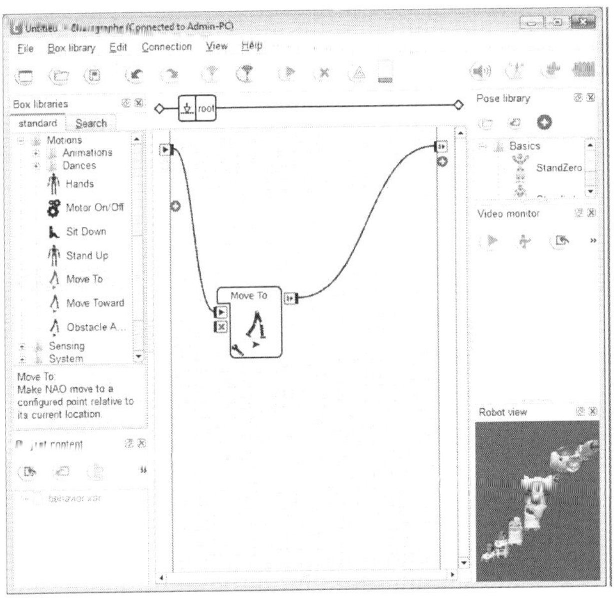

Figure 5.2. *Walk action (move to)*

Note that, before performing the movement by pressing on “play”, Nao should be controlled by clicking the control button. To create a sequence of movement in order to build an action, we must add the boxes as previously explained and connect them.

It is possible to create loops with complex conditions and thus to obtain complete and effective algorithms (see Figure 5.2). We can also create a movement manually (if it is desirable that Nao make a movement with the head or with a particular arm) by using Nao itself and by recording the positions that we want, or by choosing a virtual position through NaoQi, the simulator.

Basic actions of the Choregraphe library:

	Stand up	Control that makes Nao stand up.
	Sit down	Control that commands Nao to sit.
	Say	Control that tells Nao what the user is writing in the software.
		Control that asks Nao to walk a certain parameterizable distance.

These boxes provide a means to save time when a user wants to program a series of simple movements (stand up, walk and sit for example) and can also make it possible to simply visualize the possibilities that Nao has to offer (make a demonstration). The possibility to create the movement boxes or to insert code in Python offers the user of Choregraphe an infinity of behaviors to be created in order to make Nao livelier and more interactive:

– *Files*: it is possible to record recently created behaviors and to open others using the "Open Project" tab. However, it is important not to overwrite existing behaviors with the ongoing executing behavior.

– *Behavior*: it is possible to predict Nao's behavior. To this end, we first have to verify that it is not already in the memory of the robot (menu "Behaviors"), and then it has to be inserted using the search box for documents. The next step is to select the desired behavior file and insert it into the list. We can then click on the arrow corresponding to the selected behavior to activate it in the list. Nevertheless, beware when selecting several behaviors, Nao can get confused and then act unpredictably, without knowing what behavior he is going to adopt first.

– *Video*: there is an application that allows the user of Choregraphe to receive feedback on what Nao sees and perceives. It consists of the Camara module of the Monitor application. We can receive the data of both cameras available: one is on the front, and the second is in the mouth; it lets the user see what is happening under the feet of the robot. This second camera has been installed following the Robocup in which the robot was to kick a ball and therefore see its feet to aim correctly. The Memory module makes it possible to access and to ergonomically analyze the data that is provided by the sensors of the Nao robot.

5.2.1.4. *Telepathe and sensors*

It is a program designed to collect data provided by the sensors, generated from the robot. When the program starts, it asks whether the user wants to initiate a video capture or capture of values from different sensors. In the second case, after connecting to the robot, the list of sensors is displayed and we can select those that are of interest to us.

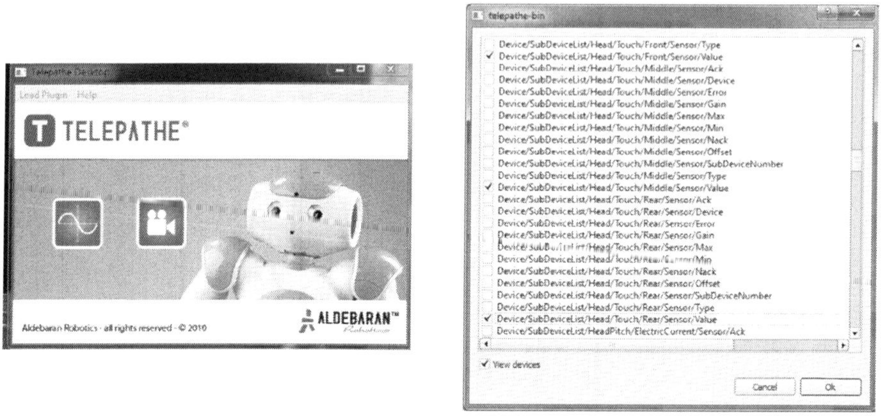

Figure 5.3. *Telepathe startup and selection of sensors for capture*

By activating the sensors to view through the selection of a refresh interval (in the menu "Subscription Mode") and then by launching the capture, we obtain curves in real time according to the values of the sensors selected, as shown in Figure 5.4.

In the case of video capture, it is possible to perform numerous color and capture mode (video, photo) adjustments as illustrated in Figure 5.5.

5.2.1.5. *Monitor*

Monitor is the new version of Telepathe, its operation remains the same.

5.2.1.6. *NaoSim (3D simulation)*

NaoSim is a software program for 3D simulation with a physical engine. It allows for the testing of Nao's programs and its behavior within an environment (outdoors, a bedroom, a house, etc.) as shown in Figure 5.7.

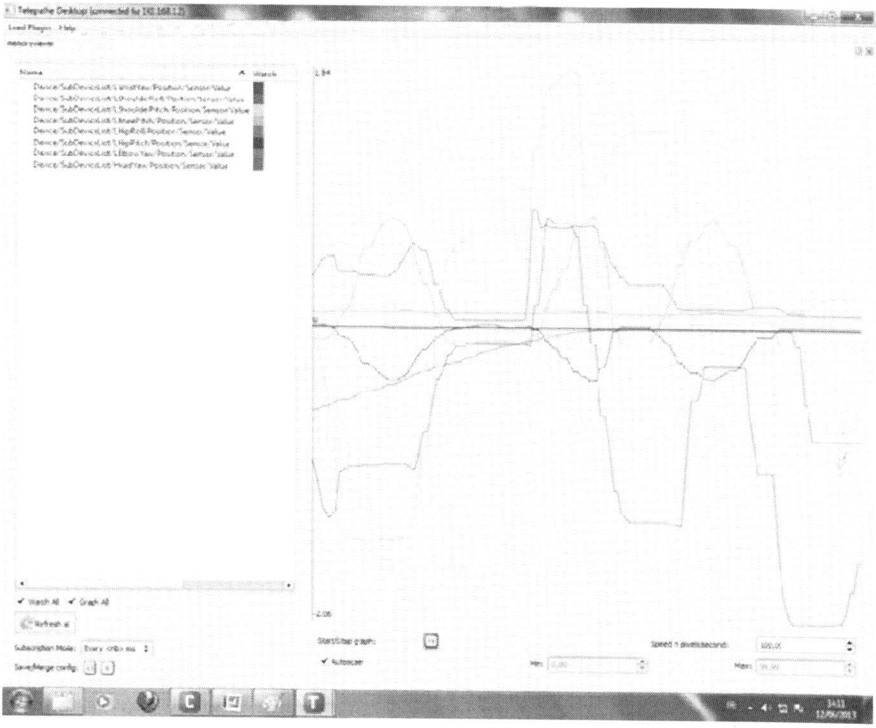

Figure 5.4. *Capture of measures with Telepathe*

There are numerous parameters and it is possible to place new objects, change their color, shape, weight, etc., and then directly connect Choregraphe to NaoSim to test the programs.

5.2.1.7. *Webots*

Webots is the new version of NaoSim; its functioning remains the same but with some improvements concerning the simulation abilities as shown in Figure 5.9.

5.2.2. *Nao Matlab SDK research*

– *What does it do*:

NaoQi-Matlab-SDK allows us to access any module of either a Nao robot or a simulated robot. Through the use of this Matlab SDK, we would be able to call functions from the computer remotely (instead of installing it into Nao directly), control Nao and access every value of the robot by Matlab.

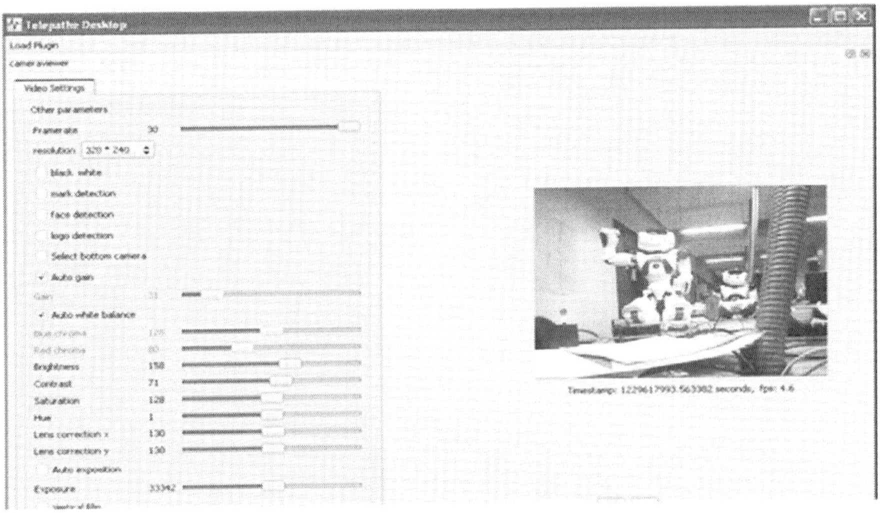

Figure 5.5. *Video capture with Telepathe*

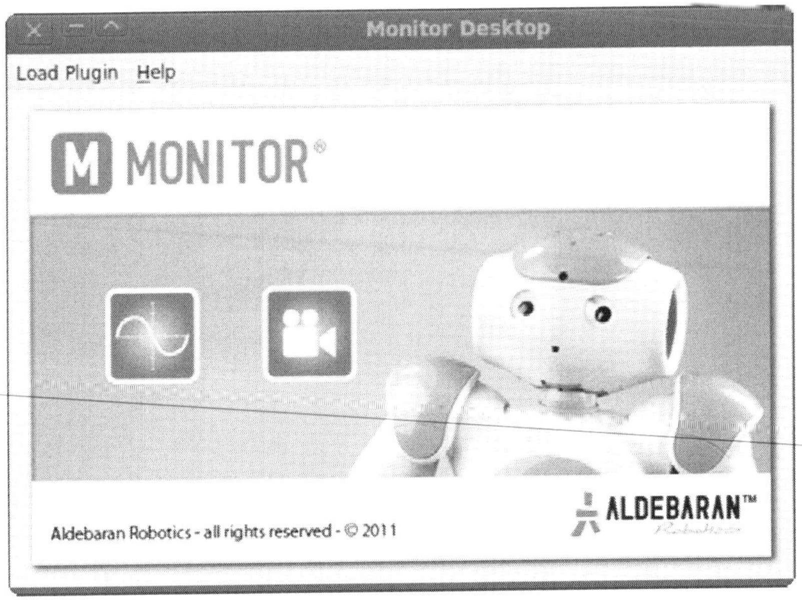

Figure 5.6. *Home screen in Monitor*

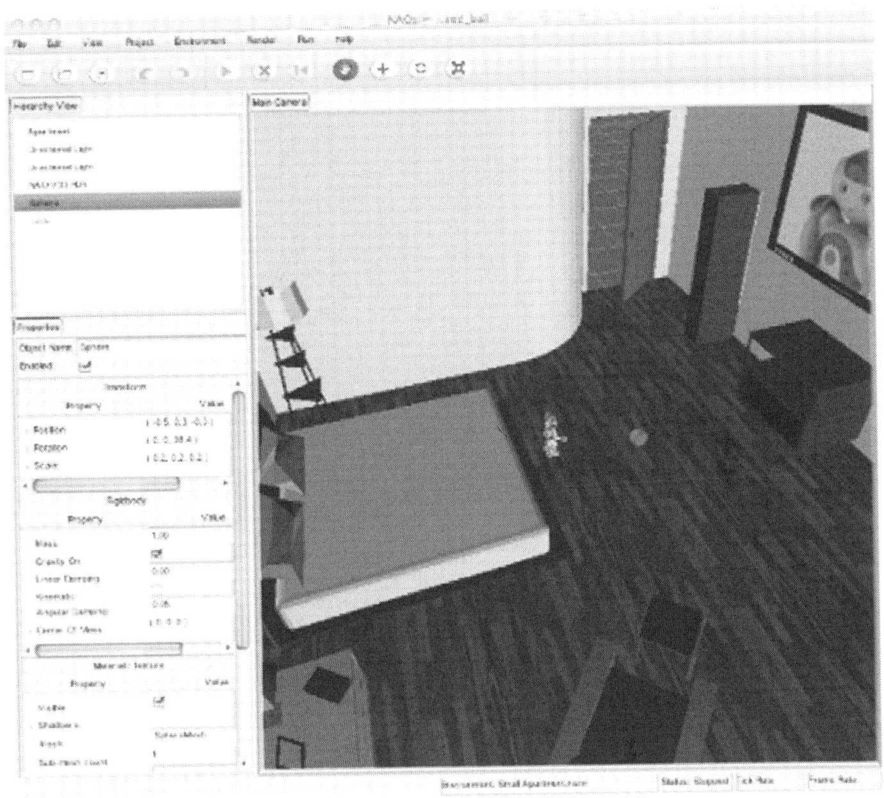

Figure 5.7. *Example of a situation in NaoSim*

– *What is NaoQi and NaoQi framework*:

NaoQi is the name of the main software that runs on the robot and controls it. The NaoQi framework is the programming framework used to program Nao. It is a cross-platform and cross-language platform.

– *Requirements*:

1) Matlab (2015 recommended)

2) NaoQi C++ SDK (source code compilation)

3) NaoQi Matlab SDK (precompiled library)

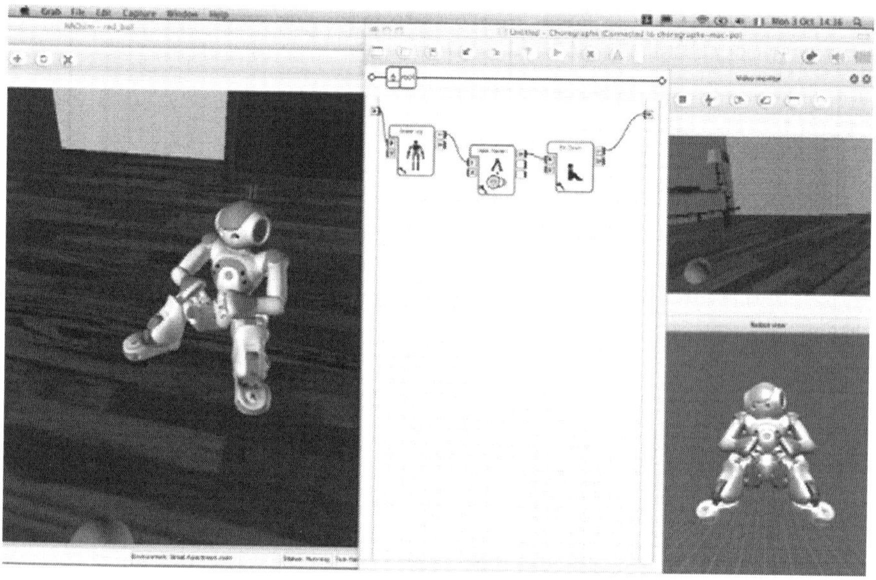

Figure 5.8. *Connection between Choregraphe and NaoSim to test a program in the virtual world*

– *What is in the Matlab library*:

1) C++ library converted to Matlab .m files

2) Several samples

– *Compilation of NaoQi-Matlab-SDK*:

Use Cmake to compile. Specify your PATH_VARIABLES if version is not 2009a.

– *CMake*.

CMake is a cross-platform free software program for managing the software build process using a compiler-independent method.

– *How to use the Matlab library*:

By pointing Matlab to the toolbox of the matlab interface so that we can access all the functions from Matlab.

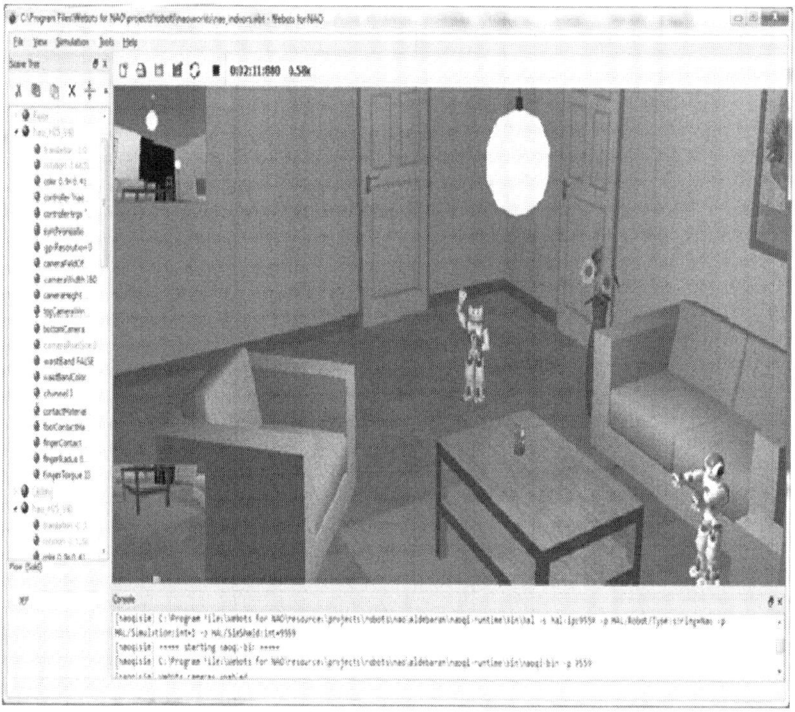

Figure 5.9. *Example of a situation in Webots*

```
z$ cd matlab
z$ mkdir build
z$ cd build
z$ cmake -DCMAKE_TOOLCHAIN_FILE=sdkpath/naoqi-sdk-1.12.0.26-linux32/toolchain-pc.cmake ..
```

Figure 5.10. *Compilation of Naoqi-Matlab-SDK*

– *Example – Get joint names from Matlab*:

Where "127.0.0.1" is the default IP address of the robot with which we specify which robot to control and access.

```
motion = ALMotionProxy('127.0.0.1');
names = motion.getAngles('Body',false);
```

Figure 5.11. *Example – Get joint names from Matlab*

By using the function getJointName of the class ALMotionProxy, which can be found in the reference of API, we can get all the joint names that we want. However, we have found that the second line should not be "getAngles" because getAngles returns Joint angles in radians that is not what we wanted. Furthermore, the image provided in the website is out of date because after searching in the library, we have found out that the function "getJointNames" is deprecated and the new function recommended by "Aldebaran" is "getBodyNames", which is a more intuitive name.

```
std::vector<std::string> ALMotionProxy::getJointNames(const std::string& name)
```

Deprecated since version 1.14: use `ALMotionProxy::getBodyNames()` instead.

param name: Name of a chain, "Body", "Chains", "JointActuators", "Joints" or "Actuators".
return: Vector of strings, one for each joint in the collection

Figure 5.12. *Deprecated function reference*

```
std::vector<std::string> ALMotionProxy::getBodyNames(const std::string& name)
```

Gets the names of all the joints in the collection.

Parameters: *name* – Name of a chain, "Body", "Chains", "JointActuators", "Joints" or "Actuators".
Returns: Vector of strings, one for each joint in the collection

Figure 5.13. *Deprecated function reference*

```
std::vector<float> ALMotionProxy::getAngles(const AL::ALValue& names, const bool& useSensors)
```

Gets the angles of the joints

Parameters: *names* – Names the joints, chains, "Body", "JointActuators", "Joints" or "Actuators".
useSensors – If true, sensor angles will be returned
Returns: Joint angles in radians.

Figure 5.14. *Deprecated function reference*

5.2.2.1. *Type conversion*

Since data types appear to be different between C++ and Matlab, we have to be careful when using the C++ library reference on the Internet as depicted in the following table.

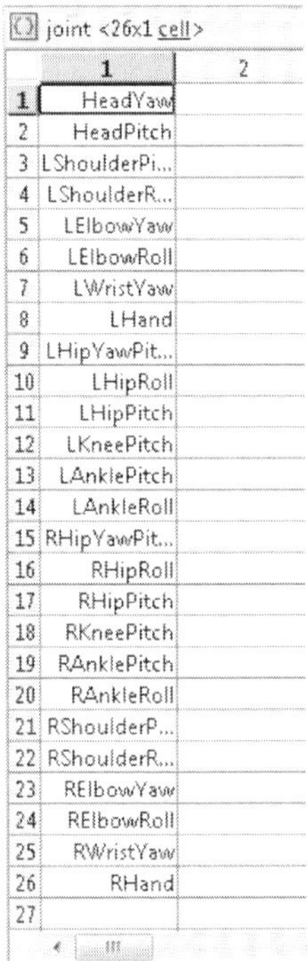

Figure 5.15. *The result returning a cell array with Strings inside*

C++	Matlab
bool	bool
int	int32
float	double
string	string
ALValue::binary	not managed
vector < float >	cells array
vector < int >	cells array
vector < string >	cells array
ALValue	cells

Every data type seems to have a correct mapping. However, we were somewhat confused about what an ALValue is but after checking, we have found out that it is just like a wrapper class. For example, a class "Integer" saving a primitive type value "int" inside so that it would have an object characteristic in JAVA. As an even better metaphor, we think it would be "NSValue" in Obj-C.

It is a wise solution providing an interface to Matlab with Nao robot because Matlab is indeed a really strong Math tool as it allows us to implement image processing, signal analysis and plenty of other things. Robots can perceive plenty of data through their cameras and their microphones. As the amount of data gets bigger and bigger, the computer inside Nao will in the end no longer satisfy users' needs. Furthermore, there are already very complete libraries provided by Matlab and other developers all over the world, so many functions would instantly become usable as long as we connect our Nao with Matlab.

We have several ideas such as, if Google Maps could be done by a robot instead of a Google Maps Van, it could provide an even better view closer to a human being. With a better height, better distance and maybe even mimic key points that human may perceive; by doing so, Google Maps could also forsake the constraint of the road width and provide more detailed information for its users. And during the information collection, the Nao robot can pass all its data into Matlab to do image processing and keyframe distraction or even collect road status (disabled friendly or not) and collect audio information of an environment (whether a place has noise pollution or not). Bearing in mind the advent of big data and more concerns about user experience and human intuitive information, we think a great robot with a combination of a strong Math tool would become a great solution.

5.2.3. *Nao and home care*

5.2.3.1. *Necessity*

Since some elderly individuals may not wish to leave their homes to move into a retirement home, home care is a service in high demand. On the other hand, it imposes very heavy requirements on home care workers or nurses and this is where robotics offers its support. This home care service falls under this context, we implement the functions that allow the Nao robot to keep company to and help an elderly person in his daily life at home.

5.2.3.2. *Context and working environment*

We define the following scenario:

– the elderly person that needs care is called Sophie;

– she lives in a single-story house;

– she is not unfit, nor suffering from a serious disease such as Alzheimer's or Parkinson's disease;

– she feels some pain, requiring weekly care such as injections by a nurse (Marie); she must take drugs at a fixed time and even suffers from some memory loss but nothing serious;

– Sophie is able to speak clearly and coherently;

– she can use a remote control and she knows the functions and how to utilize Nao (how to use the touch buttons on its head, the position of the micros, etc.).

In the tests that we are going perform, Sophie will be assimilated to a red ball (or a case) and in reality, she will wear a red scarf around the neck or on the head.

5.2.4. *The actions to be made*

Distraction

1) propose games (memory, multiplication, common knowledge questions);

2) play music;

3) read something (poetry, tales);

4) dance;

5) propose a gymnastics routine.

Care to the person

– wake up and inquire about well-being;

– follow the person;

– remind medication timetable;

– remind nurse's visit;

– call the nurse if she not coming;

– inquire about health state;

– detect if the person has fallen and call if need be (ambulance, neighbor, family).

Typical day

8 h 30: Wake-up.

9 h: Grooming.

Nao asks Sophie how she's feeling.

10 h 00: A bit of gymnastics with coach Nao.

11 h: Arrival or call from the nurse.

11 h 30: Memory game.

Nao asks Sophie how she's feeling.

12 h: Drugs reminder before the meal.

13 h: A bit of reading.

14 h 30: Upon awakening from a nap, listen to music.

15 h: Multiplication game.

Nao asks Sophie how she's feeling.

17 h: Nao dances.

20 h 30: Question game.

At any time, Sophie will be able to touch Nao urging him to distract her: memory games, multiplication, common knowledge questions, music, reading, dance or gymnastics (middle head touch button).

She can also touch Nao if she feels any problem to ask him to call the emergency services, a neighbor or family (rear head touch button).

5.3. Details of the various programs

Recurring actions.

5.3.1. *Ask for news*

Objective: Using Nao for home care aims to ensure the well-being of the person at home. Therefore, we have designed the program "Ask for news" that was meant to be as complete as possible and an attempt to anticipate all possible cases. Since this program is complex, the indication of the hierarchy at the top of each figure allows to better understand the organization of the boxes as shown in Figure 5.16.

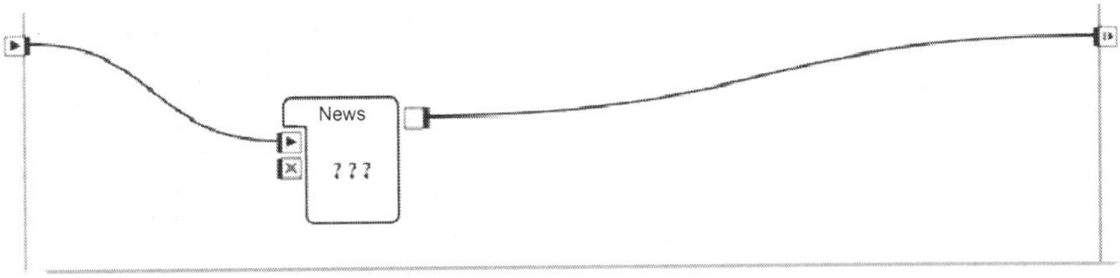

Figure 5.16. *Display of the Root*

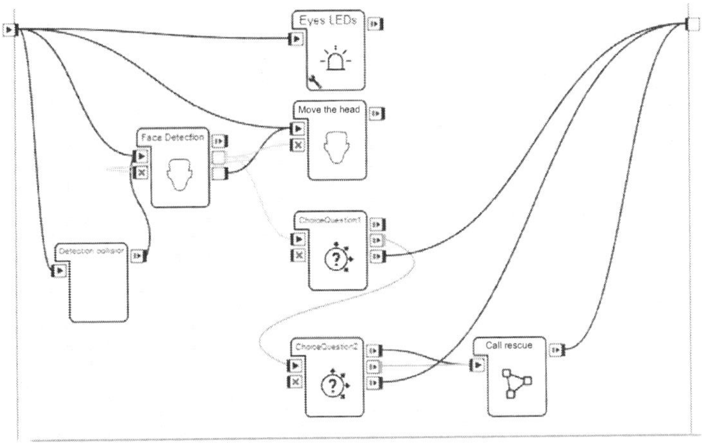

Figure 5.17. *Content of the Root*

Course :

– *Stage 1*: collision detection with the "collision detection" box allows Nao to get up in the event of a fall or to avoid obstacles. Nao has to find the person before starting the program. Face detection selecting menu "face detection" is turned on while Nao rotates his head from left to right on several levels to scan the place with his camera. Once Nao sees a face, the program enters into the box "ChoiceQuestion1" as shown in Figure 5.18.

– *Stage 2*: Nao asks the question to the person "Is everything all right?". If the answer is yes, his eyes turn blue, Nao says "Good", the signal leaves Choice and stops the whole application program asking for news. If the answer is no, the eyes become red, the signal activates another output that goes toward the box "ChoiceQuestion2" as shown in Figure 5.19.

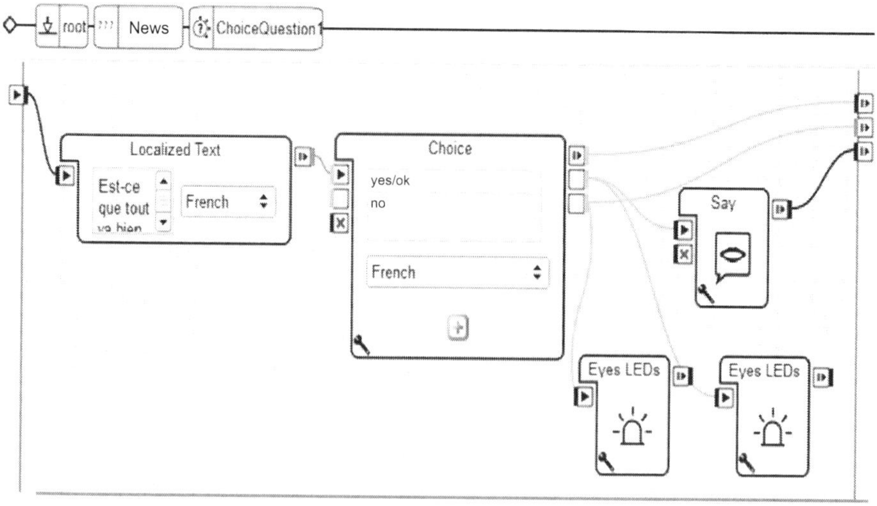

Figure 5.18. *Display of the "ChoiceQuestion" block*

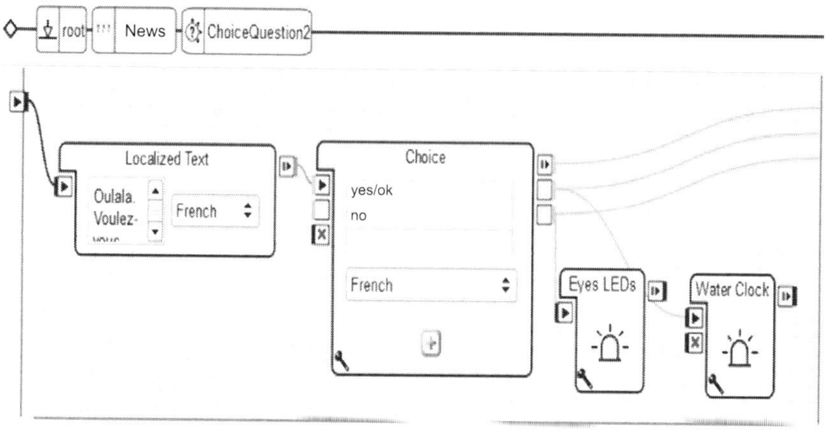

Figure 5.19. *Display of the "ChoiceQuestion2" block*

– *Stage 3*: if the person is not feeling well, Nao becomes worried and asks if she wants to call someone. If the answer is no, the eyes turn light blue and the signal comes out of the box and stops the whole program of requesting news. If the answer is yes, his eyes change color and the signal activates the box "CallHelp" as shown in Figure 5.20.

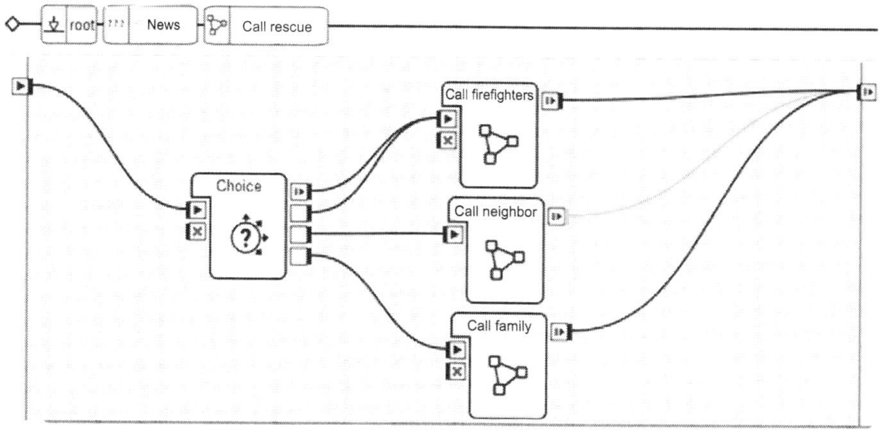

Figure 5.20. *Display of the "CallHelp" box*

– *Stage 4*: this box corresponds to everything concerning calls for help. In the box "Choice", Nao asks the person to select the person to be contacted including firefighters, the neighbor or family as depicted by Figure 5.21.

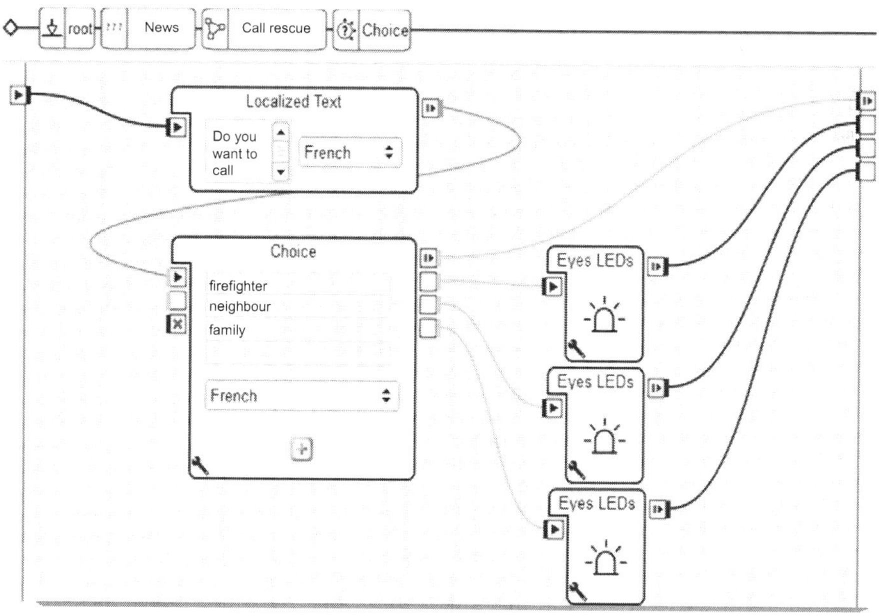

Figure 5.21. *Display of the "Choice" box in "CallHelp"*

Depending on the response, the eyes of Nao change color (red, green, yellow) and the program enters the corresponding box. If Nao does not hear any response or does not understand what the person has said, the box "CallFirefighters" is activated.

5.3.2. *CallFirefighters box*

The call to firefighters as shown in Figure 5.22 is organized as follows:

– there are two ringtones represented by the boxes "Tone";

– the firefighters answer and through the box "Firefighters1" we hear "Firefighters, hello";

– Nao explains them the emergency situation with the box "Nao". The elderly person has fallen and has injured herself. It tells them the address, the code for the door and the floor;

– the response of the firefighters saying they are coming can be heard through with the box "Firefighters2";

– then, Nao reassures the senior citizen through the box "Nao to Sophie" telling her that help is on its way;

– the final alarm allows firefighters to pinpoint the location. They will have to shut down Nao, by touching Nao's head, when they have found the person on the floor at home.

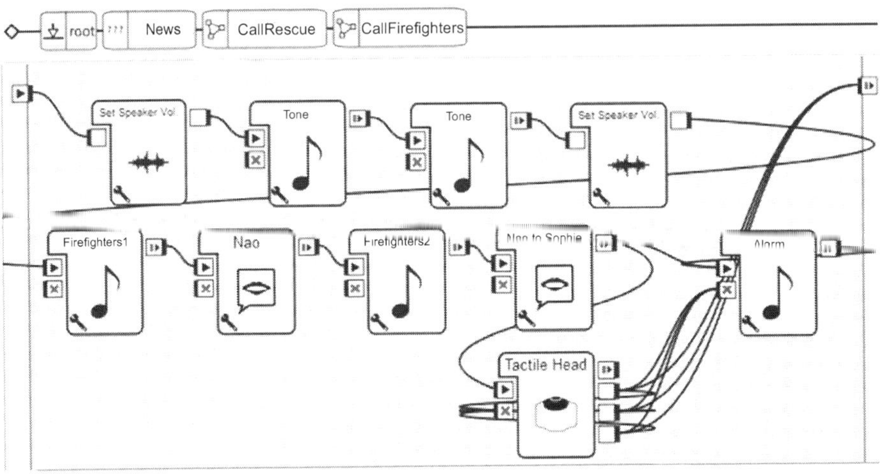

Figure 5.22. *Display of the "CallNeighbor2" block*

5.3.3. *CallNeighbor box*

As can be seen in Figure 5.23, the call to the neighbor is more complex because it may be possible that she does not answer and cannot be of help to the person in need of assistance:

– After the choice of the user to call the neighbor, Nao indicates that it is going to call. Both cases are emerging, either the neighbors respond or they do not respond. We have represented both possibilities through a function "Random" that will define what situation will be selected by assigning a random number between 0 and 1.

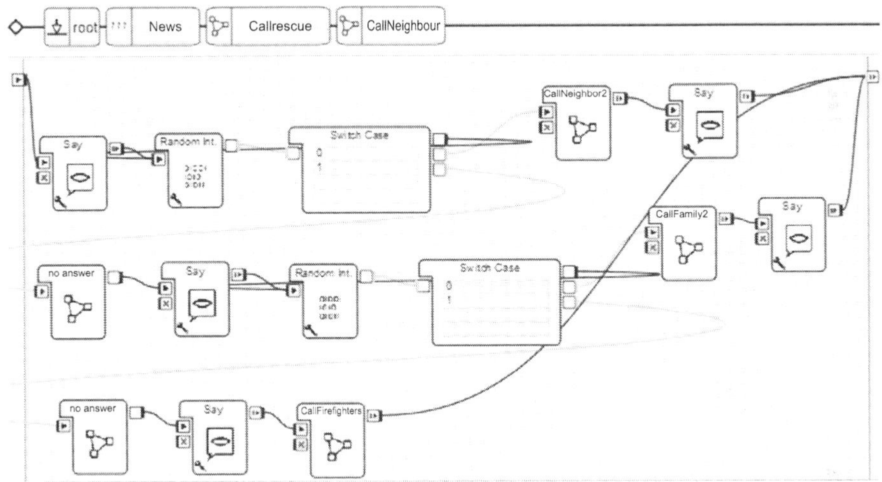

Figure 5.23. *Display of the "CallNeighbor" box*

If the number is 0, the neighbor responds and the program enters the box "CallNeighbor2" as shown in Figure 5.24.

The call to the neighbor is organized as follows:

– two ring tones;

– the neighbor picks up the phone and said "Hello?" (box "Hello");

– Nao explains the situation with the box "Nao". The elderly person has fallen and requires assistance;

– the neighbor says that she is coming over (box "Neighbor").

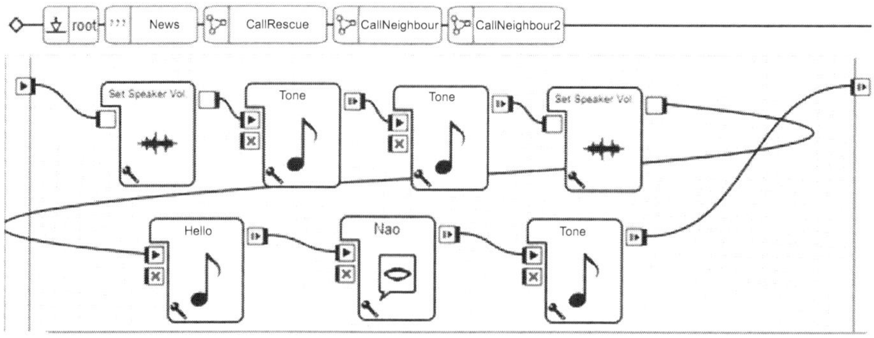

Figure 5.24. *Display of the "CallNeighbor2" box*

After this call, Nao reassures the elderly person with the box "Say" following the box "CallNeighbor2". Finally, the program stops because emergency services have been notified.

If the number is 1, the neighbor does not respond and the program enters the box "No answer" as shown in Figure 5.25.

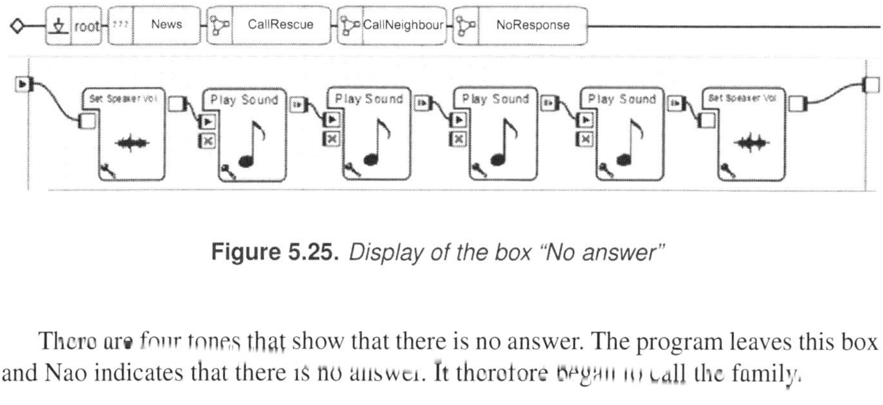

Figure 5.25. *Display of the box "No answer"*

There are four tones that show that there is no answer. The program leaves this box and Nao indicates that there is no answer. It therefore began to call the family.

The call to the family involves the same difficulties as the call to the neighbor because there is no guarantee that the person will answer. Thus in a similar fashion, a function "Random" allows making a choice between the two possible situations.

If the number is 0, the neighbor respond and the program enters the box "CallFamily2" as it can be seen in Figure 5.26.

The call to the family is organized as follows:

– two ring tones;

– the daughter of the person picks up the phone and says "Hello" ("Hello" box).

– Nao explains the situation with the box "Nao". The elderly person has fallen and requires assistance.

– the daughter says she is coming ("Family").

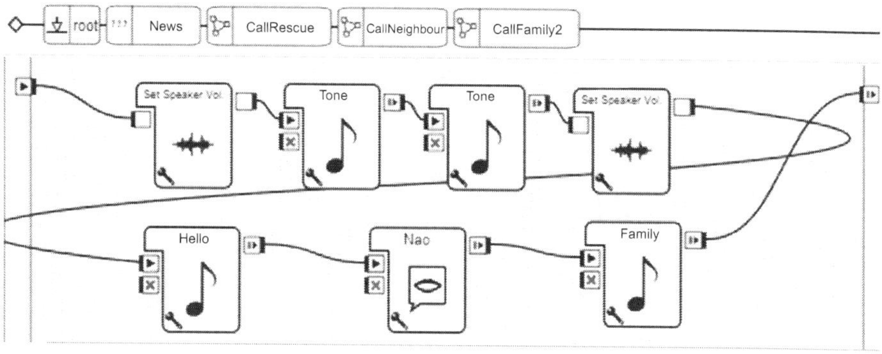

Figure 5.26. *Display of the "CallFamily2" box*

If the number is 1, the family does not answer.

As previously, the box "No answer " signals the lack of response and Nao follows on with a call to the firefighters ("CallFirefighters" box seen above) that finally answers the call for help for the senior citizen.

5.3.4. *CallFamily box*

This box behaves exactly like the second part of the call to the neighbor. If the family responds, the call takes place normally and the person will be helped by the family. If the person does not answer, Nao calls the firefighters as shown in Figure 5.27.

5.3.5. *Collision detection*

5.3.5.1. *Objective*

During the operation of the program, when Nao is standing up or if he is walking, potential collisions are detected with the "Collision detection" block. If he strikes an obstacle with his feet, this operates the bumpers, a signal stops its walk, Nao takes a few steps back, turns according to the bumper activated first and then resumes his walking. In the case of a fall, he gets up after a brief break and resumes his path.

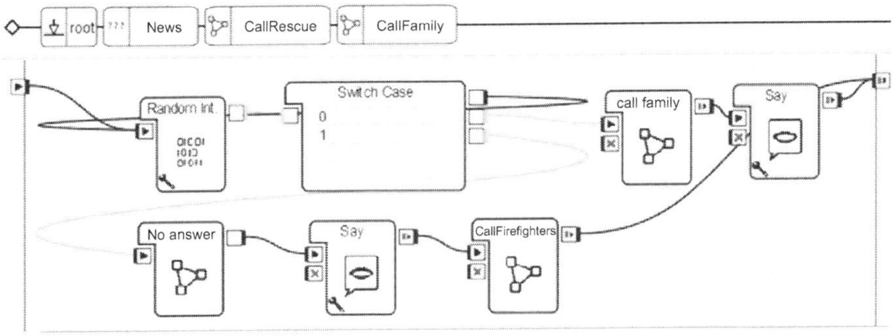

Figure 5.27. *Displaying of the "CallFamily" box*

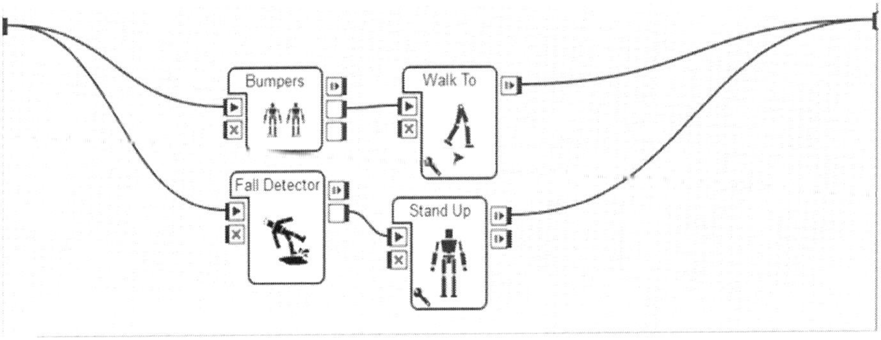

Figure 5.28. *"Collision detection"*

5.3.6. *Special actions: waking-up*

5.3.6.1. *Objective*

This application is the first of the day; it is morning, it is 8:00, Wake-up and asks her if she is feeling well and remind her of the schedule for the day.

The person is awakened by a crowing rooster. Subsequently, Nao asks the user if she has slept well:

– If yes, Nao asks the elderly person if she wants to talk about her dream.

– If no, Nao asks if everything is all right. If yes, then Nao asks if she wants to talk about her dream; if no, then Nao asks whether she wants to call someone (a neighbor, family or firefighters); otherwise, Nao asks if she wants to talk about her dream.

In the case where the elderly person decides to talk about her dream, a "Wait" of 10 min has been put in place, therefore there are two possible cases:

– the user has the possibility of interrupting the "Wait" by pressing Nao's head; as a result, Nao says "it's very interesting!".

– If the explanation of the dream lasts more than 10 min, then Nao, after this time, says "it is very interesting!".

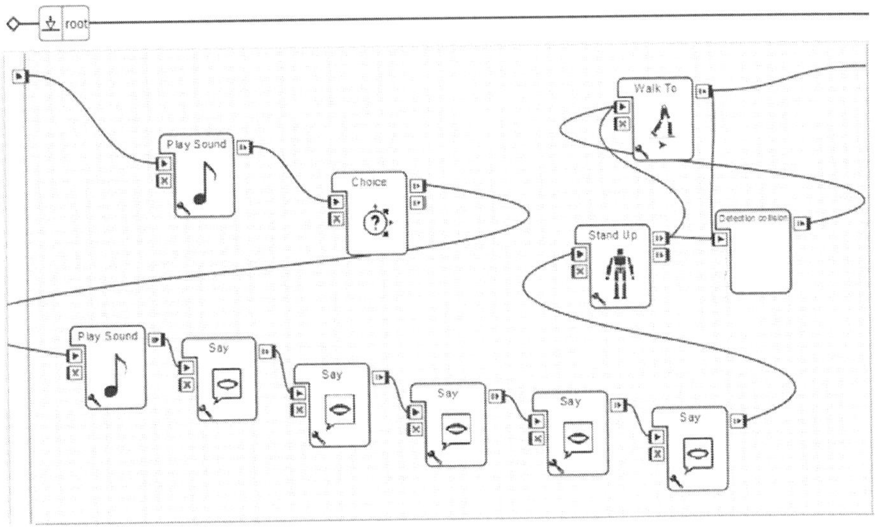

Figure 5.29. *Display of the Root*

At the end of this step, Nao reads out the schedule of the day. In the case where the elderly person decides not to talk about her dream, Nao directly reads the schedule of the day.

Finally, Nao prompts the user to have breakfast. Nao stands up and follows the person.

Description

This "Choice" as shown in Figure 5.30 corresponds to the first question posed by Nao to the senior citizen, that is to say "Good morning! Did you sleep well?". In the case in which the answer is "Yes", the signal goes to a "Choice" as illustrated in Figure 5.31 where Nao inquires "Do you want to tell me about your dream?". If Yes, then a "Wait" of 10 min is put in place so that the elderly person can talk about her dream. During this period of time, the user can interrupt it by touching one of the sensors on Nao's head by means of the "Tactile Head". If No, then the signal leaves the "Choice".

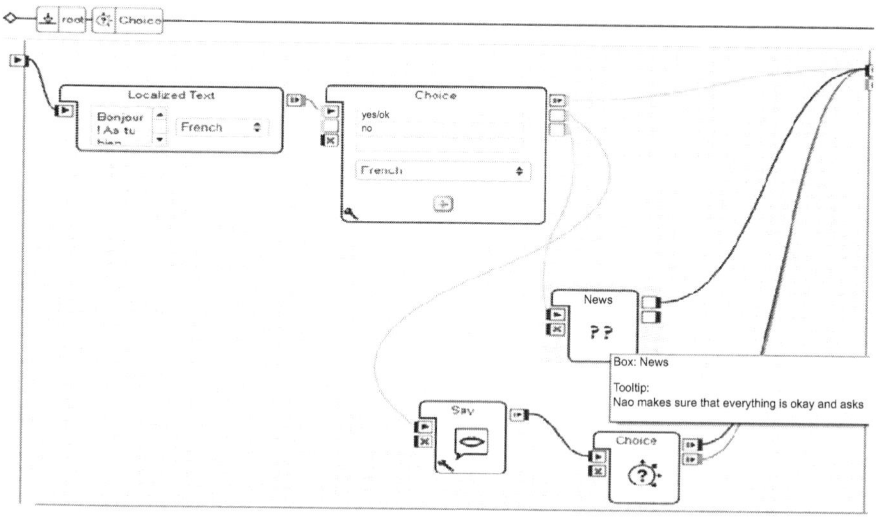

Figure 5.30. *Display of the "Sleptwell Choice"*

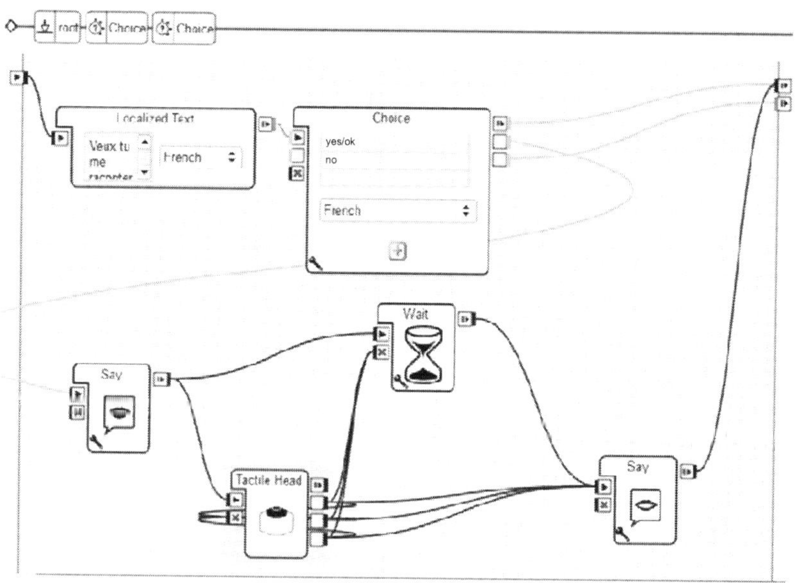

Figure 5.31. *Display of the "Choice" of the dream*

In the case in which the answer is "No", then the signal propagates to the News block below, which is explained in part 2 of "recurring action" as shown in Figure 5.32.

This is the Choice for the case where Nao asks if the user wants to talk about her dream or not.

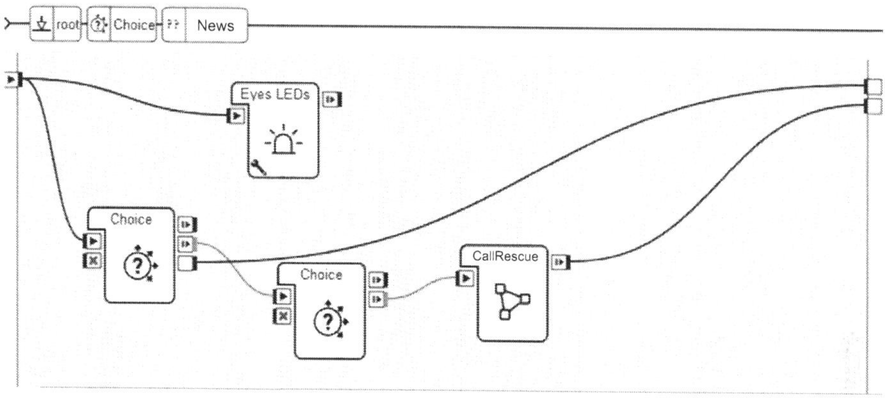

Figure 5.32. *Display with the box "News"*

When the signal leaves the Choice located in the Root, it goes to a series of "Say" that correspond to the daily schedule of the user as shown in Figure 5.33.

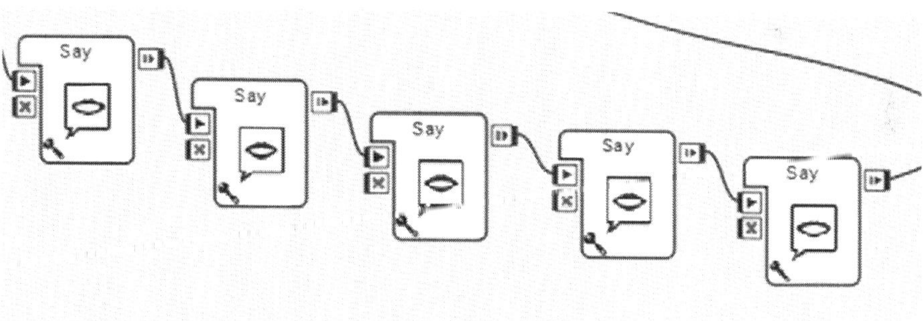

Figure 5.33. *Displaying of the user's schedule*

In the end, Nao accompanies the user to the kitchen. He gets up and walks a few steps.

5.3.7. *Morning hygiene*

5.3.7.1. *Objective*

When supporting an elderly person alone at home, it may be necessary to ensure that the person takes good care of herself. Thus, we have implemented a program reminding her of washing every morning. This program would be launched at 9:00 in the morning everyday.

5.3.7.2. *Course*

Step 1: by means of the "HygieneChoice" box, Nao asks the person if she has already washed herself as illustrated in Figure 5.35.

Step 2: if the person answers yes, Nao congratulates her. If the answer is no, the program enters the box "Wash".

Step 3: looking like a timeline, this box combines word and movements to serve as illustration. It comprised of four layers of behavior that distribute every step of the wash as shown in Figure 5.36.

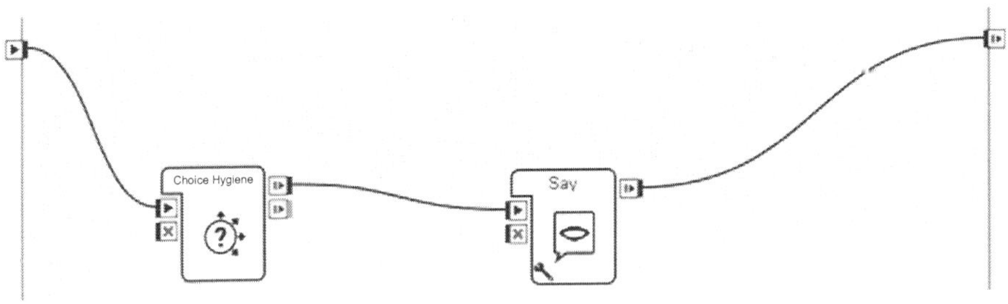

Figure 5.34. *"Root display"*

In the first layer, Nao makes gestures as if he were under the shower and says "let's get wet, let's wash up, let's scrub ourselves, let's rinse ourselves".

The second layer concerns tooth brushing and Nao illustrates once again the stated action "let's brush our teeth".

In the third layer, Nao combs its hair and invites the person to do the same "let's comb our hair". And finally the last layer illustrates the movements for dressing with the text "let's get dressed". The color change of the eyes makes it possible to make the illustrations of the movements more entertaining.

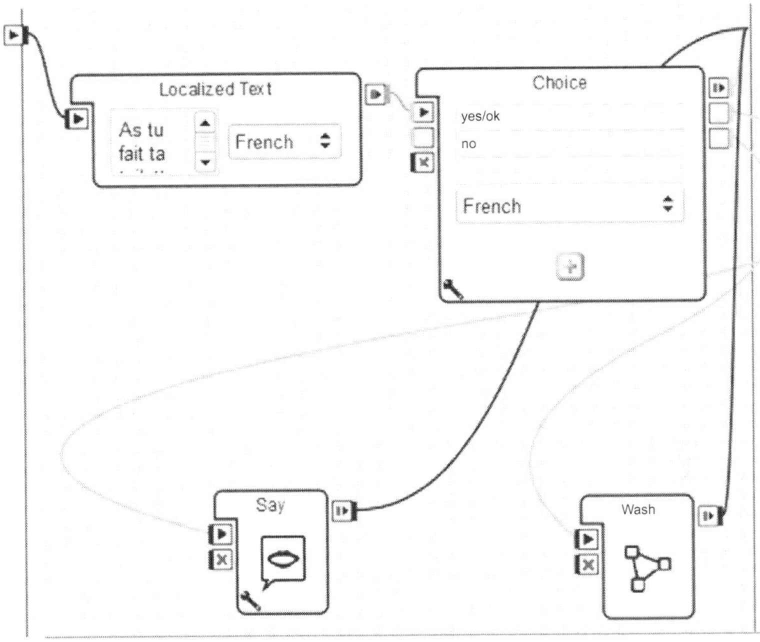

Figure 5.35. *Display of the "ChoiceHygiene" block*

5.3.8. *Gymnastics*

Aim

Make Nao a leader in gymnastics and do particular movements to the rhythm of the music.

What we have done

We started the music "Relax, take it easy" and made Nao do some slow warm ups. After several basic joints movements, we made him present some boxing. With great control, we gradually added intensity and Nao can be a great Gym coach indeed.

Difficulties

Without the help of automatic "whole body balance API", it is very difficult to control both legs without making Nao fall. For example, if we move the knee of one leg, we will have to move at least another joint of the robot of the same leg to make its feet standing flat on the floor. However, this makes the two feet of the robot differ into two altitudes and makes the robot lean or even fall down.

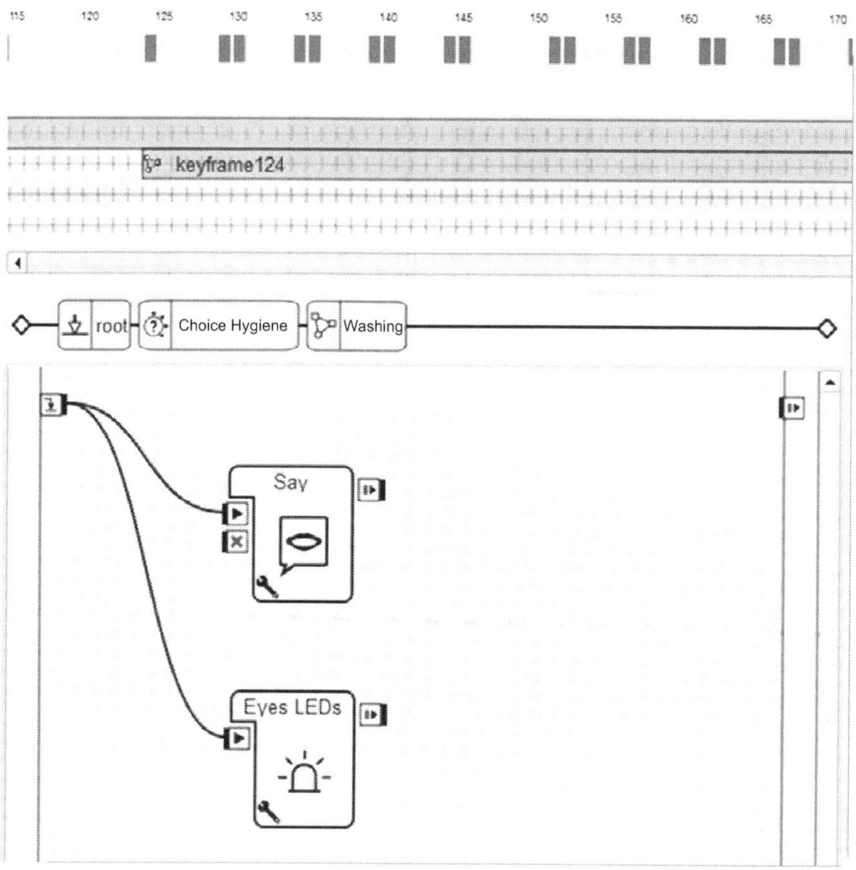

Figure 5.36. *Display of the TimeLine*

By taking a lot of gymnastic movies on Youtube as references, we can see that there are many postures that a robot cannot do: putting one hand on another while deviating to one side (with the constraint of the shoulder joint) and Nao cannot turn its neck like it is recommended in a lot of warm ups (with the constraint of the neck without the dimension of Y). The most difficult thing to implement during our development is making Nao do a "lunge". We often make the "center of mass" (COM) out of Nao's supportable region, which results in Nao falling and screaming "ouch!!". With so many constraints, there are many postures that Nao cannot do especially if there is a specific part of the body that we want to stretch, therefore we tried Nao's best in being a gymnastic leader.

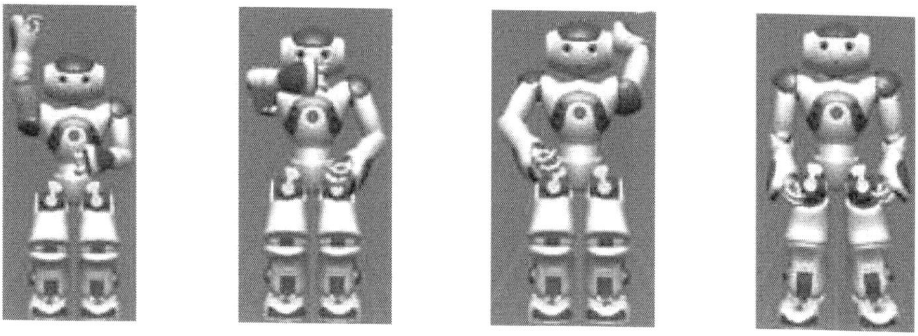

Figure 5.37. *Cleansing movements in NaoQi. Let's scrub ourselves... Let's brush our teeth... Let's comb our hair... Let's get dressed...*

Figure 5.38. *Nao does gymnastics*

Solutions

1) Moving each joint bit by bit.

2) Edit every key frame by directly moving the robot itself while it is unstiffened and then save the posture into the key frame back again.

For the first solution, it takes much time and it is not technical. The second way is rather like making a movie but seems to be the best solution when there is some posture which is hard to achieve.

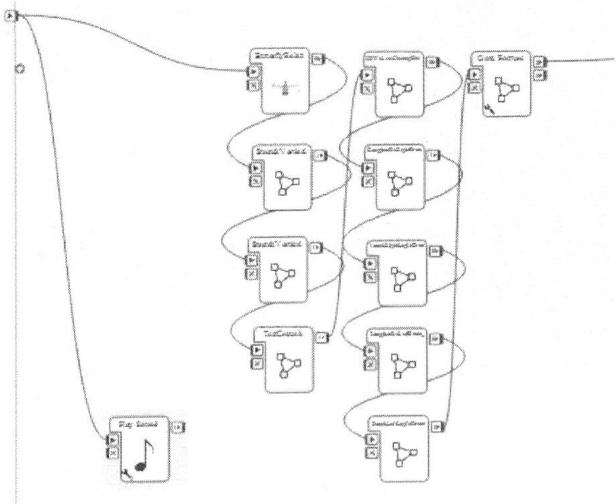

Figure 5.39. *View of the Root*

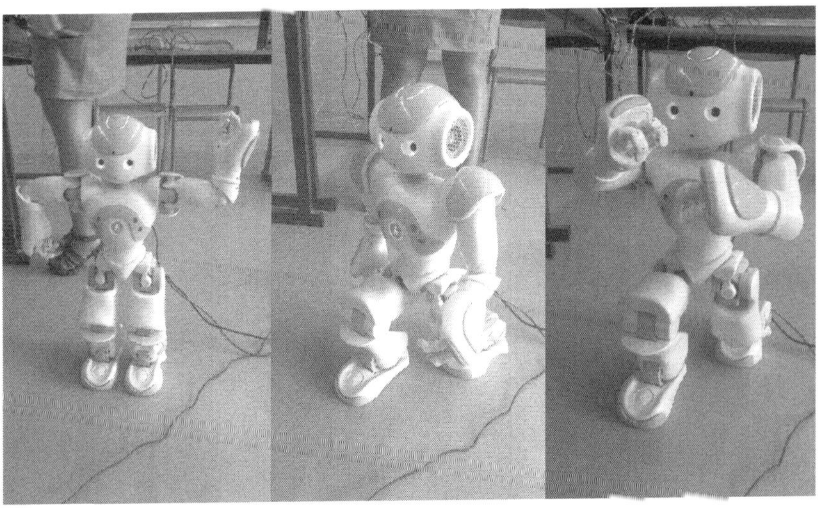

Figure 5.40. *Nao does gymnastics*

Other problems

1) Nao sometimes does not work properly: for example, sometimes we run the "toPosture" function to make the robot sit down or to "standInit" and it would sometimes fall down. We are not sure whether the problem is due to a lack of

strength in Nao's specific joints or maybe it is due to the environment in which Nao is processing.

2) The version of Choregraphe provided on the website and the one we are using are different, which make some functions incompatible.

3) Programming this kind of movement functions, we often have to waste a human resource to protect the robot from falling down again and again during the test.

4) Because there is only one Nao for each group, when we try to use the emulator to program Nao, we will not know whether it will fall down or not. It would be simpler if the emulator would be able to let us know whether it will fall or not or at least whether the COM is out of the contacted area.

5.3.9. *Nurse call*

5.3.9.1. *Objective*

This program is intended to ensure that the nurse will come see the user, in the case of a regular visit, weekly for example.

Nao will have to warn the user that the nurse is coming soon and if she does not show up at the time that was scheduled, it must be able to call her.

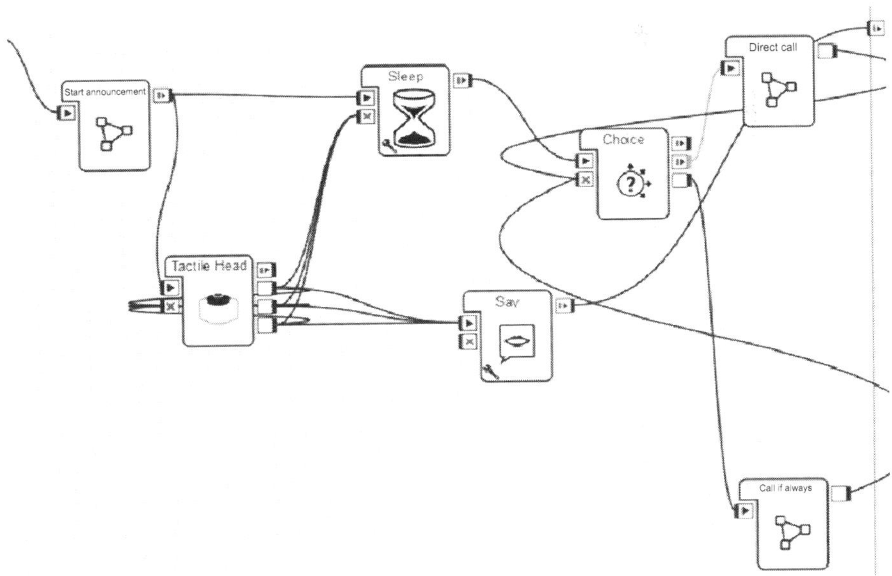

Figure 5.41. *Display of the Root*

5.3.9.2. *Course*

Step 1: Nao tells the user that the nurse is supposed to arrive in 10 min, when this is the scheduled time (10:50 in our case).

Step 2: if she is not there yet, Nao states that the nurse has not showed up and asks the user if she wants to call the nurse to find out where she is, or not. It is a box "choice" that is called as illustrated in Figure 5.42.

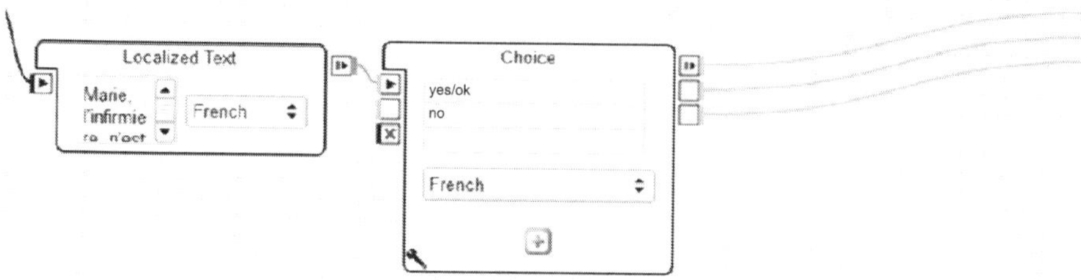

Figure 5.42. *Display of "Choice"*

If the user wishes to call the nurse, Nao makes the call. It phones Marie, has a brief conversation with her, then concludes by informing the user according to the discussion as shown in Figure 5.43.

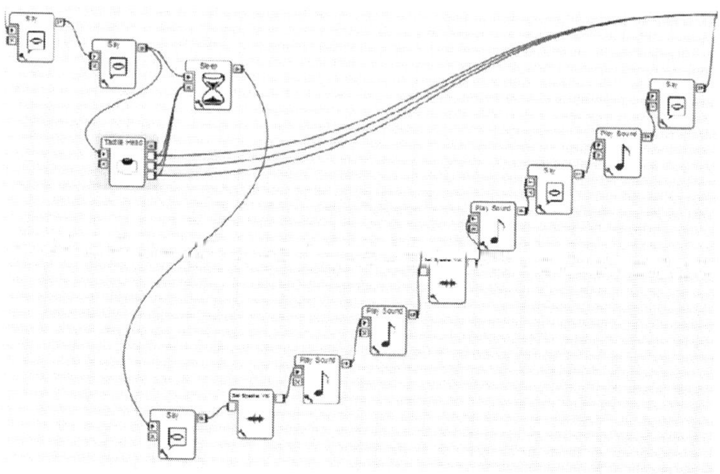

Figure 5.43. *Display of "Say", the call to Marie the nurse*

If the user does not wish to call the nurse immediately, Nao waits 10 min and then calls. The conversation remains the same as illustrated in Figure 5.44.

5.3.10. *Memory game*

5.3.10.1. *Objective*

This program is intended to distract the user, by suggesting a memory game, based on the repetition of colors.

Nao will say the name of colors and the user will have to repeat. In case of a good answer, a color will be added, and so on.

5.3.10.2. *Example*

Nao: “Red”.

Sophie: “Red”.

Nao: “Good. Red, blue”.

Sophie: “Red, blue”.

Nao: “Good. Red, blue, red”.

Sophie: “Red, blue, red”.

Nao: “Good. Red, blue, red, yellow”.

Sophie: “Red, blue, yellow”.

Nao: “Wrong! Start over”.

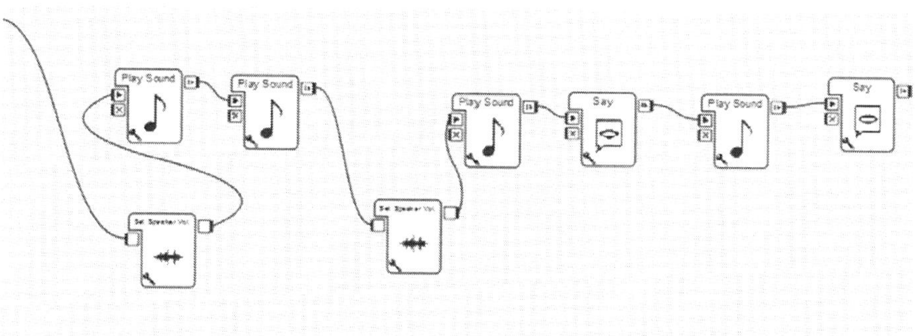

Figure 5.44. *Display with “Say2”, the call to Marie the nurse*

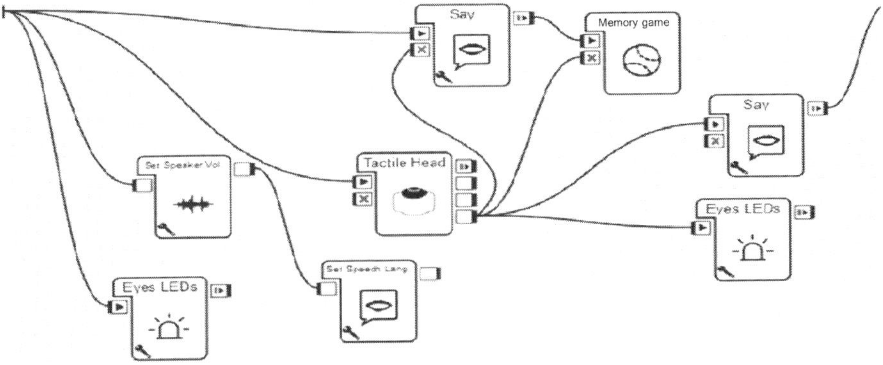

Figure 5.45. *Display of the Root*

5.3.10.3. *Course*

Step 1: Nao tells the user that it suggests her a memory game, with a first "say" box.

Step 2: Nao launches the memory game.

As shown in Figure 5.46, this block is made up of several boxes, the whole forming an infinite loop as long as the user does not indicate that she wants to stop by touching the back of Nao's head.

First of all, Nao announces that it is his turn to play. It generates a random number ("random int" box) that enters a "switch case" box, in which the number is associated with a color: 1 for blue, 2 for red, 3 for green and 4 for yellow as shown in Figure 5.47.

The result is then directed to the memory where it is stored as in a queue (first in, first out).

At the beginning of a round, Nao says "That's my turn". After announcing the level number (number of random colors stored in the "memory" block, given by the fourth output of the "memory" block), a random color is then generated and the corresponding number is added to the waiting queue in the "memory" block. The first color comes out and is read out loud. As long as there is a color remaining in the queue, Nao says its name as shown in Figure 5.48.

Nao then says "It's your turn" and it launches speech recognition. The colors in the queue are output one by one as long as the user guesses the correct answer.

The program compares the number corresponding to the color that the user has just said with the number corresponding to the expected answer.

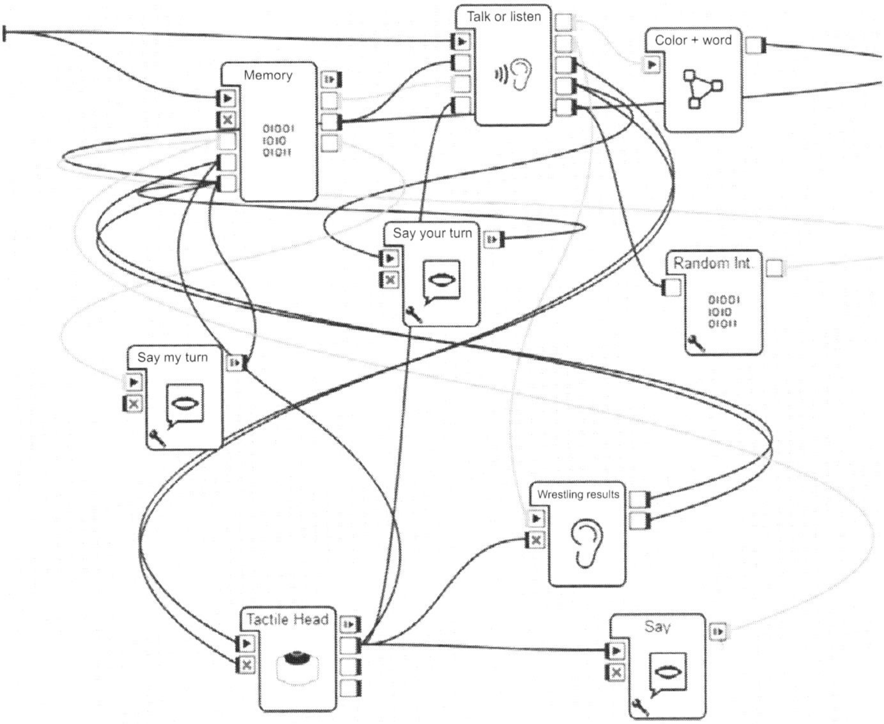

Figure 5.48. *Display of the "MemoryGame" block*

In case of a good response the program makes the sound "Tadaa", and then moves to the next color. If it happens in the end of the line, the program will return to "speak" mode where a new random color is generated.

In case of a bad answer, the program makes the sound "bongbong" and it asks the user to start again as it can be seen in Figure 5.49; it then returns to the beginning of the color file.

We have had to create the "memory" block by coding in Python in order to store the results in a queue (composed of numbers 1–4 to symbolize colors).

The third input of the block allows a new number to be added to the queue. At each addition, the number of digits in the queue (the level of the game) is returned by the fourth output of the "memory" block.

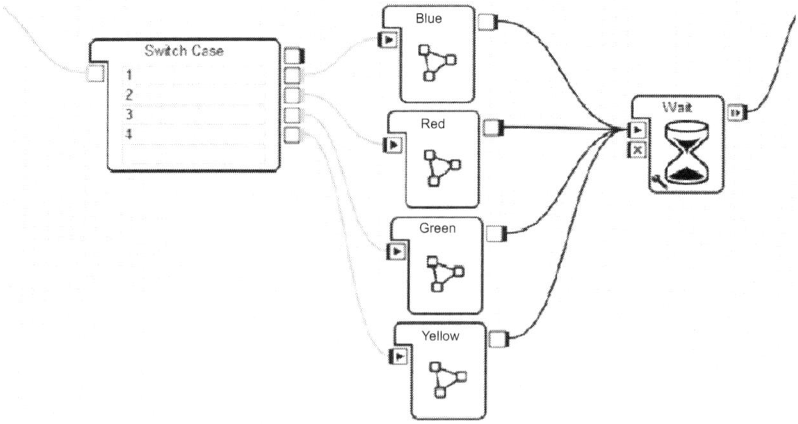

Figure 5.47. *Display of the "Color+Word" block*

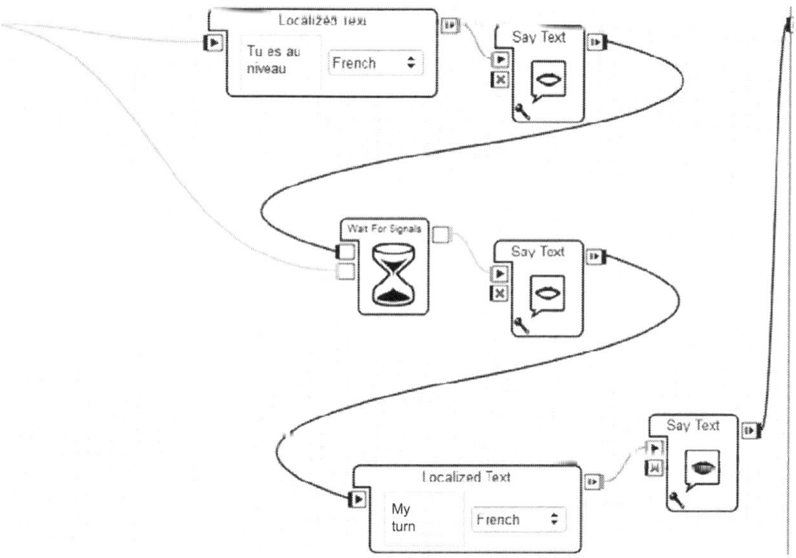

Figure 5.48. *Display of the "MyTurn" block*

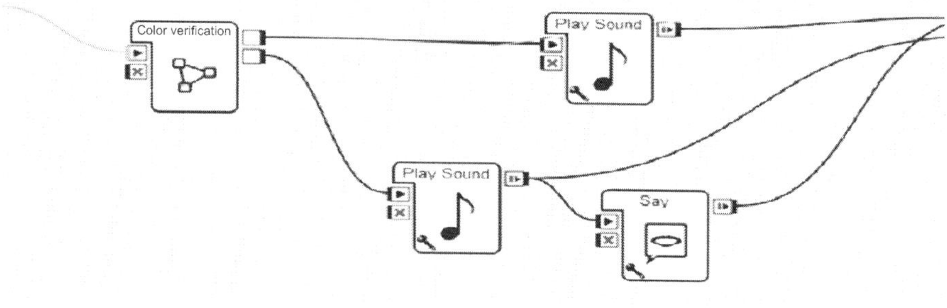

Figure 5.49. *Display of the "Wrestling Results" block*

The fourth input places the reading pointer on zero (when the user has lost or when she is asked to repeat), the fifth allows that the current number be output on the second output and the read pointer to be increased. If the reading pointer exceeds the size of the queue, output 3 is stimulated: it will thus help to switch from "talk" mode to "listen" mode (and vice versa) using the "talk or listen" block.

This last block also coded in Python acts as a rocker: in the "speaks" case, the number of the current color is going to the "color + word" block in order for the color to be uttered and "touch head" is disabled. In the "listen" case, the number is redirected toward speech recognition and the "tactile head" is activated. When shifting to "speak" mode, a new random number is generated.

Step 3: if the user wishes Nao to repeat the sequence of colors, she must touch the front of Nao's head when the latter is in "listen" mode. The "Speak or listen" block is back to "Speak" mode, the memory playback pointer is reset to zero, then the "Say" block returns the value -1 to indicate the memory not to generate any new random number.

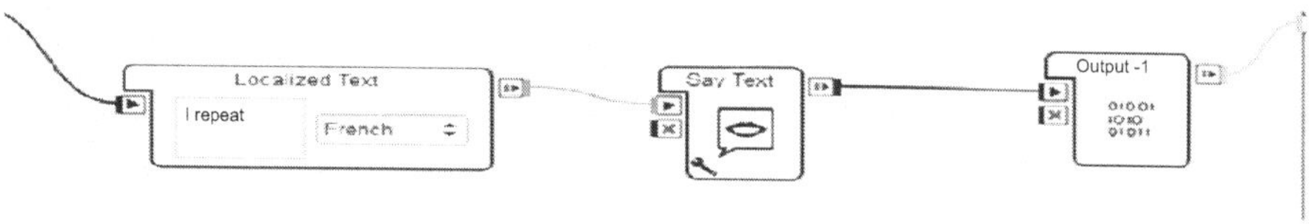

Figure 5.50. *Display of the "Say" block for repetition*

Step 4: if the user wishes to stop playing, she can touch the button on the back of Nao's tactile head. This latter leaves the "memory game" block and reads out a goodbye message.

5.3.11. *Drugs reminder*

5.3.11.1. *Objective*

This program is intended to compensate for possible memory loss from the user. If the person to whom Nao is giving assistance requires medication at a fixed time every day, it must be able to remind her, before breakfast, for example.

In the scenario that we have set up, Sophie should take her drugs at noon.

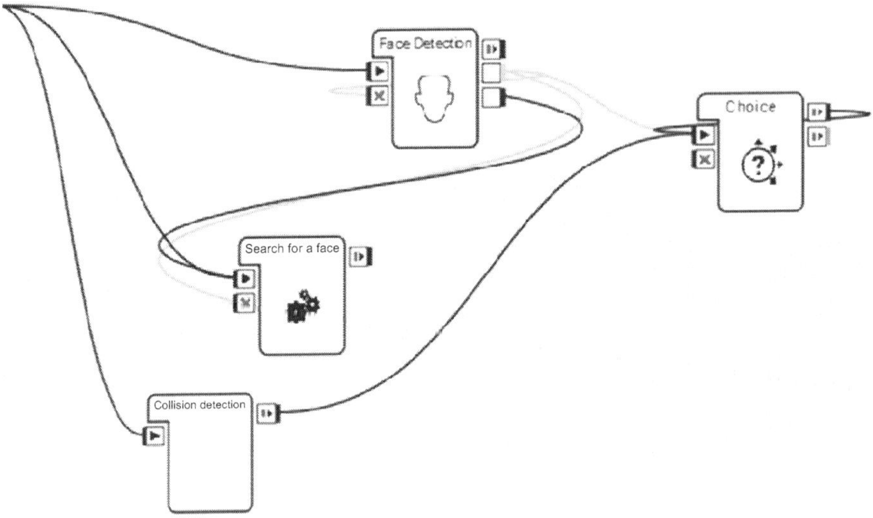

Figure 5.51. *Display with Root*

5.3.11.2. *Course*

Step 1: Nao has to detect a face in order to be able to address the user. To detect a face, he is going to search for it with the "search for a face" box. A movement of the head is programmed with the timeline to look for a face to talk to as shown in Figure 5.52.

Step 2: when Nao has detected a face and looks at its user, it says that it is time for medication before starting a question with a "choice" block to certify that the drugs have been taken. If the answer is "yes", it completes the program by saying "good", but if the answer is "no", it requires the user to take them, recalling where they are. The program ends then as shown in Figure 5.53.

Figure 5.52. *Display of the timeline and virtual Nao for the movement "search for a face"*

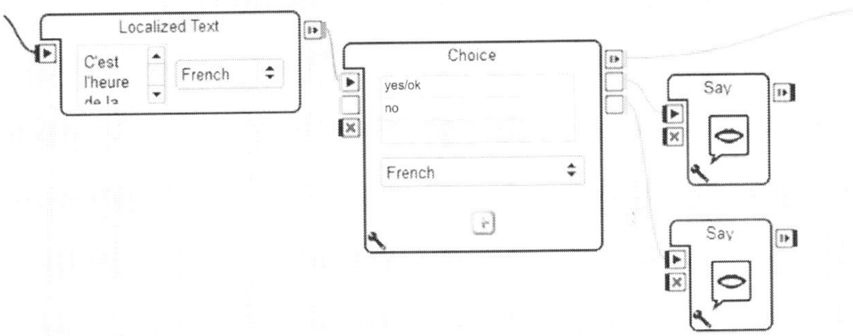

Figure 5.53. *Display of the "Choice" block*

5.3.12. *Reading*

5.3.12.1. *Objective*

To keep a senior citizen company, we have developed several programs for reading media. Here, we suggest to the person to listen to a prerecorded audiobook.

5.3.12.2. *Course*

Step 1: Nao tells the user that he is going to read her a book with the "SayIntro" box and prompts her to make her selection through the "ChoiceGenre" box. The user can make a choice between genres: tale, novel, poetry and theatre as illustrated in Figure 5.55.

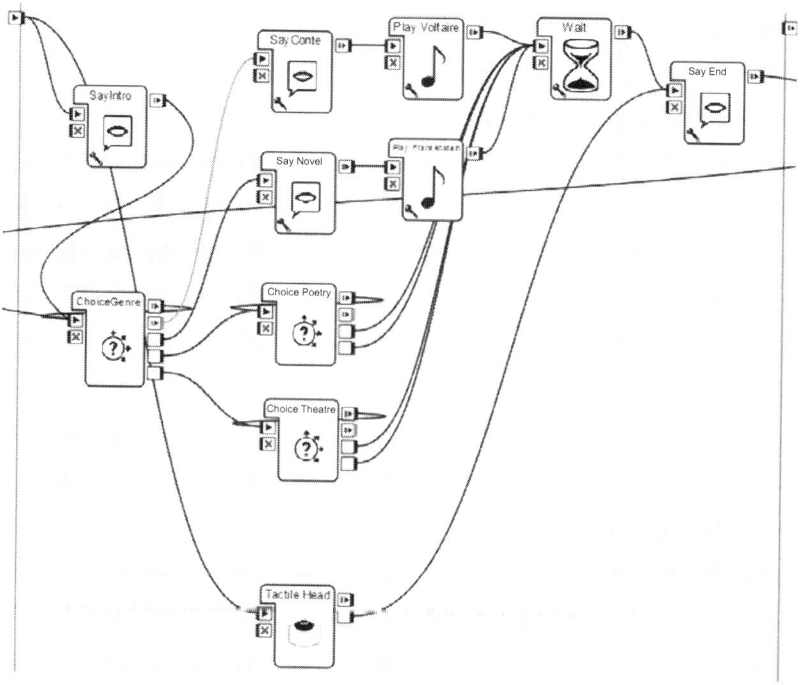

Figure 5.54. *Display of the Root*

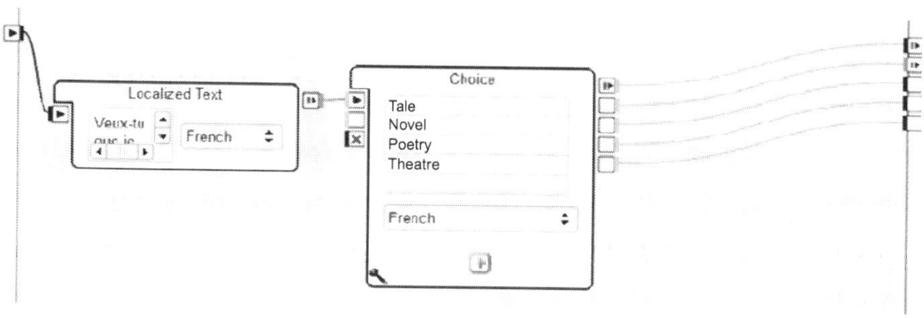

Figure 5.55. *Display of the "Choice" block for the genre of book to read*

Step 2: when the choice is made, Nao introduces the book that will be read, for genres tale and novel (because only one book of each genre is available) and the robot starts reading the book. For genres poetry and theatre, Nao proposes to the user a choice between two authors as illustrated in Figure 5.56.

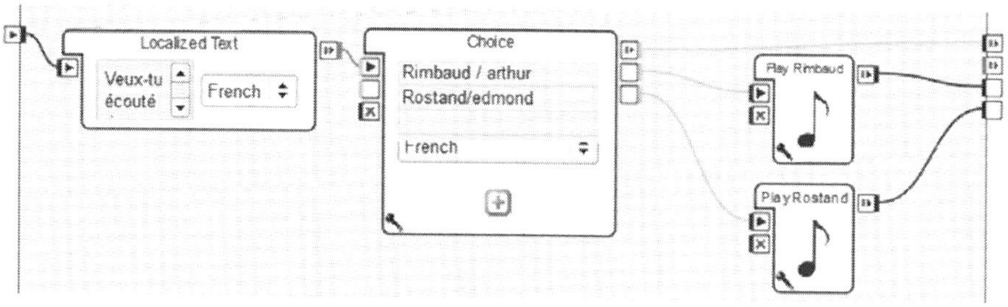

Figure 5.56. *Display of the "Choice" block of the type of book*

Here, for example, in the poetry category if the user chooses the author Rimbaud, the program starts to read the poetry "Ophelia" by Arthur Rimbaud through the box "Play Rimbaud".

Step 3: After having read the book, Nao waits 5 s and then reports the completion of the activity through the box "SayEnd" (see display of the root).

5.3.13. *Listening to music*

5.3.13.1. *Objective*

This application is located after the nap of the elderly person. It allows one to listen to music either by making a selection of the music or in a random fashion.

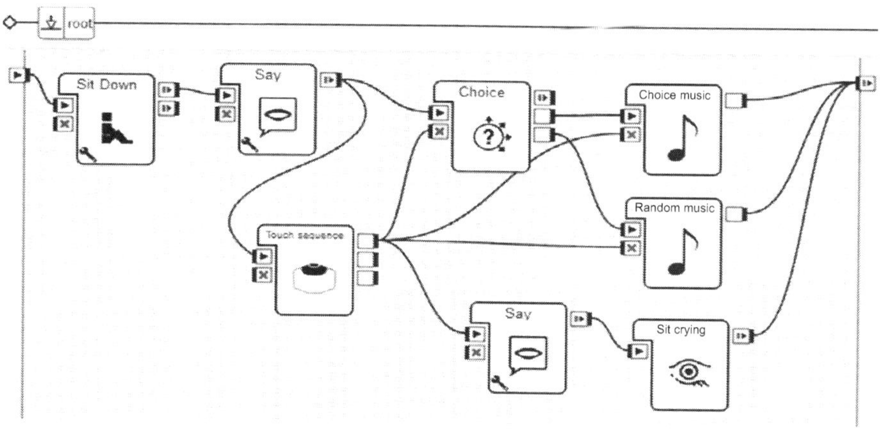

Figure 5.57. *Display of the Root*

5.3.13.2. *Course*

Nao starts by saying "I suggest that you listen to some music". Thus, the senior person has the choice between choosing a music track or listen to music randomly:

– In the first case, Nao says "I have classical, pop and rock music". Therefore, the user has the choice between several types of music (classical, pop and rock music). At the end of each track, Nao asks "Another?". If the user says yes, the loop begins again. Otherwise, the application stops.

– In the second case, a musical track is randomly selected using a random number. At the end of each track, Nao asks "Another?" If the user says yes, the loop restarts. Otherwise, the application stops.

If the elderly person wishes to stop the application, then she can touch Nao's head sensors from the front to the back at any time. In this case, Nao reads out his farewell message and cries.

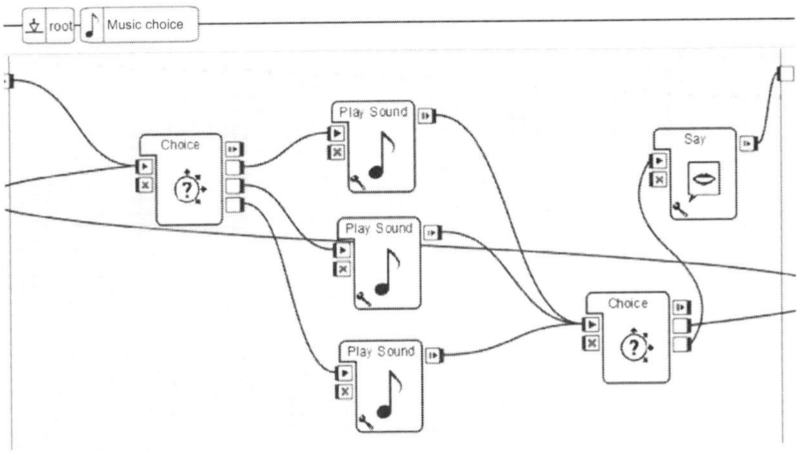

Figure 5.58. *Display of the "ChoiceMusic" block*

This Choice, as illustrated in Figure 5.58, allows the senior citizen to make a choice between listening to music in a random fashion and selecting a music track in particular.

Therefore, on output, the first is connected to the "choice music" block and the second to the "random music" block.

5.3.13.3. *Music choice*

As shown in Figure 5.59, this block allows for the selection of a music track.

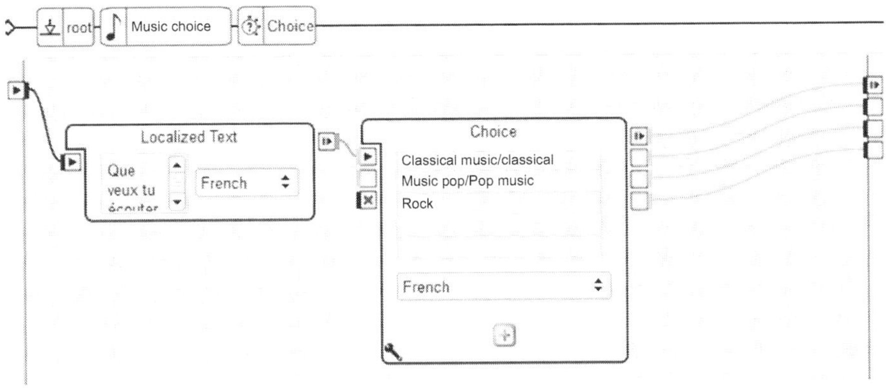

Figure 5.59. *Display of the "Choice" block in ChoiceMusic*

The first choice is connected to three tracks. One output thus corresponds to a track.

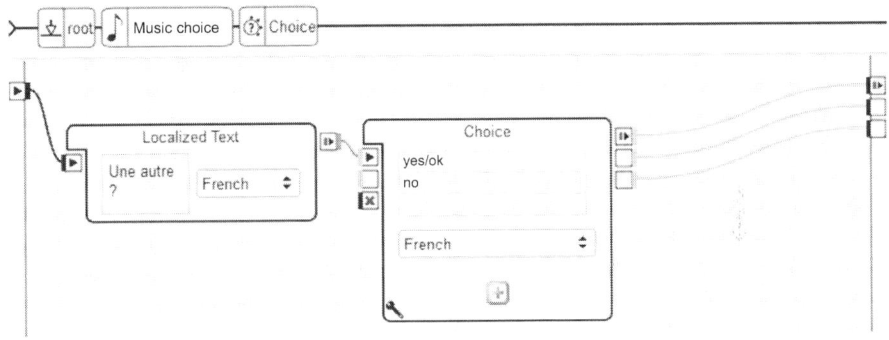

Figure 5.60. *Display of the second choice block in "ChoiceMusic"*

At the end of each track, a second choice has been implemented allowing the user to either stop the application or to listen to another track as shown in Figure 5.60.

5.3.13.4. *Random music*

When the user choses to randomly listen to music, the signal passes through a Random box. The latter generates a number that by means of the Switch Case allows for reading a music track "randomly". At the end of the track, the Choice on the right-hand side of the image above allows the user either to listen to another track or to stop the application as detailed in Figure 5.61.

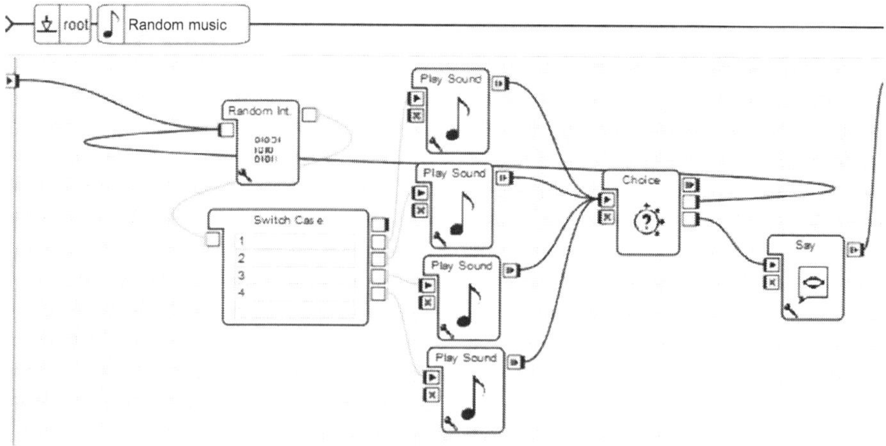

Figure 5.61. *Display of the "RandomMusic" block*

5.3.13.5 *Sitting and weeping*

We have created this block to make Nao cry when the user wants to exit this application. Sniffles have been found on the Internet corresponding to the block "ouin" and gestures using a behavior corresponding to the "arms sitting" block as shown in Figures 5.62 and 5.63.

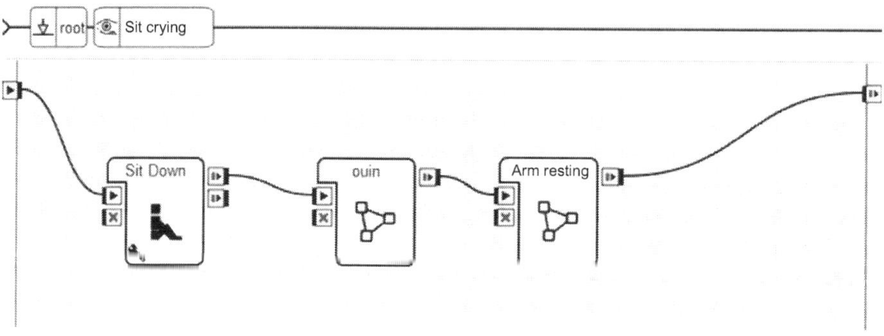

Figure 5.62. *Display of the "CrySitting" block*

5.3.13.6. *Touch sequence*

This block, created by our services, makes it possible to stop the application when touching ALL sensors on Nao's head from the front to the back. We decided to put this block in the Root to allow the user to stop the application at any time even during a music track as shown in Figure 5.64.

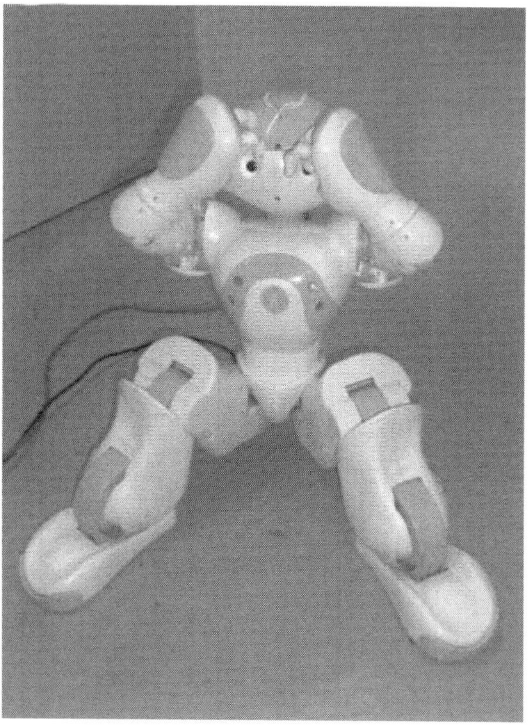

Figure 5.63. *Nao sits and cries*

5.3.14. *Multiplication game*

5.3.14.1. *Objective*

This program is intended to compensate for possible memory loss that the user may have. The intent of this game is to help the user not to forget her multiplication tables, through a series of 10 questions.

5.3.14.2. *Course*

Step 1: Nao tells the user that it proposes her a multiplication game, with a first "say" box.

Step 2: a loop is started that will iterate and propose 10 times a multiplication question.

Step 3: one enters the "Multiplication block".

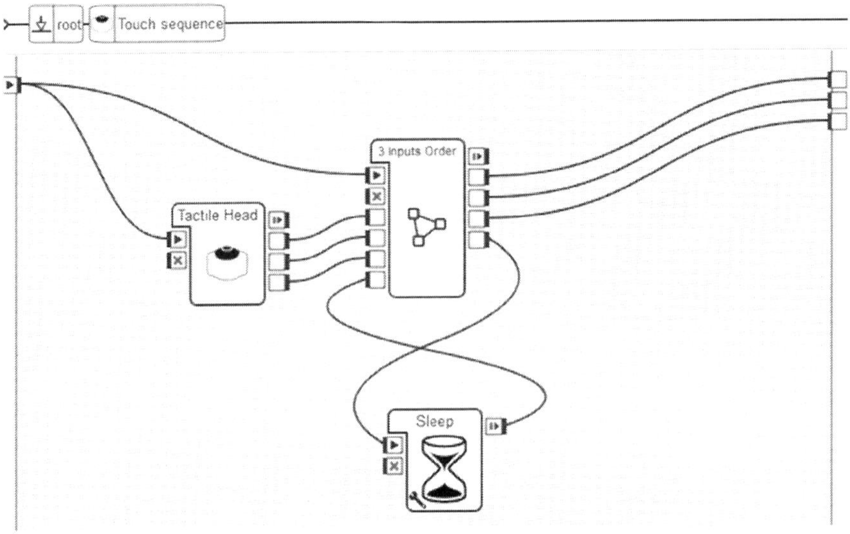

Figure 5.64. *Display of the "Touch Sequence" block*

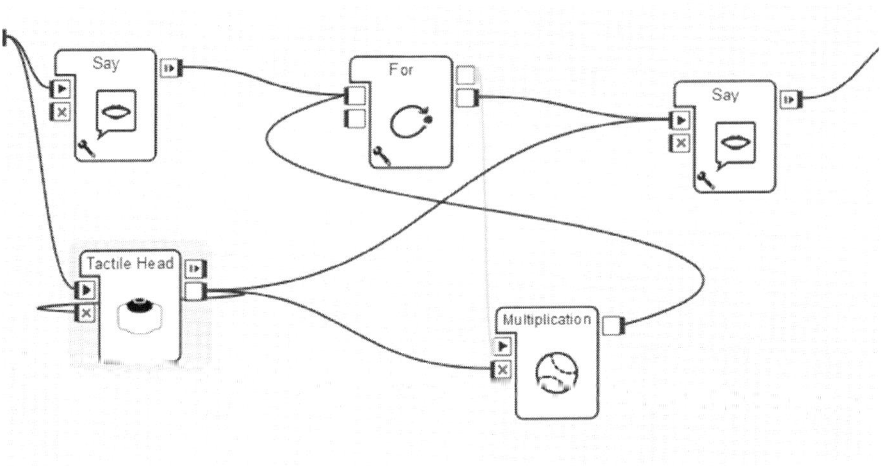

Figure 5.65. *Display of the Root*

We generate two random numbers between 1 and 9 in the box "2 numbers" and we multiply them in a first stage.

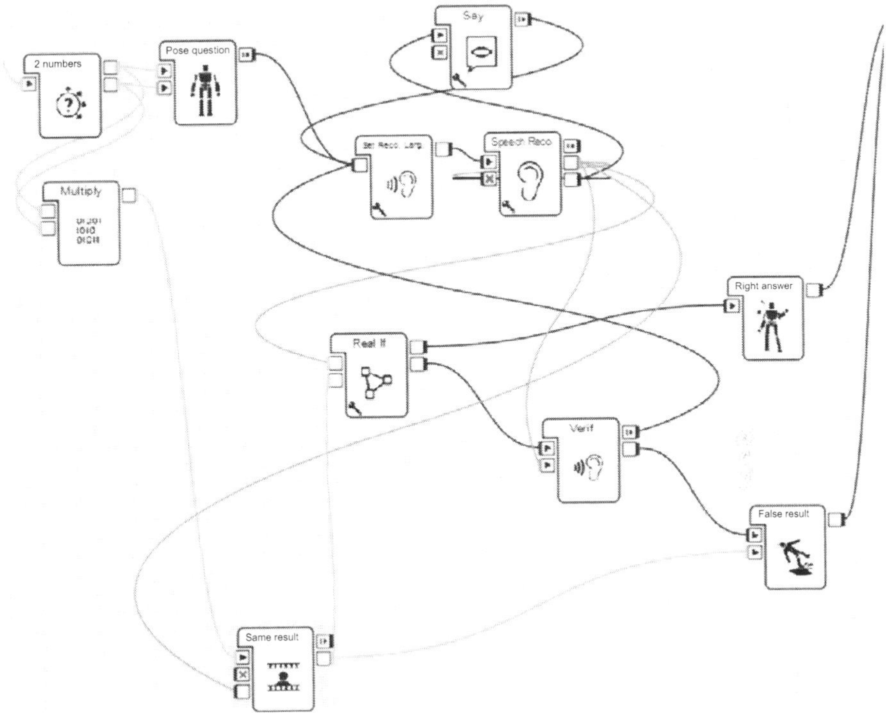

Figure 5.66. *Display of the "Multiplication" block*

At the same time as the program multiplies the two numbers, Nao will read the multiplication orally to the user. The two numbers arrive as parameters, Nao reads them by adding the text "times " in between, then ends by "equals to?" before waiting for the answer through speech recognition as pictured in Figure 5.66.

When Nao has finished asking the question, he waits for the user's answer. During the tests, we have realized that speech recognition (there are more than 80 possibilities of answers) was not perfect with numbers. Consequently, we have added a loop in order to certify that Nao has correctly understood the answer given by the user, in the case where Nao would have understood a wrong answer as shown in Figure 5.67.

In effect, if Sophie says "five", Nao may easily understand "nine" because these two numbers are phonetically close to each other as shown in Figure 5.68.

Nao therefore asks the user if she has properly pronounced the word he has heard.

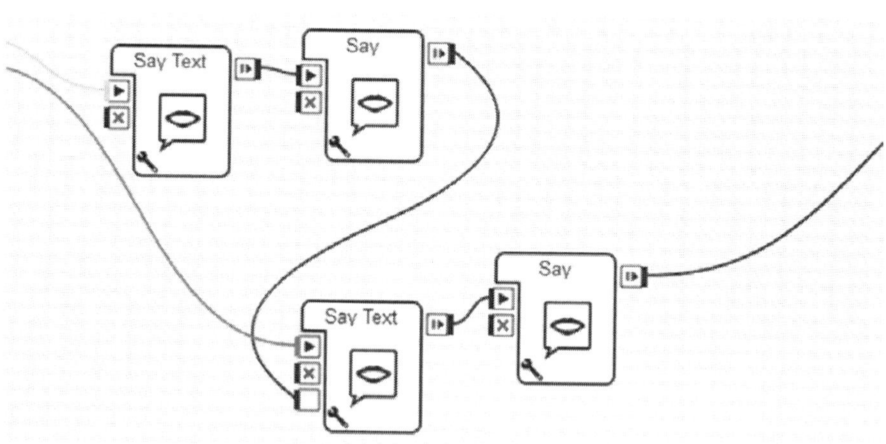

Figure 5.67. *Display of the "PoseQuestion" block*

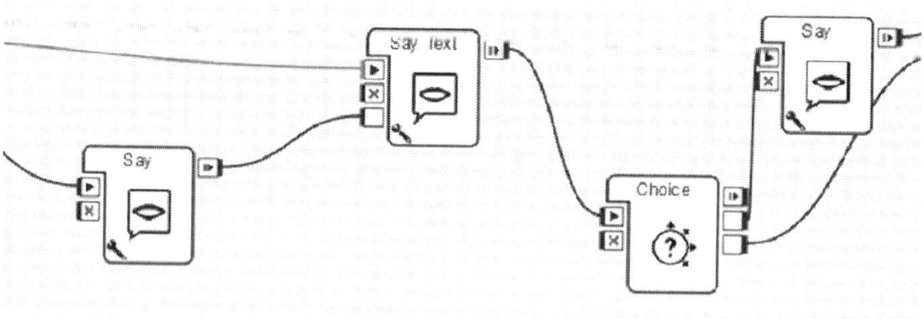

Figure 5.68. *Displaying of the "Verif" block*

5.6.14.3. *Example*

Nao: "3 times 3 equals?"

Sophie: "9".

Nao: "I have understood 33, is that what you have just said?"

If the user answers "yes" and that the result is actually false, Nao announces that it is not the good result, then tells the user what was the expected result. It is the result of the multiplication that is passed as a parameter into a simple "say text" after a regular "say" block as shown in Figure 5.69.

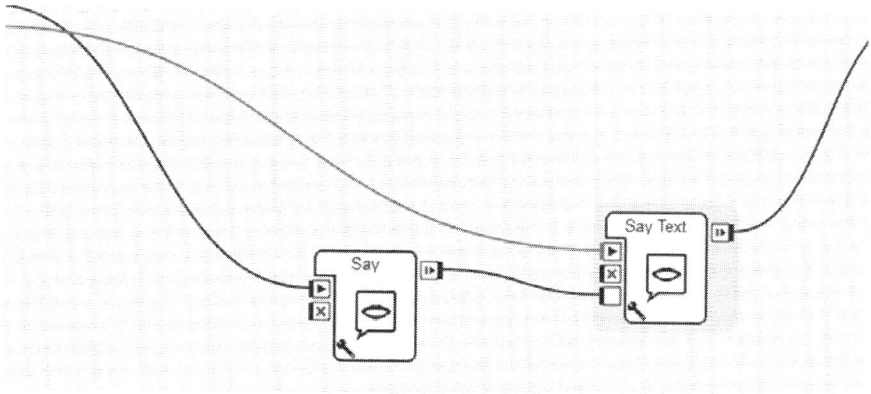

Figure 5.69. *Display of the "False Result" box*

However, if the user answers "No" (that it was not the result uttered, therefore the robot has misheard), Nao apologizes with a "say" block, in the"Verif", then goes back to listening, so that the user utters again the result of the multiplication. If Nao does not understand the result, he says "Don't get it" and listens again.

Step 4: when the 10 multiplication questions are finished, the program leaves the loop "for" and Nao thanks for playing, before saying goodbye.

If the user wishes to stop playing during the loop, she can press the touch buttons on Nao's head. This directly brings the final goodbye message.

5.3.15. *Nao's dance*

Target

By using box logics and some built-in dance function and a self-programmed "Macarena", we can ask Nao to do some dancing moves and entertain everyone.

What we have done

As the program starts, Nao would tell us that he would like to dance and Nao would initialize the program with a posture ready for dancing. Next, he would ask us what kind of dance we prefer to see. Nao knows three kind of dances: "Macarena", "Tai Chi Martial art dance" and the famous Asian pop "Gangnam Style". By specifying what kind of dance we would like to see, Nao would start happily.

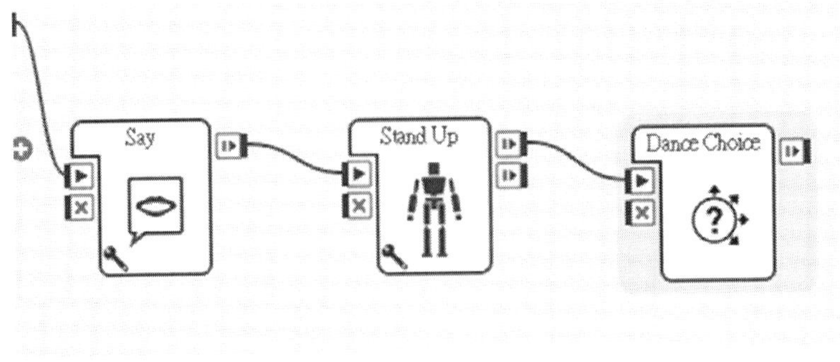

Figure 5.70. *Display of the Root*

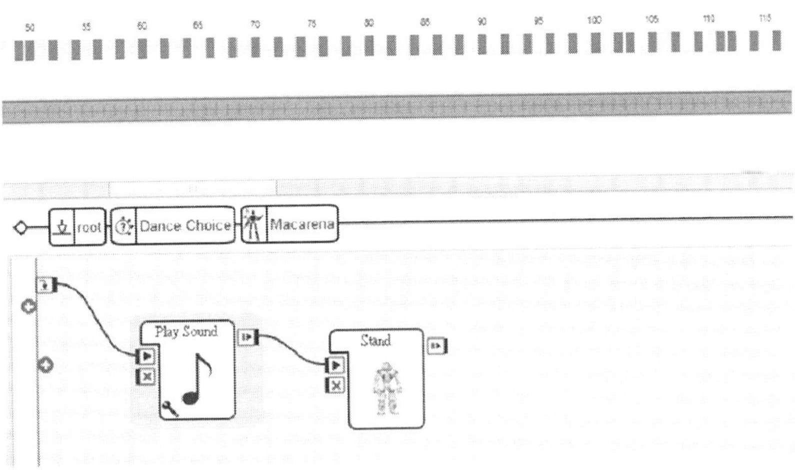

Figure 5.71. *Display of the "Macarena" block*

Difficulties:

Macarena is a dance with complicated hand and arm movements; in order to resemble the original music video, we have to rotate every joint to the right at a specific time so that it will not look like another dance that we invented. When dancing the Macarena, Nao shakes his body a lot which easily makes him lose his balance. Due to frequently bending his knees while moving his arms back and forth, up and down, we often have to protect Nao from falling. The music "Macarena" is fast-paced music so that it was a hard task to make Nao's movements fit with the music.

Solutions

– to make the dance look like the original one, we had to compare Nao with the video every three or four keyframes. By doing so, we can be sure that there were not too many differences;

– to keep Nao in good balance, we had to make sure that Nao bends his knees and stands back straight (two movements which rapidly displace its COM), we do not make too many hand and arm movements so as to cause further imbalance;

– in order to fit all the movements with the music, we had to keep adjusting every space between keyframes.

5.3.16. *Memory game*

5.3.16.1. *Objective*

We have decided to place this application after the evening meal. The goal of this program is to distract the senior person by testing her knowledge.

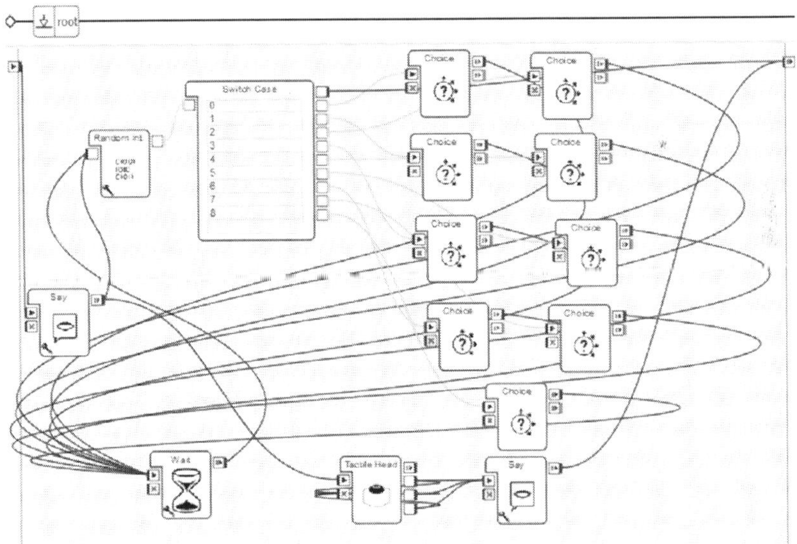

Figure 5.72. *Display of the "Root"*

5.3.16.2. *Course*

First of all, the application starts with a welcome message from Nao "Hello, welcome to Nao's question game". Then, the elderly person must answer a series of questions that will be asked to him in a random manner. In effect, a random number has been implemented so that this application does not become redundant by always asking these questions in the same order.

To each raised question, the user has two chances to tell the right answer. Therefore, if she says the right answer at the first attempt, then Nao says “Bravo!”. Otherwise, he says “this is not the right answer” and the user has a second chance.

At any time, the user can stop the application by touching one of the sensors in his head. In this case, Nao says “all right, you don’t want play with me anymore”.

Randomly asked question

To avoid that Nao asks all questions always in the same order, we have decided to put a Random. Therefore, the Switch case makes it possible to associate a number to each question and the Random by generating a number randomly, which enables the game to stay interactive as shown in Figure 5.73.

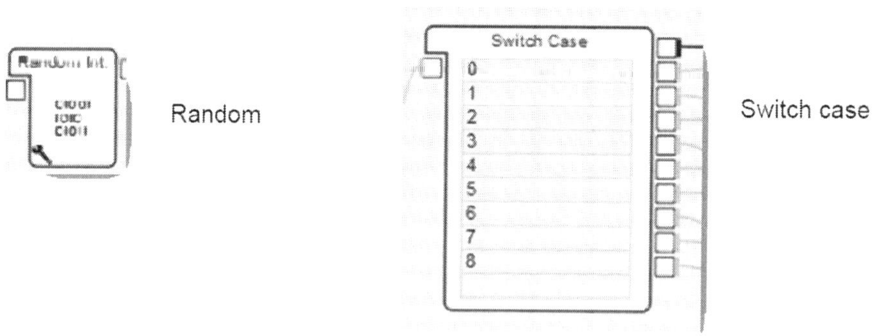

Figure 5.73. *“Random” and“Switch case” boxes*

Example

Step 1: a number is generated by Random, for instance 0.

Step 2: through the Switch case this number corresponds to a question. In our case, as Figure 5.74 illustrates it.

Here, Nao asks the question “Who has invented the electric light bulb?”. If the elderly person says the right answer, then Nao says “Bravo!”. Otherwise, the user has a second chance, then Nao says “This is not the right answer, try again!”.

So if the user says the right answer, then Nao says “Bravo!”. Otherwise, Nao says the right answer as shown in Figure 5.75.

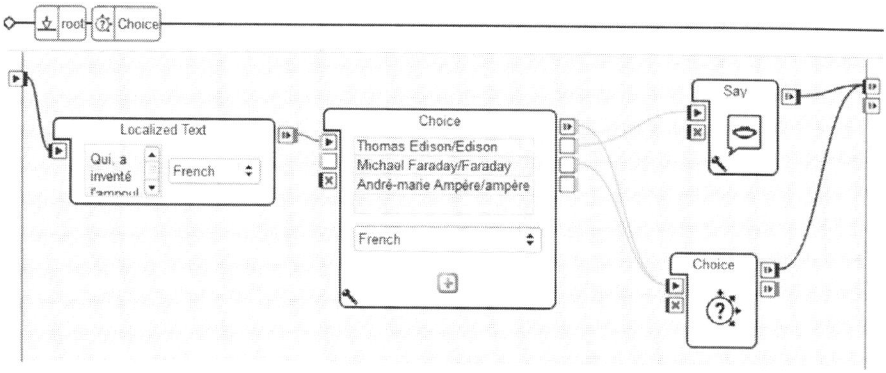

Figure 5.74. *Example of a question in the game*

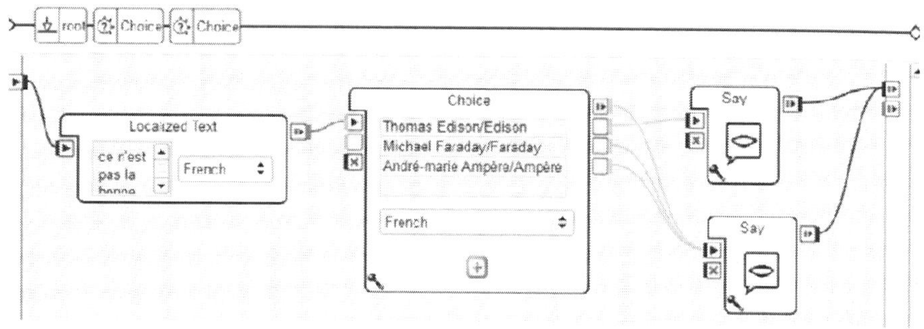

Figure 5.75. *If the answer is wrong*

In this way, the program runs and generates random questions until the user decides to stop the program by touching Nao's head.

5.3.17. *Detect person on the ground*

5.3.17.1. *Objective*

The goal of this program is to follow the person (represented by a red ball) with the eyes, to turn the head in all directions and to make Nao spin to find her if the person is no longer in his field of vision, to recognize if the person has fallen.

5.3.17.2. *Course*

Step 1: Nao announces to the user who is willing to follow and the "touch sequence" block is activated.

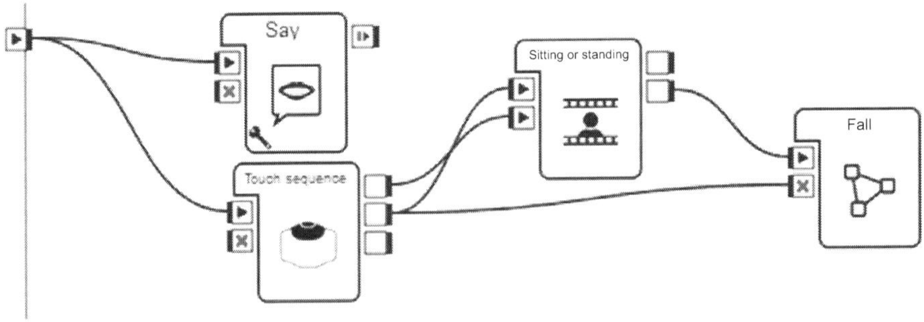

Figure 5.76. *Display of Root*

Step 2: when the user slides her finger on Nao's head from his forehead to the back of his neck, Nao gets up and then the fall recognition program is activated. When the user slides her finger on Nao's head from his neck toward his forehead, the fall recognition program ends and Nao sits down.

Step 3: while the Chute block is active, Nao turns his head and then makes quarter turns on himself. Once Nao has spotted the user (represented by a red ball attached to the person's chest), he stops searching and starts to follow her. If the red ball is at a low height (at knee height) Nao interpretates that the user has fallen, he asks her if everything is alright and offers to call for help if need be.

Blocks detail:

– *"Sitting or standing" block*

In the block "sitting or standing", the first input is requested, the robot sits and then a signal comes out of the first output. If it is the second input, the robot gets up and then a signal comes out from the second output.

– *"Fall" block*

In this block, the "seek and walk" box is first activated: the robot seeks for the red ball representing the user.

Case 1: the first output of the "seek and walk" block is requested when Nao has lost the user and could not find her despite its search movements. The program stops, she will have to reactivate Nao by pressing the touch buttons of his head from his forehead to his neck.

Case 2: the second output of the "seek and walk" block is requested when the red ball has been found at the ground level. Nao then asks the person to clap her hands to check if she is still conscious. Otherwise, Nao calls the neighbors and waits before returning to the person search block. In case the person responds,

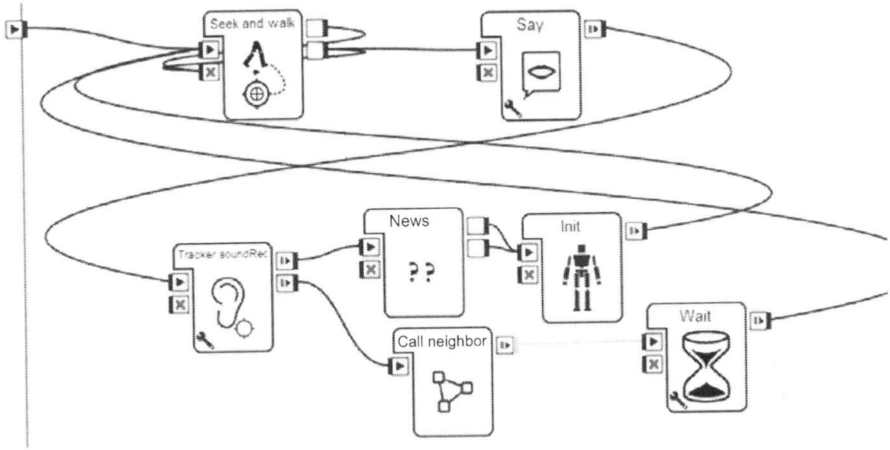

Figure 5.77. *Display of the "Fall" box*

Nao asks for news and calls for help if necessary. Regardless of whether an emergency call has been made or not, Nao is restarted to his original position and continues to track the person.

"Call" block

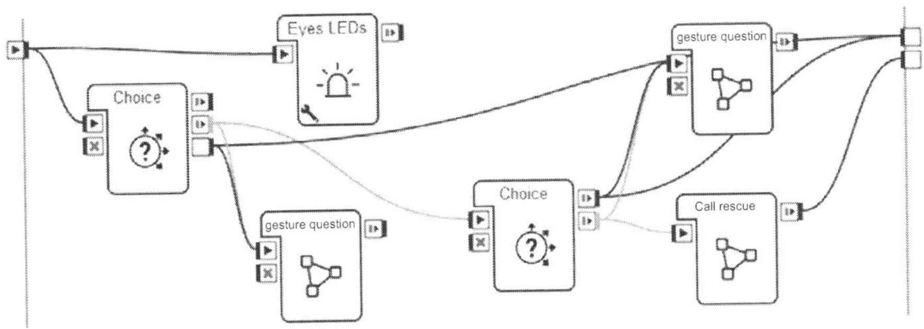

Figure 5.78. *Display of the "Call" block*

Nao asks the user if she is okay; if yes, the first output of the block is solicited. If no, Nao asks if the user wishes to call for help. In the negative case, the first output of the block is requested. In the positive case, the "CallHelp" block previously described is called and then the second output of the block is solicited.

"Seek and walk" block

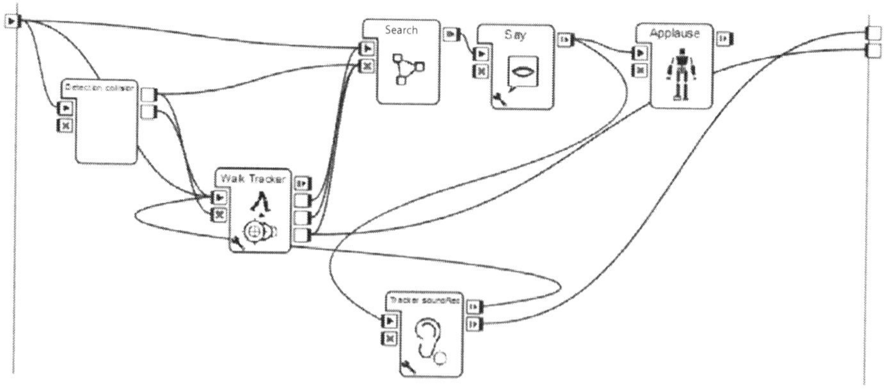

Figure 5.79. *Display of the "seek and walk" block*

When launching this block, collision detection is also initiated, Nao starts searching for the red ball symbolizing the person and the "walk tracker" is activated. The "collision detection" block is similar to that of section 4.5.2 except that a signal is emitted on the first output when detecting a fall or strike, then a signal is emitted on the second output when the robot got back on his feet after falling or has avoided the obstacle. As a result, the search is stopped when detecting a fall or a strike and then it is restarted once the problem solved.

When the "walk tracker" block finds the ball (signal on output 2), the search is stopped and the internal functions of the "walk tracker" allow Nao to track the ball. A block coded in Python inside of the "walk tracker" makes it possible to know if Nao's head is oriented at more than 0.2 radian (10°) downwards and therefore if the red ball (the user) is on the ground. If this is the case, a signal is emitted on output 4 of the "walk tracker" block, the search is stopped and the block emits a signal on its second output.

If the ball is lost, output 3 of the "walk tracker" block initiates the search. The search block is a succession of blocks where Nao rotates the head in all directions and then completes a quarter turn. Once Nao has completed a full circle, if the search block has not been interrupted, that is if the red ball has not been found, the output of the block "search" is stimulated. Nao then ask the user to clap his hands to try to find him, the "sound/red tracker" block is activated. If the latter block allows the red ball to be found, the "walk tracker" is once more activated. If after some time the ball has still not been found, the second output of the "tracker sound/red " block is stimulated, the signal comes out on the first output of the block indicating that the red ball has been lost and that the program is terminated.

"Tracker sound/red" block

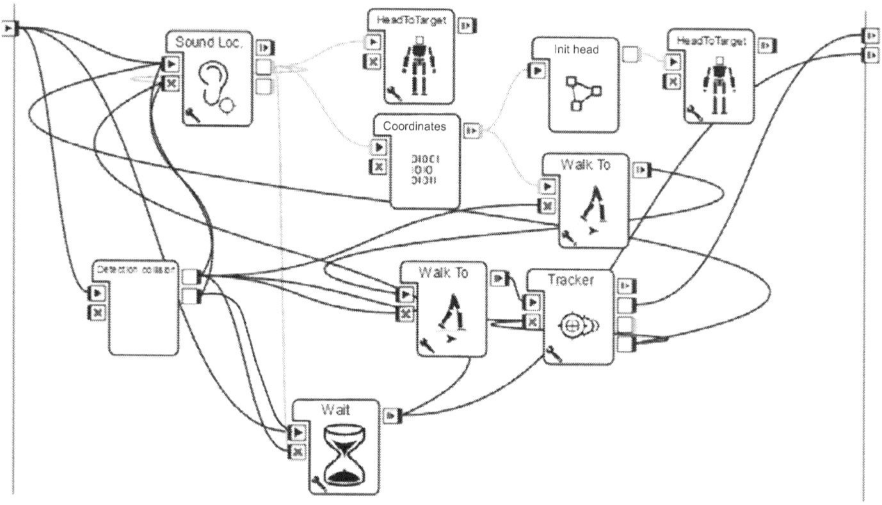

Figure 5.80. *Display of the "Tracker sound/red" block*

This block is rather complex: collision detection is activated at the same time as sound localization and a timer.

Case 0: if a strike or a collision are detected, the timer and sound localization are stopped, the robot avoids the obstacle or stands up and then they are restarted.

Case 1: if no sound has been found in the time allotted, the "wait" emits a signal that flows to the second output of the block, indicating "timeout".

Case 2: As sound is detected, Nao rotates his head toward the sound and then faces the point of origin of the sound and replaces his head back to its initial position; finally, he travels a distance of 50 cm toward the point of origin of the sound. A red ball tracker is then activated, if the ball is found, the block is terminated with a signal on its first output: the ball is found again. If the ball has still not been detected, the signal goes back to the "sound tracker" and the cycle starts over.

5.3.18. *At any time*

5.3.18.1. *Objective*

The goal of this program is to provide activities to the user at any time of the day when the latter makes a sequence on the touch buttons on Nao's head. It is possible to stop the activity at any time through the reverse sequence on the buttons.

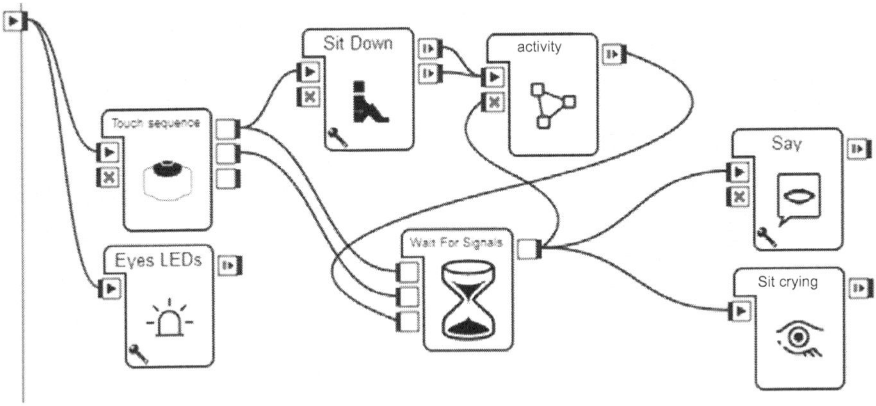

Figure 5.81. *Display of the Root*

Sequence and block details

When the user slides her finger on Nao's head from his forehead toward his neck, he sits and proposes to do an activity.

When the user slides her finger on Nao's head from his forehead toward his forehead while an activity has been launched, this latter terminates and Nao starts crying because they were having a good time.

"Wait for signals" block

When a signal arrives on the 1st input, input 2 is "open". If a signal arrives on the third input, input 2 is shut.

When a signal arrives on the second input while it is open, a signal is created on output of the block and the second input is closed.

This block controls that Nao cries if only an activity has been started.

"Cry sitting" block

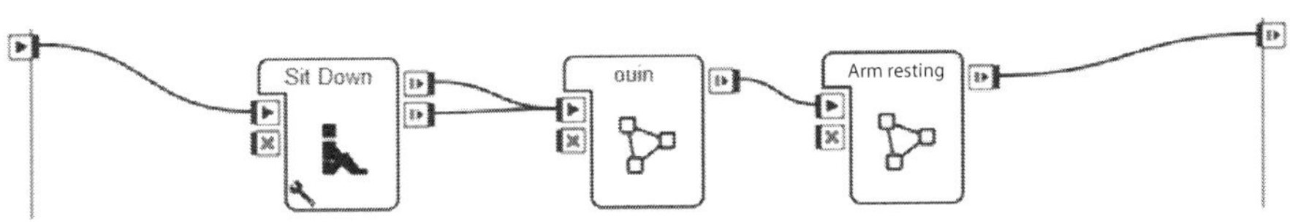

Figure 5.82. *Display of the "CrySitting" block*

This block makes Nao sit if he was standing up during the activity, then makes him do movements to wipe his eyes and produces a sniffing sound to express his sadness. Nao's bars are then put back into original position.

"Activities" block

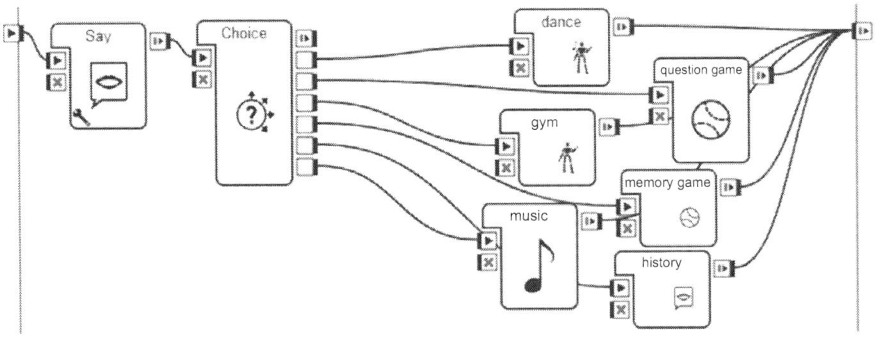

Figure 5.83. *Display of the "Activity" block*

In this block, Nao asks the user what activity does she want to do and launches the corresponding activity block.

5.4. Conclusion

5.4.1. *Nao's limitations and possible improvements*

After having studied the Nao robot, we have confirmed that this project for home care can be included within an approach for the future.

Nao is limited by many aspects:

– *His size*: 54 cm, Nao cannot reach all the objects in a home, its user will have to bend down to touch his head, or just lower her head to look at the robot.

The solution will be the robot Romeo, with his 140 cm, he can reach objects and act at another height level.

– *His motor capabilities (his strength)*: Nao weighs 4.5 kg, and his engines are limited and therefore he cannot carry heavy loads, without altering his movements or even his balance.

– *His autonomy*: during this work, we have observed that Nao's autonomy reached 90 min when he was recharged for 3 h in DC without use. This is a short period of time and it is clear that during a 24h/24 daily use, a system should be devised so that the robot be able to recharge himself on his own, without the user having to plug his cable in the back.

We can imagine a pedestal for recharging into which Nao could plug himself, such that he may recharge himself when he sits on top of it and is still able to keep his attention on the person he is caring for.

– *His limited speed*: to keep his balance, Nao cannot exceed 10 cm/s, which greatly limits his movements, if it is desirable that he follows a person walking at a normal speed (for a human being between 3 and 5 km/h).

To overcome this problem, the user has to adapt himself to the speed of the robot. We can also picture a system with a skateboard so that Nao can move faster, perhaps pulled by his user, why not.

– *His speech recognition*: Nao's understanding is limited to an integrated dictionary. He can only recognize expected words or words included in each program such as "cancel", "repeat" or other.

If the user wants to ask him a question then Nao has to be prepared for it and must be programmed to answer it in order to satisfy the user. The artificial intelligence programmed by developers must therefore be as complete as possible and ideally, Nao should be able to integrate on his own made-up words or frequently used by his user.

– *His inaccurate gestural abilities*: Nao has 25 degrees of freedom, it is certainly a lot for a "machine", but very little to perform precise movements. In addition, he only has three fingers, it is enough to make basic gestures or to catch an object with an adapted form but this makes it difficult for more complicated gestures such as pressing on a TV remote control, grabbing a pill to hand it to a patient, catching a fragile object such as a footed glass, etc.

In addition, Nao is very sensitive to irregularities of the ground, he may lose his balance even if the floor is at a very slight slope.

– *His cost*: Nao currently costs 4,500 euros (in 2009 it costed 12,000 euros). This price makes this technology unaffordable to most homes but the maintenance and development of a customized program for a user make this tool/robot far remote from common use nowadays. It might be necessary to consider a reimbursement by health insurance for robots destined to support dependent persons or individuals with disabilities.

To conclude, Nao (54 cm robot) would not be able to help an elderly person of 60 kg if she was unconscious. He would only be able to call for more reliable help.

This study has enabled the development of a humanoid in a comprehensive manner. We have made advances compared to the humanoid robotics sector and brought solutions to problems that belong to the future.

6

Application Problems with Solutions

This chapter explores car suspension, electromechnical systems, identification and state–space representation observation, and control of nonlinear systems.

6.1. Exercise 6.1: car suspension

Modeling of the suspension of a quarter of car whose simplified diagram is shown in Figure 6.1.

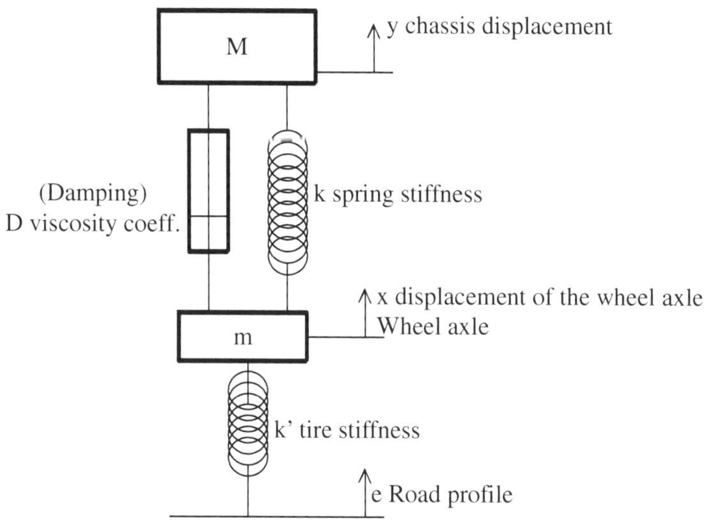

Figure 6.1. *Modeling of the suspension of a quarter of car*

6.1.1. *Modeling*

– Static equations of the system:

When the system (x_0, y_0) is in equilibrium, the static behavior is governed by the following equations:

$$\begin{cases} Mg = -k(y_0 - x_0) \\ mg = -k'(x_0 - e_0) - k(x_0 - y_0). \end{cases} \qquad [6.1]$$

– Dynamic equations of the system:

In dynamics, displacements are achieved around the equilibrium (static) and for that purpose we define:

$$\begin{aligned} Y &= y_0 + y \\ X &= x_0 + x \\ E &= e_0 + e. \end{aligned} \qquad [6.2]$$

Dynamic equations are thus given by:

$$\begin{cases} M\ddot{Y} = -D(\dot{Y} - \dot{X}) - k(Y - X) - Mg \\ m\ddot{X} = -D(\dot{X} - \dot{Y}) - k(X - Y) - k'(X - E) - mg. \end{cases} \qquad [6.3]$$

If quantities X, Y and E are replaced by their values given by [6.2], we have the following:

$$\begin{cases} M\ddot{y} = -D(\dot{y} - \dot{x}) - k(y - x) - k(y_0 - x_0) - Mg \\ m\ddot{x} = -D(\dot{x} - \dot{y}) - k(x - y) - k'(x - e) \\ \qquad\quad -k(x_0 - y_0) - k'(x_0 - e_0) - mg. \end{cases} \qquad [6.4]$$

After simplification using [6.1], we get the following model:

$$\begin{cases} M\ddot{y} = -D(\dot{y} - \dot{x}) - k(y - x) \\ m\ddot{x} = -D(\dot{x} - \dot{y}) - k(x - y) - k'(x - e). \end{cases} \qquad [6.5]$$

By rearranging both equations, we obtain:

$$\begin{cases} M\ddot{y} + D\dot{y} + kx = D\dot{x} + kx \\ m\ddot{x} + D\dot{x} + (k + k')x = D\dot{y} + ky + k'e. \end{cases} \qquad [6.6]$$

Assuming zero initial conditions and using the properties of the Laplace transform, [6.6] can be transformed into the following:

$$\begin{cases} (Mp^2 + Dp + k)Y(p) = (Dp + k)X(p) \\ (mp^2 + Dp + (k + k'))X(p) = (Dp + k)Y(p) + k'E(p). \end{cases} \qquad [6.7]$$

After rearrangement, we have:

$$Y(p) = \frac{Dp + k}{Mp^2 + Dp + k} X(p) \qquad [6.8]$$

$$X(p) = \frac{Dp + k}{mp^2 + Dp + (k + k')} Y(p) + \frac{k'}{mp^2 + Dp + (k + k')} E(p).$$

Hence, the transfer function between the system input $E(p)$ and output $Y(p)$ is given by:

$$H(p) = \frac{Y(p)}{E(p)} = \frac{k'(Dp + k)}{(Mp^2 + Dp + k)(mp^2 + k') + Mp^2(Dp + k)}. \qquad [6.9]$$

Neglecting the elasticity of the tire ($x = e$), we have the following transfer function:

$$H(p) = \frac{Dp + k}{Mp^2 + Dp + k}. \qquad [6.10]$$

6.1.2. *Analysis*

The transfer function $H(p)$ is of the second order; it can be expressed in the form of:

$$H(p) = \frac{\frac{D}{M}p + \frac{k}{M}}{p^2 + \frac{D}{M}p + \frac{k}{M}} = \frac{2\xi\omega_n p + \omega_n^2}{p^2 + 2\xi\omega_n p + \omega_n^2} \qquad [6.11]$$

with

$$\begin{cases} 2\xi\omega_n = \frac{D}{M} \\ \omega_n^2 = \frac{k}{M} \end{cases} \Longrightarrow \begin{cases} \omega_n = \sqrt{\frac{k}{M}} \\ \xi = \frac{D}{2\sqrt{Mk}} \end{cases} \qquad [6.12]$$

– If $\xi \geq 1 \Leftrightarrow D \geq 2\sqrt{Mk}$, the transfer function has two real poles.

– If $\xi < 1 \Leftrightarrow D < 2\sqrt{Mk}$, the transfer function has two complex conjugate poles.

– The static gain is $H(0) = 1$.

The index response (motion of the chassis when driving over the sidewalk) is based on parameters M, D and k. It is represented for different values of these parameters, as shown in Figure [6.2].

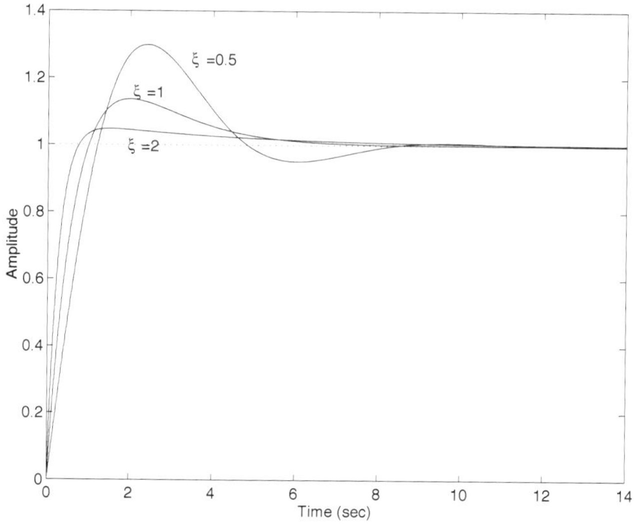

Figure 6.2. *Step responses (suspension)*

The asymptotic Bode diagram is shown in Figure 6.3.

In the Bode plot, we can observe that there is always resonance regardless of the value of the damping coefficient ξ. This is due to the presence of the zero in the transfer function of the system. In the case where $\xi \geq 1$, the system does not have dominant poles because the zero z_1 is closer to the imaginary axis than the pair of complex conjugate poles (p_1, p_2).

$$
\begin{aligned}
z_1 &= -\frac{\omega_n}{2\xi}, \qquad [6.13]\\
p_1 &= -\xi\omega_n + \omega_n\sqrt{1-\xi^2},\\
p_2 &= -\xi\omega_n - \omega_n\sqrt{1-\xi^2}.
\end{aligned}
$$

However, resonance is all the more important when the damping factor is small ($\xi < 1$: system with dominant poles).

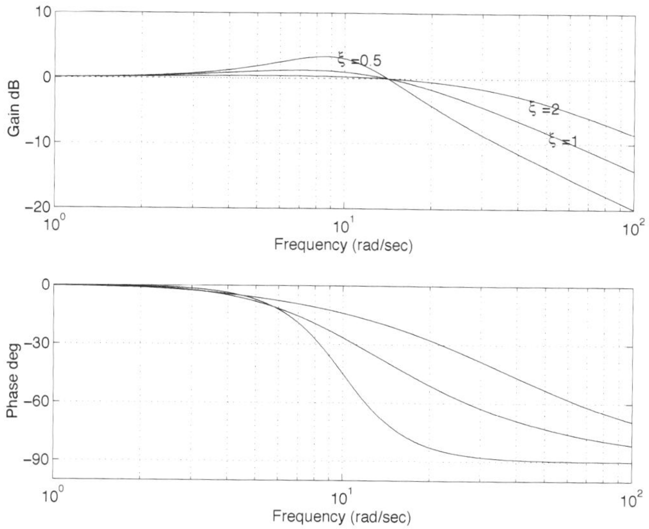

Figure 6.3. *Asymptotic Bode plot*

Regarding the phenomenon of vibration felt in some cars around 100 km/h, it can actually be explained using Bode plots. The road can actually be considered as presenting a periodic phenomenon, for example a sine wave over a few meters, thus for any velocity of the car, the same car sees a profile of the road with sinusoidal amplitude and different periods. It can therefore deducted that if the velocity varies, the period of the sinusoidal profile of the road varies. This relates directly to the definition itself of the harmonic behavior of a system (Bode plot). A sinusoidal input with variable period of a system causes a sinusoidal output of the same frequency but with amplitude and phase that represent the Bode plot.

The vibration phenomenon is explained by the fact that at 100 km/h, the frequency of the profile of the road is close to the resonance frequency seen in the Bode plot. It can also be noted that when exceeding this velocity, we have the impression that the car is glued to the road. This also can be observed in the Bode plot, it can be seen that the modulus decreases enormously far beyond the resonance frequency.

6.2. Exercise 6.2: electromechanical system

The purpose of this exercise is the modeling and analysis of the behavior of a DC electric motor controlled by the voltage of the armature. The electromechanical diagram of the system is shown in Figure 6.4.

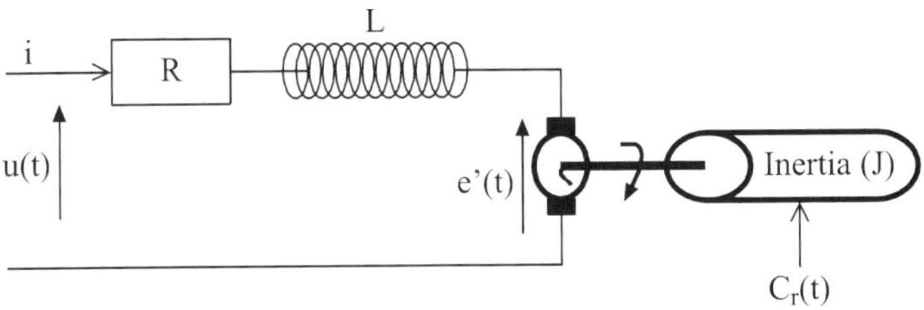

Figure 6.4. *Electromechanical diagram of the system*

Where

– L and R are the inductance and the resistance of the armature;

– J is the total moment of inertia of the load time and the motor shaft;

– f represents the viscous friction of the rotating parts;

– the motor torque is given by:

$$C_m(t) = k \cdot i(t) \qquad [6.14]$$

– k is the torque constant $(\mathrm{Nm/A})$;

– $i(t)$ is the current in the armature (A);

– the backelectromotive force $e'(t)$ is given by

$$e'(t) = k\Omega_m(t) \qquad [6.15]$$

– $\Omega_m(t)$ is the angular velocity of the motor shaft;

– $C_r(t)$ is the resistive torque external to the load.

6.2.1. *Modeling*

The system thus constituted is a multivariable system with the pair $(u(t), C_r(t))$ as input and the pair $(i(t), \Omega_m(t))$ as output. The model is obtained by writing the electric equations of the motor and mechanical load, with the torque that is ensured by the relationship between torque and current (equation [6.14]).

– The electrical equation is written as:

$$\begin{aligned} u(t) &= Ri(t) + L\frac{di(t)}{dt} + e'(t) \\ &= Ri(t) + L\frac{di(t)}{dt} + k\Omega_m(t). \end{aligned} \tag{6.16}$$

– The mechanical equation at the engine level is written as:

$$C_m(t) = J_m\frac{d\Omega_m(t)}{dt} + f_m\Omega_m(t) + \frac{C_l(t)}{N}. \tag{6.17}$$

– The mechanical equation at the load level is given by:

$$C_l(t) = J_l\frac{d\Omega_l(t)}{dt} + f_l\Omega_l(t) + C_r(t) \tag{6.18}$$

and

$$\Omega_l(t) = \frac{\Omega_m}{N} \qquad (N \text{ and the reduction ratio of the reducer}) \tag{6.19}$$

from which the mechanical equation becomes

$$\begin{aligned} C_m(t) &= (J_m + \frac{J_l}{N^2})\frac{d\Omega_m(t))}{dt} + (f_m + \frac{f_l}{N^2})\Omega_m(t) + \frac{C_r(t)}{N} \\ &= J\frac{d\Omega_m(t))}{dt} + f\Omega_m(t) + \frac{C_r(t)}{N}. \end{aligned} \tag{6.20}$$

– The torque equation is given by equation [6.14]:

$$C_m(t) = k \cdot i(t). \tag{6.21}$$

By combining the equations, we get the following system:

$$\begin{cases} u(t) = Ri(t) + L\frac{di(t)}{dt} + k\Omega_m(t) \\ C_m(t) = k \cdot i(t) \\ C_m(t) = J\frac{d\Omega_m(t)}{dt} + f\Omega_m(t) + \frac{C_r(t)}{N}. \end{cases} \tag{6.22}$$

By applying the Laplace transform to this equation system, we get:

$$\begin{cases} U(p) = (R + Lp)I(p) + k\Omega_m(p) \\ C_m(p) = k \cdot I(p) = (Jp + f)\Omega_m(p) + \frac{C_r(t)}{N} \end{cases} \quad [6.23]$$

wherefrom it can be derived that:

$$\begin{cases} I(p) = \frac{U(p) - k\Omega_m(p)}{Lp+R} \\ \Omega_m(p) = \frac{C_m(p) - \frac{C_r(t)}{N}}{Jp+f}. \end{cases} \quad [6.24]$$

Based on these equations, we propose the following block diagram:

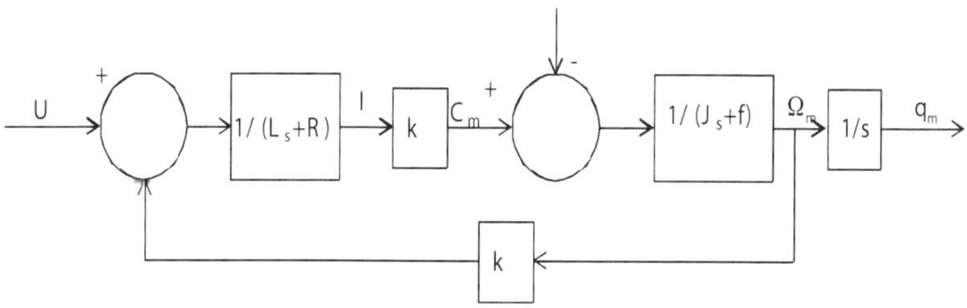

Figure 6.5. *Block diagram of the motor*

The input/output relation is given by:

$$\begin{bmatrix} I(p) \\ \Omega_m(p) \end{bmatrix} = [H(p)] \begin{bmatrix} U(p) \\ C_r(p) \end{bmatrix} \quad [6.25]$$

from which the transfer function is given by:

$$[H(p)] = \begin{bmatrix} \frac{Jp+f}{(Jp+f)(R+Lp)+k^2} & -\frac{1}{N}\frac{k}{(Jp+f)(R+Lp)+k^2} \\ \frac{k}{(Jp+f)(R+Lp)+k^2} & -\frac{1}{N}\frac{Lp+R}{(Jp+f)(R+Lp)+k^2} \end{bmatrix}. \quad [6.26]$$

6.2.2. *Analysis*

The system pictured by the block diagram (Figure 6.5) represents the motor controlled by voltage $u(t)$ and is subjected to a resistive torque (disruption) $C_r(t)$. The outputs are the angular velocity of the motor shaft. We are going to study the responses $(i(t), \Omega_m(t))$ for step inputs $(u(t), C_r(t))$.

– Step voltage $u(t)$, and $C_r(t) = 0$.

The transfer functions between $U(p)$ and $I(p)$ and between $U(p)$ and $\Omega_m(p)$ are given by:

$$\begin{aligned} H_1(p) &= \frac{I(p)}{U(p)} = \frac{Jp+f}{(Jp+f)(R+Lp)+k^2}, \\ H_2(p) &= \frac{\Omega_m(p)}{U(p)} = \frac{k}{(Jp+f)(R+Lp)+k^2}. \end{aligned} \quad [6.27]$$

– Resistive torque step function $C_r(t)$, $u(t) = 0$.

The transfer functions between $C_r(p)$ and $I(p)$ and between $C_r(p)$ and $\Omega_m(p)$ are given by:

$$\begin{aligned} H_1(p) &= \frac{I(p)}{C_r(p)} = -\frac{1}{N}\frac{k}{(Jp+f)(R+Lp)+k^2}, \\ H_2(p) &= \frac{\Omega_m(p)}{C_r(p)} = -\frac{1}{N}\frac{Lp+R}{(Jp+f)(R+Lp)+k^2}. \end{aligned} \quad [6.28]$$

6.3. Exercises: identification and state–space representation

6.3.1. *Exercise 6.3*

We are going to analyze an economic system that is represented by a model linking prices $q(t)$, revenues $r(t)$ and wages $w(t)$. This model is described by the following differential equations:

$$\begin{aligned} \frac{dr(t)}{dt} &= -a \cdot r(t) + c \cdot w(t), \\ \frac{dq(t)}{dt} &= -d \cdot q(t) + b \cdot r(t) + f \cdot w(t). \end{aligned} \quad [6.29]$$

The transfer function between wages and revenues is obtained by applying the Laplace transform to the first equation, which gives:

$$pR(p) = -aR(p) + cW(p) \Longrightarrow \frac{R(p)}{W(p)} = \frac{c}{p+a} = \frac{c/a}{1+1/a \cdot p}. \quad [6.30]$$

The first time constant is

$$\tau_1 = \frac{1}{a}. \qquad [6.31]$$

The transfer function between prices and wages is obtained by applying the Laplace transform to the second differential equation, from which we obtain:

$$\begin{aligned} pQ(p) &= -dQ(p) + bR(p) + fW(p) \Longrightarrow \frac{Q(p)}{W(p) + b/f \cdot R(p)} \\ &= \frac{f}{p+d} = \frac{f/d}{1 + 1/d \cdot p}. \end{aligned} \qquad [6.32]$$

The second time constant is

$$\tau_2 = \frac{1}{d}. \qquad [6.33]$$

Finally, by combining both results we have the relation between wages and prices:

$$\begin{aligned} pQ(p) &= -dQ(p) + \frac{bc}{p+a}W(p) + fW(p) \Leftrightarrow (p+d)Q(p) \\ &= \frac{bc + fp + fa}{p+a}W(p). \end{aligned} \qquad [6.34]$$

Hence, the transfer function

$$\frac{Q(p)}{W(p)} = H(p) = \frac{fp + bc + fa}{(p+a)(p+d)}. \qquad [6.35]$$

Numerical application ($a = 1/4$, $b = 1/6$, $c = 1/5$, $d = 1/8$, $f = 2/15$):

$$H(p) = \frac{2/15p + 1/30 + 1/30}{(p+1/4)(p+1/8)} = \frac{64/15p + 32/15}{(1+4p)(1+8p)}. \qquad [6.36]$$

Time constants are

$$\tau_1 = 4 \text{ weeks}, \qquad \tau_2 = 8 \text{ weeks}. \qquad [6.37]$$

6.3.2. *Exercise 6.4*

The three state–space representations of the process modeled by

$$H(p) = \frac{1}{p^3 + 10p^2 + 27p + 18} \qquad [6.38]$$

are given as follows:

– Companion form, written in the temporal domain:

$$x_1 = y, \qquad [6.39]$$

$$x_2 = \dot{y} = \dot{x}_1, \qquad [6.40]$$

$$x_3 = \ddot{y} = \dot{x}_2, \qquad [6.41]$$

$$\dot{x}_3 = -10\ddot{y} - 27\dot{y} - 18y + u = -10x_3 - 27x_2 - 18x_1 + u. \qquad [6.42]$$

In matrix form, we have,

state equation: $$\begin{bmatrix} \dot{x}_1 \\ \dot{x}_2 \\ \dot{x}_3 \end{bmatrix} = \begin{bmatrix} 0 & 1 & 0 \\ 0 & 0 & 1 \\ -18 & -27 & -10 \end{bmatrix} \begin{bmatrix} x_1 \\ x_2 \\ x_3 \end{bmatrix} + \begin{bmatrix} 0 \\ 0 \\ 1 \end{bmatrix} u, \qquad [6.43]$$

output equation: $$y = \begin{bmatrix} 1 & 0 & 0 \end{bmatrix} \begin{bmatrix} x_1 \\ x_2 \\ x_3 \end{bmatrix}. \qquad [6.44]$$

– Modal form (Jordan form) or parallel decomposition, written in the temporal domain:

$$H(p) = \frac{Y(p)}{U(p)} = \frac{1}{p^3 + 10p^2 + 27p + 18} = \frac{1}{(p+1)(p+3)(p+6)}. \qquad [6.45]$$

The transfer function $H(p)$ can be rewritten in simple elements:

$$H(p) = \frac{Y(p)}{U(p)} = \frac{1}{10}\frac{1}{p+1} - \frac{1}{6}\frac{1}{p+3} + \frac{1}{15}\frac{1}{p+6}. \qquad [6.46]$$

$$\begin{cases} X_1(p) = \frac{1}{p+1}U(p) & \longrightarrow & \dot{x}_1 = -x_1 + u \\ X_2(p) = \frac{1}{p+3}U(p) & \longrightarrow & \dot{x}_2 = -3x_2 + u \\ X_3(p) = \frac{1}{p+6}U(p) & \longrightarrow & \dot{x}_3 = -6x_3 + u. \end{cases} \quad [6.47]$$

$$Y(p) = \frac{1}{10}X_1(p) - \frac{1}{6}X_2(p) + \frac{1}{15}X_3(p) \longrightarrow y = \frac{1}{10}x_1 - \frac{1}{6}x_2 + \frac{1}{15}x_3. \quad [6.48]$$

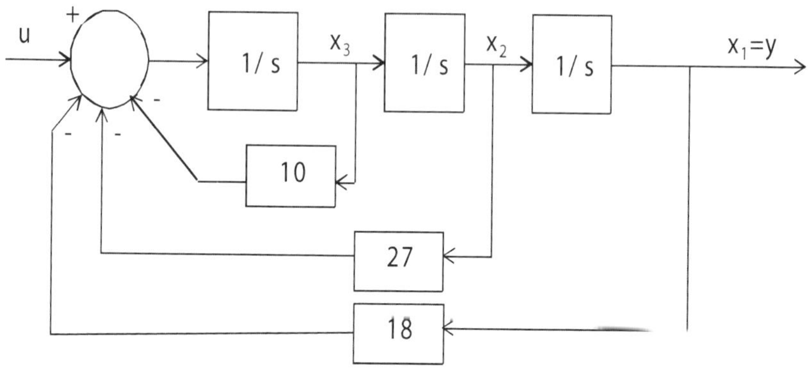

Figure 6.6. *Block diagram of the system companion form*

In matrix form,

$$\text{state equation:} \quad \begin{bmatrix} \dot{x}_1 \\ \dot{x}_2 \\ \dot{x}_3 \end{bmatrix} = \begin{bmatrix} -1 & 0 & 0 \\ 0 & -3 & 0 \\ 0 & 0 & -6 \end{bmatrix} \begin{bmatrix} x_1 \\ x_2 \\ x_3 \end{bmatrix} + \begin{bmatrix} 1 \\ 1 \\ 1 \end{bmatrix} u, \quad [6.49]$$

$$\text{output equation:} \quad y = \begin{bmatrix} \frac{1}{10} & -\frac{1}{6} & \frac{1}{15} \end{bmatrix} \begin{bmatrix} x_1 \\ x_2 \\ x_3 \end{bmatrix}. \quad [6.50]$$

– Cascade form or series decomposition, written in the time domain,

$$H(p) = \frac{1}{p^3 + 10p^2 + 27p + 18} = \frac{1}{(p+1)(p+3)(p+6)}. \quad [6.51]$$

$$\begin{cases} X_1(p) = \frac{1}{p+1}U(p) & \longrightarrow \quad \dot{x}_1 = -x_1 + u \\ X_2(p) = \frac{1}{p+3}X_1(p) & \longrightarrow \quad \dot{x}_2 = -3x_2 + x_1 \\ X_3(p) = \frac{1}{p+6}X_2(p) & \longrightarrow \quad \dot{x}_3 = -6x_3 + x_2 \end{cases} \qquad [6.52]$$

$$Y(p) = X_3(p) \quad \longrightarrow \quad y = x_3. \qquad [6.53]$$

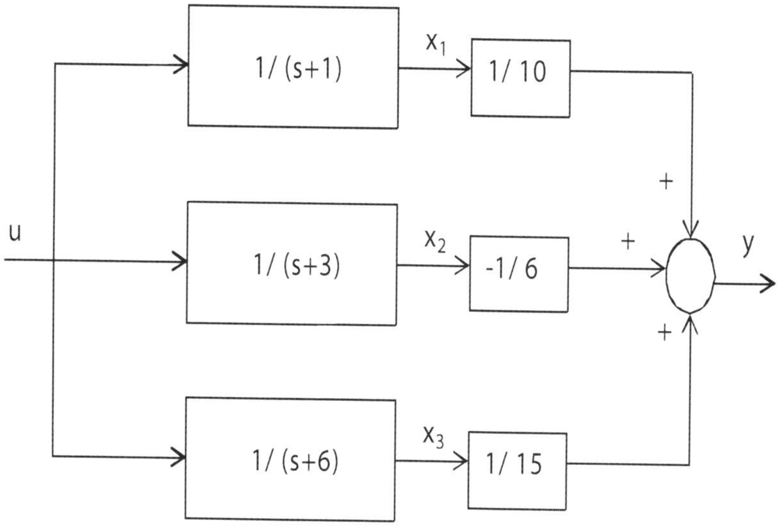

Figure 6.7. *Block diagram of the system, Jordan form*

In matrix form,

$$\text{state equation:} \quad \begin{bmatrix} \dot{x}_1 \\ \dot{x}_2 \\ \dot{x}_3 \end{bmatrix} = \begin{bmatrix} -1 & 0 & 0 \\ 1 & -3 & 0 \\ 0 & 1 & -6 \end{bmatrix} \begin{bmatrix} x_1 \\ x_2 \\ x_3 \end{bmatrix} + \begin{bmatrix} 1 \\ 0 \\ 0 \end{bmatrix} u; \qquad [6.54]$$

$$\text{output equation,} \quad y = \begin{bmatrix} 0 & 0 & 1 \end{bmatrix} \begin{bmatrix} x_1 \\ x_2 \\ x_3 \end{bmatrix}. \qquad [6.55]$$

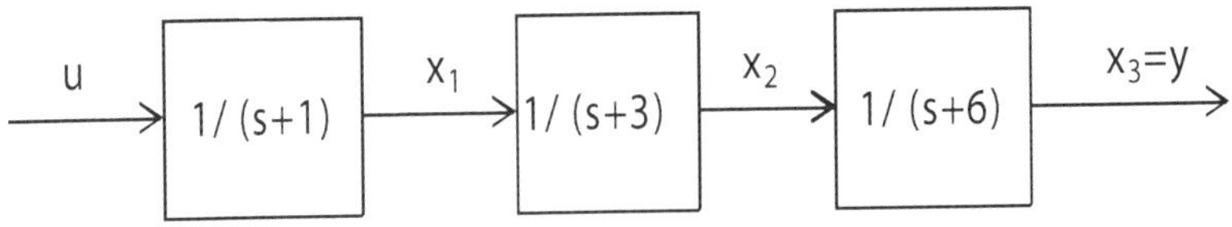

Figure 6.8. *Block diagram of the system cascade form*

6.3.3. *Exercise 6.5*

The state vector is given by

$$x(t) = e^{At}x(0^+) + \int_0^t e^{A(t-\tau)}Bu(\tau)d\tau, \qquad [6.56]$$

where

$$x(t) = L^{-1}\{[pI - A]^{-1}\}x(0^+) + L^{-1}\{[pI - A]^{-1}Bu(p)\}\ (L^{-1} = (T.L)^{-1}). \qquad [6.57]$$

There are several methods for calculating e^{At}:

– Laplace transformation:

$$e^{At} = L^{-1}\{[pI - A]^{-1}\}. \qquad [6.58]$$

– Diagonalization of matrix A (if$\lambda_i \neq \lambda_j$):

$$e^{At} = Pe^{\Lambda t}P^{-1} \qquad P \text{ is the eigenvectors matrix.} \qquad [6.59]$$

– Cayley–Hamilton's theorem:

Each matrix satisfies its own characteristic equation. If matrix A is $(n \times n)$ and admits as characteristic equation:

$$\Delta(\lambda) = \lambda^n + a_{n-1}\lambda^{n-1} + a_{n-2}\lambda^{n-2} + \ldots + a_1\lambda + a_0 = 0 \qquad [6.60]$$

then, we can write:

$$\Delta(A) = A^n + a_{n-1}A^{n-1} + a_{n-2}A^{n-2} + \ldots + a_1A + a_0I = 0. \qquad [6.61]$$

The last equation shows that any matrix A^N, $N \geq n$, can be written as a linear combination of $I, A, A^2, \ldots, A^{n-1}$. The same happens for any power series of A. We can thus write:

$$f(A) = \sum_{k=0}^{\infty} b_k A^k = \sum_{k=0}^{n-1} \alpha_k A^k. \qquad [6.62]$$

Coefficients α_i ($i = 0, \ldots, n-1$) can be computed from the equations obtained by making use of the eigenvalues of matrix A, because it can be written that:

$$f(\lambda_i) = \sum_{k=0}^{n-1} \alpha_k \lambda_i^k \qquad (\lambda_i \text{ are the eigenvalues of } A). \qquad [6.63]$$

If $\lambda_i \neq \lambda_j$, we then have n equations with n unknowns and all the coefficients α_k can be determined. In the case of multiple eigenvalues, the method must be amended.

Let λ_i be an eigenvalue of multiplicity m.

$$\Delta(\lambda) = (\lambda - \lambda_i)^m \Delta_1(\lambda). \qquad [6.64]$$

Differentiating this equation yields:

$$\begin{aligned} \Delta'(\lambda) &= m(\lambda - \lambda_i)^{m-1}\Delta_1(\lambda) - (\lambda - \lambda_i)^m \Delta_1'(\lambda) \\ &= (\lambda - \lambda_i)^{m-1}\Delta_2(\lambda). \end{aligned} \qquad [6.65]$$

Therefore, we see that:

$$\Delta(\lambda_i) = \Delta'(\lambda_i) = \Delta''(\lambda_i) = \cdots = \Delta^{m-1}(\lambda_i) = 0, \qquad [6.66]$$

hence the function $f(\lambda_i) = \sum_{k=0}^{n-1} \alpha_k \lambda_i^k$ can be differentiated to give $(m-1)$ independent equations, which can be used to determine coefficients α_k.

We apply the Cayley–Hamilton theorem to calculate e^{At}. Matrix A has two distinct eigenvalues ($\lambda_0 = -1$, $\lambda_1 = -2$); we can then write:

$$e^{At} = C_0(t)I + C_1(t)A. \qquad [6.67]$$

This yields two equations with two unknowns:

$$\begin{cases} e^{-t} = C_0(t) - C_1(t) \\ e^{-2t} = C_0(t) - 2C_1(t) \end{cases} \quad \Longrightarrow \quad \begin{cases} C_0(t) = 2e^{-t} - e^{-2t} \\ C_1(t) = e^{-t} - e^{-2t} \end{cases} \qquad [6.68]$$

from which:

$$e^{At} = \begin{bmatrix} 2e^{-t} - e^{-2t} & e^{-t} - e^{-2t} \\ -2e^{-t} + 2e^{-2t} & -e^{-t} + 2e^{-2t} \end{bmatrix} \quad [6.69]$$

and

$$\int_0^t e^{A(t-\tau)} Bu(\tau) d\tau = \begin{bmatrix} \frac{1}{2} - e^{-t} + \frac{1}{2} e^{-2t} \\ e^{-t} - e^{-2t} \end{bmatrix}. \quad [6.70]$$

6.3.4. *Exercise 6.6*

6.3.4.1. *Modeling*

Based on the step response shown in Figure 6.9, we are able to propose a second-order model structure:

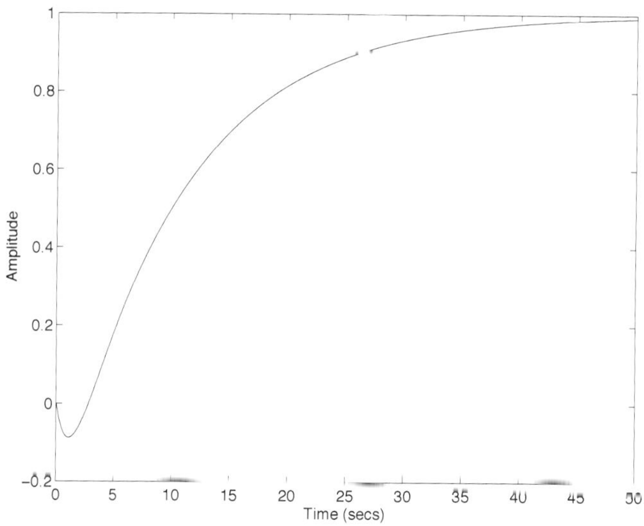

Figure 6.9. *Step response*

1) The derivative at the origin is finite (a degree of difference between the numerator and the denominator) and there is an inflection point, we thus propose the

second order. The zero has a real positive part because the derivative at the origin is negative, hence the following transfer function:

$$H(p) = \frac{K(1 - T_1 p)}{(1 + T_2 p)(1 + T_3 p)}. \qquad [6.71]$$

2) The impulse response (see Figure 6.10) is given by the derivative of the step response:

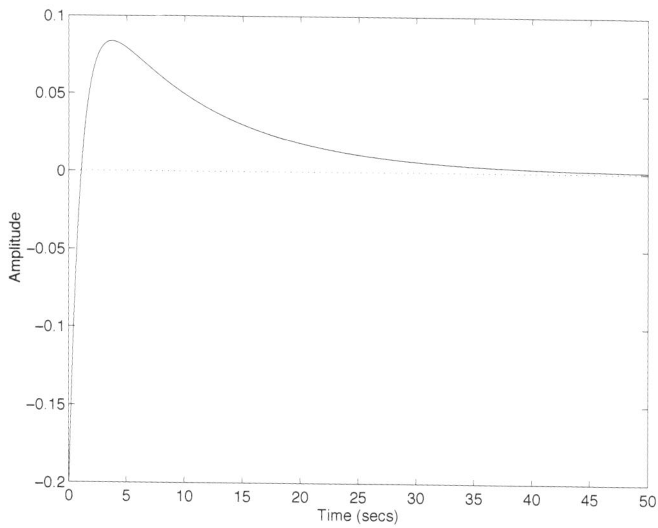

Figure 6.10. *Impulse response*

3) State–space representation of the system using the modes method (Jordan form):

$$\begin{aligned} H(p) &= \frac{Y(p)}{U(p)} = \frac{K(1 - T_1 p)}{(1 + T_2 p)(1 + T_3 p)} \Rightarrow Y(p) \\ &= \frac{K}{T_2 - T3}[\frac{T_2 + T_1}{1 + T_2 p} - \frac{T_1 + T_3}{1 + T_3 p}]U(p). \end{aligned} \qquad [6.72]$$

Taking as state variables

$$X_1(p) = \frac{U(p)}{1 + T_2 p} \quad \text{and} \quad X_2(p) = \frac{U(p)}{1 + T_3 p}, \qquad [6.73]$$

it follows that:

$$\begin{cases} \dot{x}_1 = -\frac{1}{T_2}x_1 + \frac{1}{T_2}u \\ \dot{x}_2 = -\frac{1}{T_3}x_2 + \frac{1}{T_3}u \\ y = \frac{K}{T_2 - T3}[(T_1 + T_2)x_1 - (T_1 + T_3)x_2] \end{cases}$$

$$\Rightarrow \begin{cases} A = \begin{bmatrix} -\frac{1}{T_2} & 0 \\ 0 & -\frac{1}{T_3} \end{bmatrix}; B = \begin{bmatrix} \frac{1}{T_2} \\ \frac{1}{T_3} \end{bmatrix} \\ C = K \begin{bmatrix} \frac{T_1+T_2}{T_2-T_3} & -\frac{T_1+T_3}{T_2-T_3} \end{bmatrix}. \end{cases} \quad [6.74]$$

Companion form is given by

$$H(p) = \frac{Y(p)}{U(p)} = \frac{K(1 - T_1 p)}{(1 + T_2 p)(1 + T_3 p)}. \quad [6.75]$$

This allows us to obtain the following equation

$$Y(p)(1 + (T_2 + T_3)p + T_2 T_3 p^2) = K(1 - T_1 p)U(p). \quad [6.76]$$

By reordering the equation, it yields that

$$Y(p) + (T_2 + T_3)pY(p) + T_2 T_3 p(\frac{KT_1}{T_2 T_3}U(p) + pY(p)) = KU(p). \quad [6.77]$$

We choose as state vector

$$X_1(p) = Y(p); \qquad X_2(p) = \frac{KT_1}{T_2 T_3}U(p) + pY(p). \quad [6.78]$$

By transforming equations [6.77] and [6.78] into the time domain, we can write:

$$x_1(t) + (T_2 + T_3)\dot{x}_1 + (T_2 T_3)\dot{x}_2(t) = Ku(t); \qquad \frac{KT_1}{T_2 T_3}u(t) + \dot{x}_1 = x_2, \quad [6.79]$$

hence the following state equation:

$$\begin{cases} \dot{x}_1 = x_2 - \frac{KT_1}{T_2T_3}u(t) \\ \dot{x}_2 = -\frac{1}{T_2T_3}x_1(t) - \frac{T_2+T_3}{T_2T_3}x_2 + \frac{K}{T_2T_3}[1 + T_1\frac{T_2+T_3}{T_2T_3}]u(t) \\ y = x_1. \end{cases} \quad [6.80]$$

The state–space representation matrices are given by:

$$\begin{cases} A = \begin{bmatrix} 0 & 1 \\ -\frac{1}{T_2T_3} & -\frac{T_2+T_3}{T_2T_3} \end{bmatrix}; \quad B = \begin{bmatrix} -\frac{KT_1}{T_2T_3} \\ \frac{K}{T_2T_3}[1 + T_1\frac{T_2+T_3}{T_2T_3}] \end{bmatrix} \\ C = \begin{bmatrix} 1 & 0 \end{bmatrix} \end{cases} \quad [6.81]$$

4) The expression of the step response $y(t)$:

$$\begin{aligned} Y(p) &= H(p)U(p)\mid_{U(p)=\frac{1}{p}} = \frac{K(1-T_1p)}{(1+T_2p)(1+T_3p)} \cdot \frac{1}{p} \\ &= \frac{K}{T_3 - T2}[\frac{T_3 - T2}{p} + \frac{T_1 + T_2}{p + \frac{1}{T_2}} - \frac{T_1 + T_3}{p + \frac{1}{T_3}}]. \end{aligned} \quad [6.82]$$

By calculating the inverse Laplace transform, we get:

$$\begin{aligned} y(t) &= K[1 + \frac{T_1 + T_2}{T_3 - T2}e^{-\frac{t}{T_2}} - \frac{T_1 + T_3}{T_3 - T2}e^{-\frac{t}{T_3}}], \\ y'(0) &= \lim_{p\to\infty} p^2 Y(p) = -\frac{KT_1}{T_3T2}. \end{aligned} \quad [6.83]$$

6.3.4.2. *Identification*

1) When $t \gg \max(T_1, T_2)$, the expression of $y(t)$ becomes

$$y(t) \simeq K[1 - \frac{T_1 + T_3}{T_3 - T2}e^{-\frac{t}{T_3}}], \quad [6.84]$$

from where the expression of $\alpha(t)$ is given by:

$$\alpha(t) = \text{Log}\frac{K - y(t)}{K} = \text{Log}\frac{T_1 + T_3}{T_3 - T2} - \frac{t}{T_3}. \quad [6.85]$$

We choose different values of α at different times:

$$\left.\begin{array}{l}\left.\begin{array}{l} t = 30\,\mathrm{s}\ \ \alpha = -2.70 \\ t = 40\,\mathrm{s}\ \ \alpha = -3.7 \end{array}\right\} \Rightarrow T_3 = 10\,\mathrm{s} \\ \qquad\qquad\qquad\qquad t = 40\,\mathrm{s}\ \ \alpha = -3.70 \end{array}\right\}$$

$$\Rightarrow 1.34 = \frac{T_1 + 10}{10 - T_2}. \qquad [6.86]$$

The derivative at the origin is given by:

$$y'(0) = \lim_{p \to \infty} p^2 Y(p) = -\frac{KT_1}{T_2 T_3} = 0.2 \quad \Rightarrow \quad T_1 = 2T_2. \qquad [6.87]$$

Therefore, the values of T_1 and T_2 can be obtained:

$$T_2 \simeq 1.05\,\mathrm{s} \quad \text{and} \quad T_1 \simeq 2.1\,\mathrm{s}. \qquad [6.88]$$

2) The system is associated according to the plot shown in Figure 6.11, $C(p) = k$.

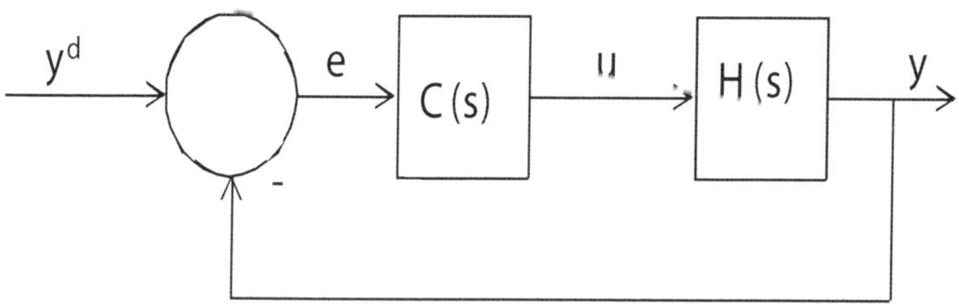

Figure 6.11. *Closed-loop system*

3) The transfer function of the closed-loop (CL) system is given by

$$F(p) = \frac{kK(1 - T_1 p)}{(1 + T_2 p)(1 + T_3 p) + kK(1 - T_1 p)} \qquad [6.89]$$

$$= \frac{kK(1 - T_1 p)}{T_2 T_3 p^2 + (T_2 + T_3 - kKT_1)p + 1 + kK}.$$

If k is large $\Rightarrow$ Complex poles $\Rightarrow$ oscillating system.

The transfer function can be written as:

$$F(p) = K_1 \cdot \frac{1 - T_1 p}{\frac{1}{\omega_0^2} p^2 + \frac{2\xi}{\omega_0} p + 1}, \qquad [6.90]$$

with

$$\omega_0^2 = \frac{1 + kK}{T_3 T2}; \quad \xi = \frac{T_2 + T_3 - kKT_1}{1 + kK} \quad \text{and} \quad K_1 = \frac{kK}{1 + kK}. \qquad [6.91]$$

The shape of the step response of the control when k is large enough ($k = 4$) is shown in Figure 6.12.

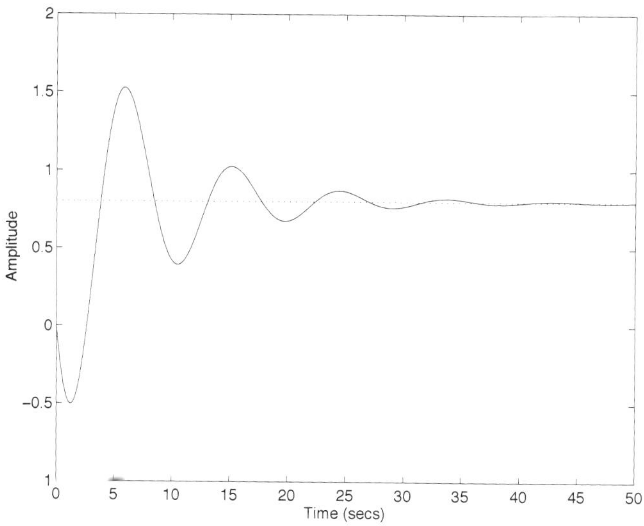

Figure 6.12. *Step system response in CL*

4) Based on the experimental measurements of this response, we can obtain the characteristic parameters of the system (K_1, T_1, ω_0 and ξ) from:

- the pseudo-period $T_p \Rightarrow \omega_p = \omega_0 \sqrt{1 - \xi^2}$;

- ξ is calculated using the logarithmic decrement;

- the final value is given by

$$y(\infty) = K_1; \qquad [6.92]$$

- the derivative at the origin:

$$y'(0) = -\frac{kKT_1}{T_3T2}. \qquad [6.93]$$

A system of four equations with four unknowns.

6.3.5. *Exercise 6.7*

The state–space representation of the process modeled by the recurrence equation

$$y((k+2)T) + 0.2y((k+1)T) - 0.63y(kT) = u(kT) \qquad [6.94]$$

is given as follows:

$$y_k = y(kT) = x_{1k} = x_1(kT) \qquad [6.95]$$

$$y_{k+1} = x_{1(k+1)} = x_{2k}$$

$$y_{k+2} = x_{2(k+1)} = -0.2x_{2k} + 0.63x_{1k} + u_k.$$

– In matrix form, we have,

state equation: $$\begin{bmatrix} x_{1(k+1)} \\ x_{2(k+1)} \end{bmatrix} = \begin{bmatrix} 0 & 1 \\ 0.63 & -0.2 \end{bmatrix} \begin{bmatrix} x_{1k} \\ x_{2k} \end{bmatrix} + \begin{bmatrix} 0 \\ 1 \end{bmatrix} u_k, \qquad [6.96]$$

output equation: $$y = \begin{bmatrix} 1 & 0 \end{bmatrix} \begin{bmatrix} x_{1k} \\ x_{2k} \end{bmatrix} \qquad [6.97]$$

- The stability of the system is ensured if the eigenvalues of the evolution matrix A are inside the unit circle.

$$det(\lambda I - A) = 0 \Longrightarrow \begin{cases} \lambda_1 = 0.7 \\ \lambda_2 = -0.9 \end{cases} \qquad [6.98]$$

The system is stable because:

$$|\lambda_1| < 1 \quad \text{and} \quad |\lambda_2| < 1. \qquad [6.99]$$

– The unit step response:

From the recurrence equation, we calculate the Z-transform,

$$(z^2 + 0.2z - 0.63)Y(z) = U(z). \qquad [6.100]$$

The transfer function $H(z)$ is given by:

$$H(z) = \frac{Y(z)}{U(z)} = \frac{1}{z^2 + 0.2z - 0.63} = \frac{1}{(z+0.9)(z-0.7)}. \qquad [6.101]$$

The step response:

$$Y(z) = H(z)U(z) \quad \text{with} \quad U(z) = \frac{z}{z-1} \qquad [6.102]$$

from which

$$Y(z) = \frac{z}{(z+0.9)(z-0.7)(z-1)}. \qquad [6.103]$$

We carry out the simple elements decomposition of the following expression:

$$\begin{aligned}\frac{Y(z)}{z} &= \frac{1}{(z+0.9)(z-0.7)(z-1)} \\ &= \frac{25}{4}\frac{1}{z+0.9} + \frac{50}{3}\frac{1}{z-0.7} + \frac{100}{57}\frac{1}{z-1},\end{aligned} \qquad [6.104]$$

wherefrom

$$Y(z) = \frac{25}{4}\frac{z}{z+0.9} + \frac{50}{3}\frac{z}{z-0.7} + \frac{100}{57}\frac{z}{z-1}. \qquad [6.105]$$

The response is given by:

$$y(nT) = (\frac{25}{4}(-0.9)^n + \frac{50}{3}(0.7)^n + \frac{100}{57})u(nT). \qquad [6.106]$$

6.4. Exercises: observation and control of nonlinear systems

6.4.1. *Exercise 6.8*

The unit step response is given by:

$$Y(p) = H(p) \cdot \frac{1}{p} = \frac{1}{p^2(p+1)} = -\frac{1}{p} + \frac{1}{p^2} + \frac{1}{p+1}, \qquad [6.107]$$

from which, by calculating the inverse Laplace transformation, it follows that:

$$y(t) = (t - 1 + e^{-t})\Gamma(t) \quad (\Gamma(t) \text{ is the Heaviside function}). \qquad [6.108]$$

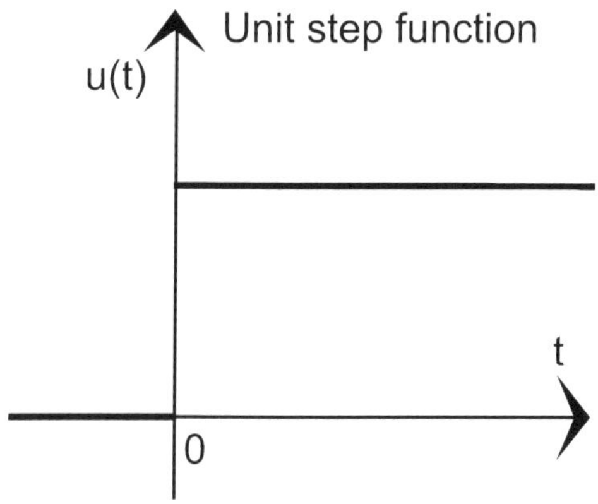

Figure 6.13. *Step response of* $H(s) = \frac{1}{s(s+1)}$

The system is controlled according to the function $F(p)$.

Let $C(s) = K$, the transfer function of the CL system is given as follows:

$$F(p) = \frac{KH(p)}{1 + KH(p)} = \frac{K}{p^2 + p + K} = \frac{\omega_0^2}{p^2 + 2\xi\omega_0 p + \omega_0^2} \qquad [6.109]$$

with the following characteristic parameters:

$$\omega_0 = \sqrt{K} \quad \text{and} \quad \xi = \frac{1}{2\sqrt{K}}. \qquad [6.110]$$

It can be observed that if $K \nearrow \Longrightarrow \omega_0 \nearrow$ and $\xi \searrow$.

The permanent position ε_p and velocity ε_v errors are given as follows:

$$\begin{aligned}\varepsilon_p &= \lim_{p\to 0} p\varepsilon(p)\mid_{u(p)=1/p} = \lim_{p\to 0} p\cdot\frac{1}{1+C(p)H(p)}\cdot\frac{1}{p}\\ \Rightarrow\quad \varepsilon_p &= \frac{1}{1+C(0)H(0)} = 0.\end{aligned} \quad [6.111]$$

The system presents an integration; therefore, the position error is $\varepsilon_p = 0$.

$$\begin{aligned}\varepsilon_v &= \lim_{p\to 0} p\varepsilon(p)\mid_{u(p)=1/p^2} = \lim_{p\to 0} p\cdot\frac{1}{1+C(p)H(p)}\cdot\frac{1}{p^2}\\ \Rightarrow\quad \varepsilon_v &= \lim_{p\to 0}\frac{1}{pC(p)H(p)} = 1/K.\end{aligned} \quad [6.112]$$

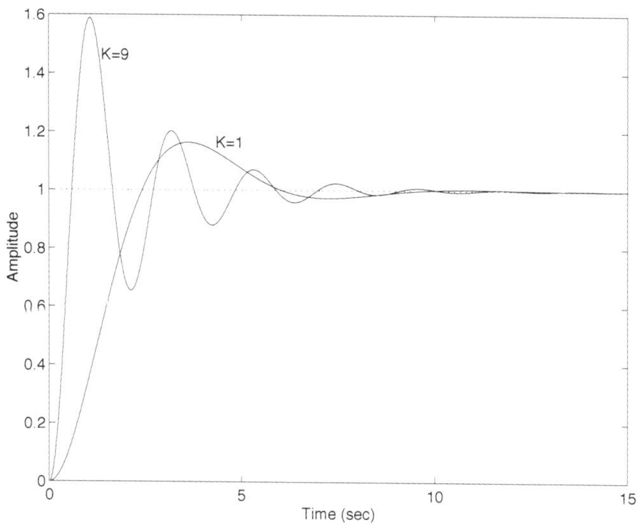

Figure 6.14. $K = 1 \Rightarrow (\xi = 0.5;\ \omega_0 = 1;\ \varepsilon_p = 0;\ \varepsilon_v = 1)$ *and* $K = 9 \Rightarrow (\xi = 1/6;\ \omega_0 = 3;\ \varepsilon_p = 0;\ \varepsilon_v = 1/9)$

The system shows a single integration; therefore, the velocity error is $\varepsilon_v = cte.$

The step response of the system with this loop is given by ($\xi < 1$):

$$y(t) = (1 - \frac{e^{-\xi\omega_0 t}}{\sqrt{1-\xi^2}} \cdot sin(\omega_0\sqrt{1-\xi^2}t + \phi))\Gamma(t)$$

$$\text{with } \begin{cases} \sin(\phi) = \sqrt{1-\xi^2} \\ \cos(\phi) = \xi \end{cases} \qquad [6.113]$$

In conclusion, $K \nearrow \Longrightarrow \varepsilon_v \searrow$ but $y(t)$ badly damped (compromise between accuracy and stability).

6.4.2. *Exercise 6.9*

The open-loop (OL) transfer functions $H_i(p)$ of the processes below are all functions of the parameter K. As a result, the general aspect of Bode and Nyquist plots are given for $H_i(p)/K$, and the stability is then studied according to this parameter K.

Knowing that $P = 1$, then:

i) and iii) $\overset{\frown}{N} = 0 \Longrightarrow$ Unstable system in CL, because $Z = 1$.

ii) $\overset{\frown}{N} = -1 \Longrightarrow$ Stable system in CL, since $Z = 0$.

Knowing that $P = 0$, then:

i) $\overset{\frown}{N} = 2 \Longrightarrow$ Unstable system in CL, because $Z = 2$.

ii) $\overset{\frown}{N} = 1 \Longrightarrow$ Unstable system in CL, because $Z = 1$.

Knowing that $P = 1$, then:

i) $\overset{\frown}{N} = 1 \Longrightarrow$ Unstable system in CL, because $Z = 2$.

ii) $\overset{\frown}{N} = 0 \Longrightarrow$ Unstable system in CL, because $Z = 1$.

Knowing that $P = 1$, then:

i) and iii) $\overset{\frown}{N} = 0 \Longrightarrow$ Unstable system in CL, because $Z = 1$.

ii) $\overset{\frown}{N} = -1 \Longrightarrow$ Stable system in CL, since $Z = 0$.

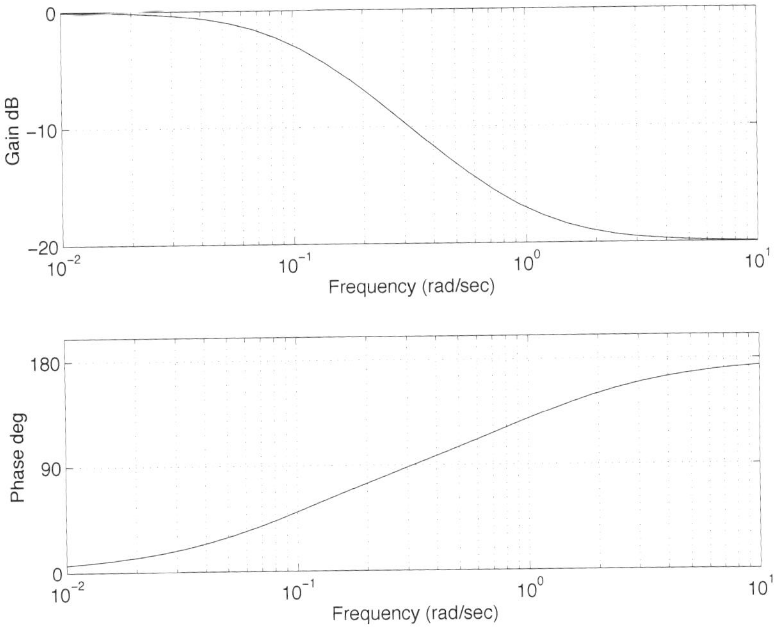

Figure 6.15. *Bode plot of* $\frac{H_1(s)}{K} = \frac{1+\tau s}{1-Ts}$

Knowing that $P = 0$, then:

ii) and v) $\overset{\curvearrowright}{N} = 0 \Longrightarrow$ Stable system in CL, since $Z = 0$.

ii) $\overset{\curvearrowright}{N} = 2 \Longrightarrow$ Unstable system in CL, because $Z = 2$.

iii) $\overset{\curvearrowright}{N} = 3 \Longrightarrow$ Unstable system in CL, because $Z = 3$.

iv) $\overset{\curvearrowright}{N} = 1 \Longrightarrow$ Unstable system in CL, because $Z = 1$.

Using the Routh criterion, verify the stability of the CL system.

The transfer function of the CL system no. 1:

$$F_1(p) = \frac{K(1+\tau p)}{(K\tau - T)p + K + 1}. \qquad [6.114]$$

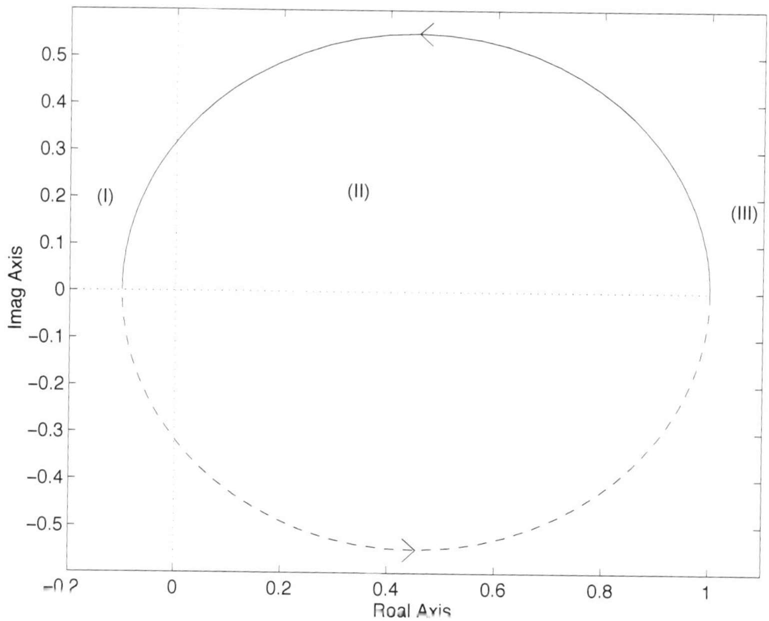

Figure 6.16. *Nyquist plot of* $\frac{H_1(s)}{K} = \frac{1+\tau s}{1-Ts}$

For the CL system no. 1 to be stable, it is necessary that:

$$K\tau - T > 0 \Longrightarrow K > \frac{T}{\tau}. \qquad [6.115]$$

The transfer function of the CL system no. 2:

$$F_2(p) = \frac{K(1+\tau p)}{Tp^3 + p^2 + K\tau p + K}. \qquad [6.116]$$

p^3	T	$K\tau$
p^2	1	K
p^1	$K(\tau - T)$	0
p^0	K	0

Table 6.1. *Routh table system no. 2 results*

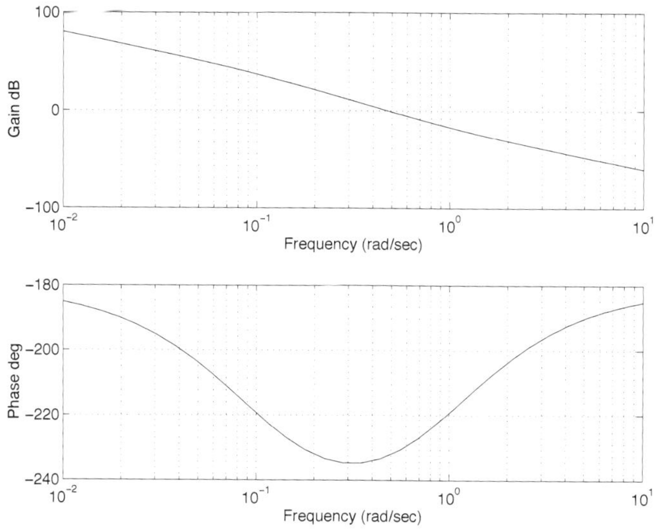

Figure 6.17. *Bode plot of* $\frac{H_2(s)}{K} = \frac{1+\tau s}{s^2(1+Ts)}$

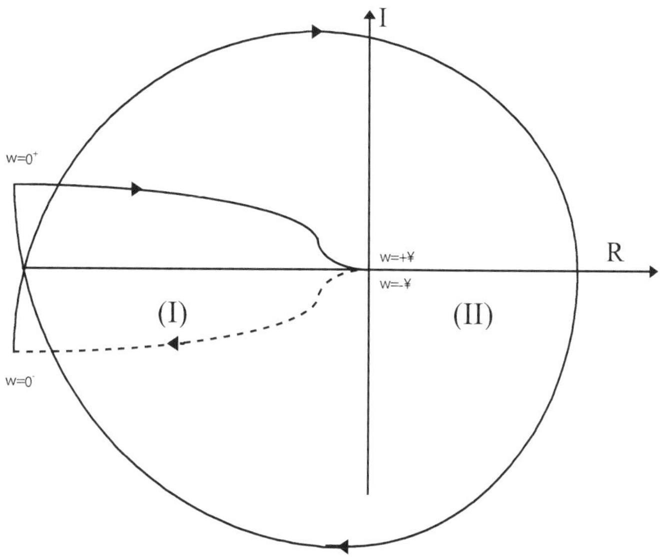

Figure 6.18. $\frac{H_2(s)}{K} = \frac{1+\tau s}{s^2(1+Ts)}$

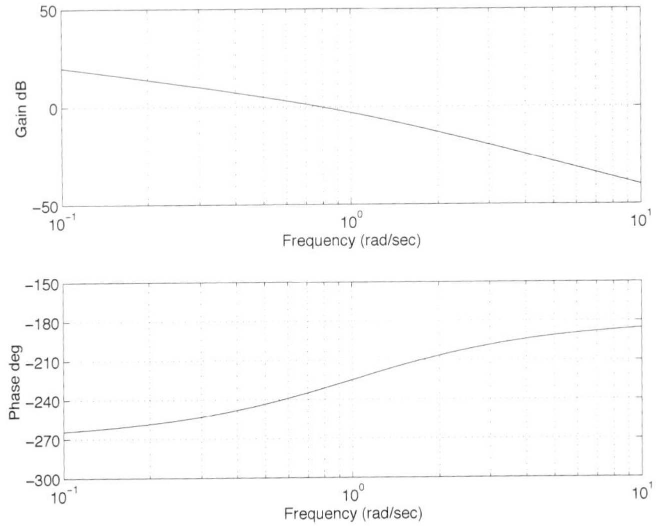

Figure 6.19 *Bode plot of* $\frac{H_3(s)}{K} = \frac{1}{s(s-1)} = \frac{1}{-s(1-s)}$

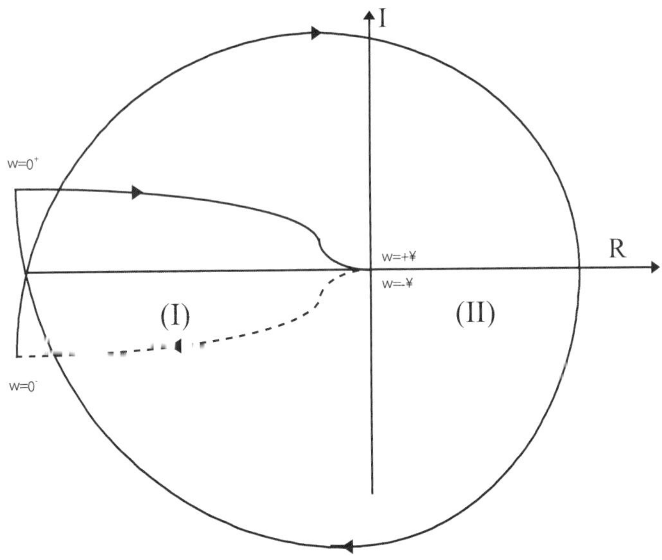

Figure 6.20. *Nyquist plot of* $\frac{H_3(s)}{K} = \frac{1}{s(s-1)} = \frac{1}{-s(1-s)}$

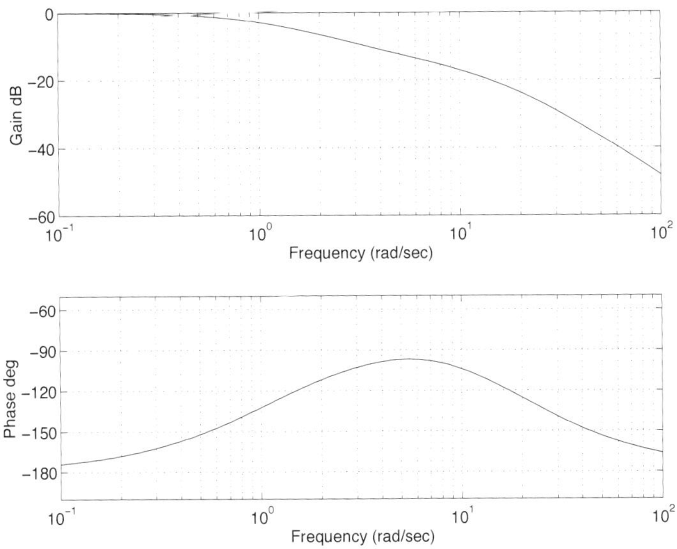

Figure 6.21. *Bode plot of* $\frac{H_4(p)}{K} = \frac{1+0.2s}{-(1+0.05s)(1+0.1s)(1-s)}$

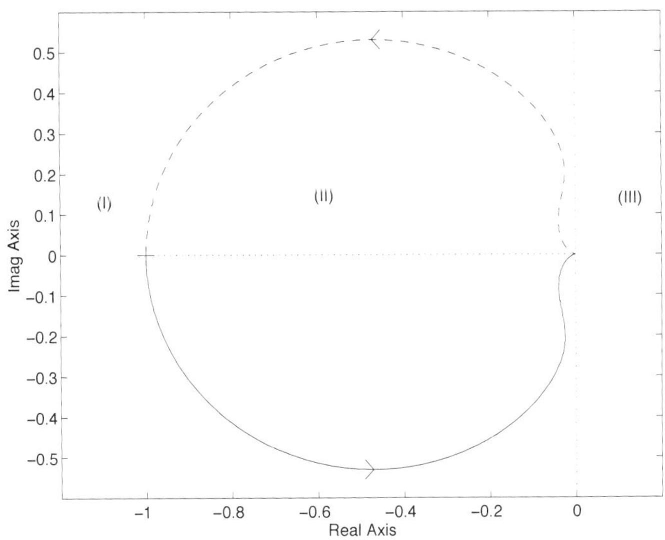

Figure 6.22. *Nyquist plot of* $\frac{H_4(s)}{K} = \frac{1+0.2s}{-(1+0.05s)(1+0.1s)(1-s)}$

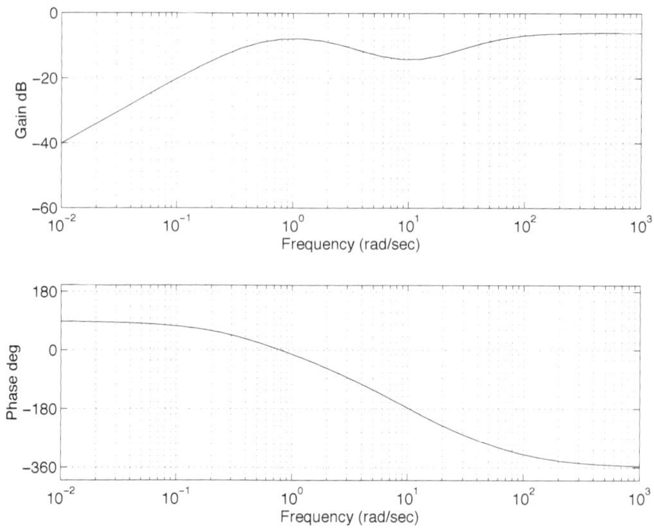

Fiqure 6.23. *Bode plot of* $\frac{H_5(s)}{K} = \frac{s(1-0.1s)^2}{(1+2s)(1+0.5s)(1+0.02s)}$

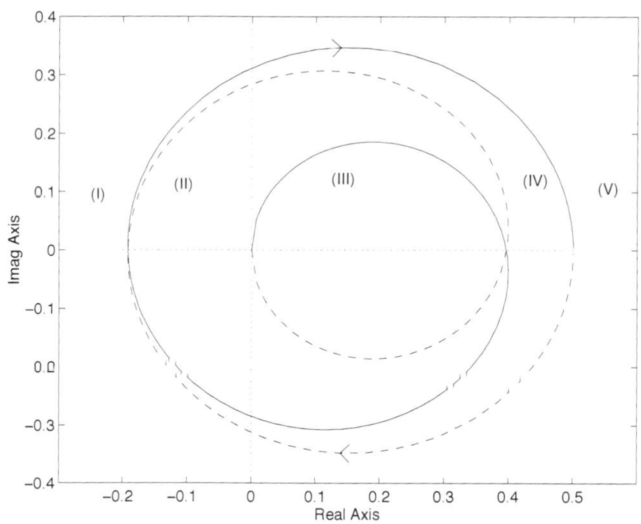

Figure 6.24. *Nyquist plot of* $\frac{H_5(s)}{K} = \frac{s(1-0.1s)^2}{(1+2s)(1+0.5s)(1+0.02s)}$

The CL system no. 2 is still unstable because there are two sign changes due to the fact that

$$(\tau - T) < 0 \text{ and } K > 0. \qquad [6.117]$$

The transfer function of the third CL system:

$$F_3(p) = \frac{K}{p^2 - p + K}. \qquad [6.118]$$

The CL system no. 3 is still unstable because all the coefficients do not have the same sign.

The transfer function of the CL system no. 4:

$$F_4(p) = \frac{K(1 + 0.2p)}{0.005p^3 + 0.145p^2 + (0.85 + 0.2K)p + K - 1}. \qquad [6.119]$$

p^3	0.005	$0.85 + 0.2K$
p^2	0.145	$K - 1$
p^1	$\frac{.1282-.024K}{0.145}$	0
p^0	$K - 1$	0

Table 6.2. *Routh table system no. 4 results*

The CL system no. 4 is stable if (first column of the same sign):

$$K \in]1, \quad 5.34[. \qquad [6.120]$$

The transfer function of the CL system no. 5:

$$F_5(p) = \frac{Kp(1 - 0.1p)^2}{(0.02 + 0.01K)p^3 + (1.05 - 0.2K)p^2 + (2.52 + K)p + 1}. \qquad [6.121]$$

p^3	$0.02 + 0.01K$	$2.52 + K$
p^2	$1.05 - 0.2K$	1
p^1	$\frac{2.626+.536K-.2K^2}{1.05-0.2K}$	0
p^0	1	0

Table 6.3. *Routh table system no. 5 results*

The CL system no. 5 is stable if

$$K \in]-0.48, \quad 5.25[. \qquad [6.122]$$

6.4.3. *Exercise 6.10*

The transfer function in OL of the system is given by:

$$H(p) = \frac{K(1+5p)}{p^2(1+0.8p+4p^2)}. \qquad [6.123]$$

The characteristic polynomial of the CL system is obtained by writing:

$$1 + H(p) = 0 \Longleftrightarrow p^2(1+0.8p+4p^2) + K(1+5p) = 0. \qquad [6.124]$$

Hence, the characteristic polynomial:

$$4p^4 + 0.8p^3 + p^2 + 5Kp + K = 0. \qquad [6.125]$$

The Routh table is presented in Table 6.4.

p^4	4	1	K
p^3	0.8	$5K$	0
p^2	$1-25K$	K	
p^1	$K\frac{4.2-125K}{1-25K}$		
p^0	K		

Table 6.4. *Routh table OL system results*

The system in CL is stable if:

$$\left.\begin{array}{l} 1-25K > 0 \quad\;\; \Rightarrow K < 0.04 \\ 4.2-125K > 0 \Rightarrow K < 0.0336 \\ \qquad\quad K > 0 \end{array}\right\} \Longrightarrow 0 < K < K_c = 0.0336. \qquad [6.126]$$

The transfer function of the OL system $H(p)$ is of the form:

$$H(p) = \frac{K(1+5p)}{p^2(1+2\frac{\xi}{\omega_0}p+\frac{1}{\omega_0^2}p^2)} \quad \text{with} \begin{cases} \omega_0^2 = \frac{1}{4} \Rightarrow \omega_0 = 0.5 \\ 2\frac{\xi}{\omega_0} = 0.8 \Rightarrow \xi = 0.2. \end{cases} \quad [6.127]$$

The Bode plot is given as follows:

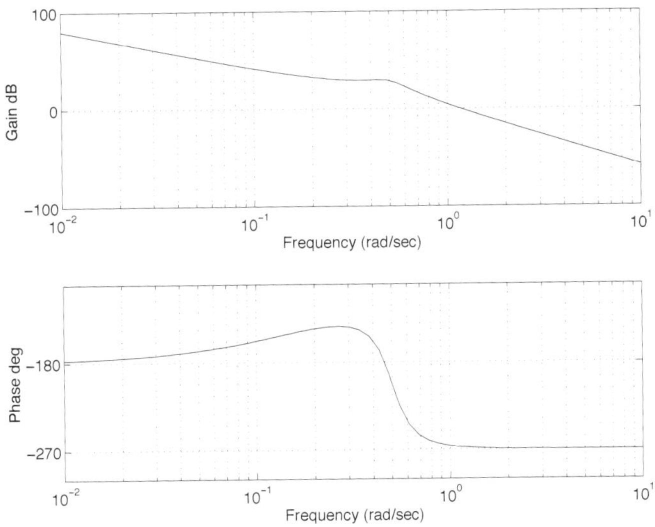

Figure 6.25. *Bode plot of* $\frac{H(p)}{K} = \frac{1+5s}{s^2(1+0.8s+4s^2)}$

Derived from the Bode plot, the Nyquist plot is thus:

We have $P = 0$, therefore for the system to be stable N has to be equal to zero. For this purpose, it is necessary that

$$-\frac{1}{K} < A. \quad [6.128]$$

The critical point A is such that $\phi(\omega = \omega_c) = -180°$. Looking for the ω_c that corresponds to this:

$$\phi(\omega_c) = -180° \Longrightarrow -180° + arctg(5\omega_c) - arctg(\frac{0.8\omega_c}{1-4\omega_c^2}) = -180°. \quad [6.129]$$

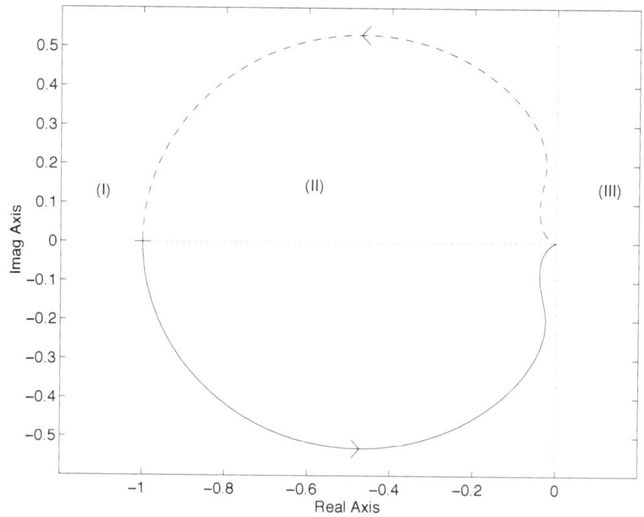

Figure 6.26. *Nyquist plot of* $\frac{H(s)}{K} = \frac{1+5s}{s^2(1+0.8s+4s^2)}$

This yields:

$$arctg(5\omega_c) = arctg(\frac{0.8\omega_c}{1-4\omega_c^2}) \Longrightarrow 20\omega_c^2 = 4.2. \qquad [6.130]$$

Let:

$$\omega_c = 0.458 \text{ rad/s}. \qquad [6.131]$$

The system is stable if:

$$\frac{1}{K} > \frac{1}{\omega_c^2}\sqrt{\frac{1+25\omega_c^2}{(1-4\omega_c^2)^2+0.64\omega_c^2}} = 29.76 \Longrightarrow K < K_c = 0.0336. \qquad [6.132]$$

The result can be verified using the auxiliary equation:

$$(1-25K_c)p^2 + K_c = 0 \Longleftrightarrow 0.16p^2 + 0.0336 = 0 \Rightarrow p = \pm j\omega_c = \pm j0.458. [6.133]$$

6.4.4. *Exercise 6.11*

The transfer function can be expressed in the form:

$$H(p) = \frac{0.64}{p(p^2 + 2\xi\omega_0 p + \omega_0^2)} \qquad \text{with} \begin{cases} \omega_0^2 = 1 \Rightarrow \omega_0 = 1 \\ 2\xi\omega_0 = 1 \Rightarrow \xi = 0.5 \end{cases} \qquad [6.134]$$

The Bode and Black–Nichols plots are given as follows:

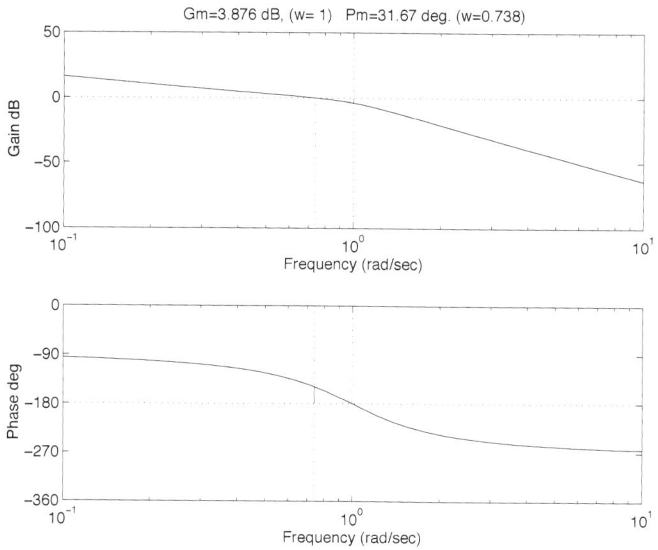

Figure 6.27. *Bode plot of* $H(p) = \frac{0.64}{s(s^2+s+1)}$

The second-order ξ can be approximated by

$$\xi = \frac{\Delta\Phi}{100} \qquad (\Delta\Phi \text{ phase margin in degrees}). \qquad [6.135]$$

This approximation remains valid for a higher order system *dominated* by a pair of complex poles. In the Black–Nichols nomograph, gain and phase margins can be read:

$$|H(j\omega)| = 0 \; db \Longrightarrow \Phi(\omega) = -150° \Longrightarrow \Delta\Phi = 30°. \qquad [6.136]$$

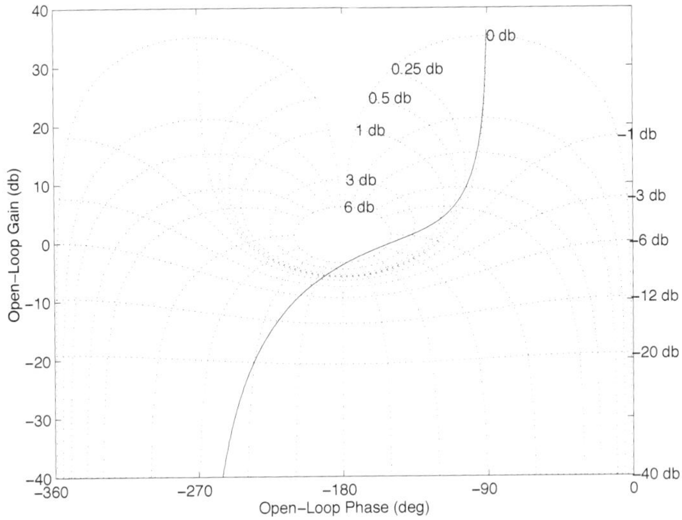

Figure 6.28. *Black–Nichols plot of* $H(s) = \frac{0.64}{s(s^2+s+1)}$

Therefore, the ξ based on the phase margin is given by

$$\xi = \frac{\Delta\Phi}{100} = 0.3. \qquad [6.137]$$

The maximum modulus in the CL system $\left|\frac{H}{1+H}\right|_{\max}$ occurs at $\omega_r = 0.88$ rad/s and is equal to $+9$ db, from which:

$$20\log_{10}(M_r) = 9 \text{ db} \Longrightarrow M_r = 2.8. \qquad [6.138]$$

With the help of the curves of the second-order $M_r = f(\xi)$, we can estimate that ξ is equivalent to

$$\xi = 0.175. \qquad [6.139]$$

The characteristic equation is given by:

$$(p + 0.77)(p^2 + 0.225p + 0.826) = 0 \Leftrightarrow (p + 0.77)(p^2 + 2\xi\omega_0 p + \omega_0^2) = 0. \qquad [6.140]$$

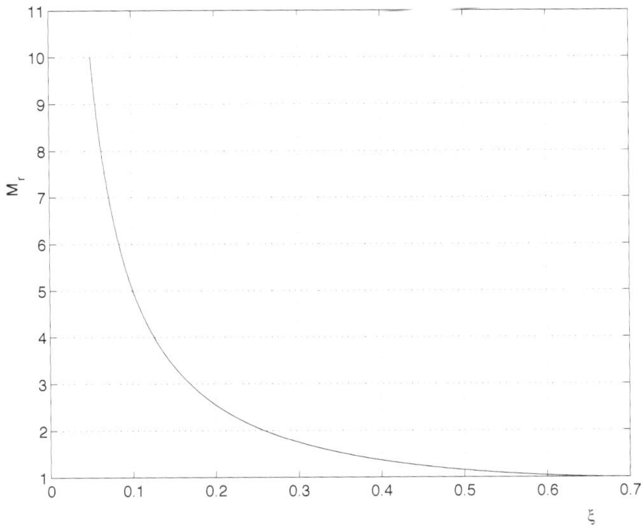

Figure 6.29. *Resonance maximum amplitude M_r according to ξ*

This allows us to obtain

$$\xi = 0.124. \quad [6.141]$$

We can conclude that the complex roots of this CL system do not dominate and the real root adds a little damping. This is a system that does not have dominant poles (the real pole is close to the pair of complex conjugate poles). It is therefore advisable to choose the lowest value of ξ obtained by $\Delta\Phi$ and M_r.

6.4.5. *Exercise 6.12*

6.4.5.1. *Observable canonical form*

This is the so-called Horner's method; we are going to present it with an example. Given the transfer function of Exercise 6.1,

$$H(p) = \frac{Y(p)}{U(p)} = \frac{K(1 - T_1 p)}{(1 + T_2 p)(1 + T_3 p)} = \frac{-\frac{KT_1}{T_2 T_3} p + \frac{K}{T_2 T_3}}{p^2 + \frac{T_2 + T_3}{T_2 T_3} p + \frac{1}{T_2 T_3}}. \quad [6.142]$$

Let n be the higher order of the transfer function. We then divide the denominator of the transfer function by p^n, and get for our example:

$$\Rightarrow H(p) = \frac{Y(p)}{U(p)} = \frac{-\frac{KT_1}{T_2T_3}p^{-1} + \frac{K}{T_2T_3}p^{-2}}{1 + \frac{T_2+T_3}{T_2T_3}p^{-1} + \frac{1}{T_2T_3}p^{-2}}. \qquad [6.143]$$

By expanding:

$$Y(p)\left(1 + \frac{T_2+T_3}{T_2T_3}p^{-1} + \frac{1}{T_2T_3}p^{-2}\right) = U(p)\left(-\frac{KT_1}{T_2T_3}p^{-1} + \frac{K}{T_2T_3}p^{-2}\right) \qquad [6.144]$$

$$Y(p) = U(p)\left(-\frac{KT_1}{T_2T_3}p^{-1} + \frac{K}{T_2T_3}p^{-2}\right) - Y(p)\left(\frac{T_2+T_3}{T_2T_3}p^{-1} + \frac{1}{T_2T_3}p^{-2}\right) \qquad [6.145]$$

$$Y(p) = -\frac{KT_1}{T_2T_3}p^{-1}U(p) + \frac{K}{T_2T_3}p^{-2}U(p)\frac{T_2+T_3}{T_2T_3}p^{-1}Y(p) - \frac{1}{T_2T_3}p^{-2}Y(p). \qquad [6.146]$$

Finally,

$$Y = (p) = p^{-1}\left[-\frac{KT_1}{T_2T_3}U(p) - \frac{T_2+T_3}{T_2T_3}Y(p) + p^{-1}\left[\frac{K}{T_2T_3}U(p) - \frac{1}{T_2T_3}Y(p)\right]\right]. \qquad [6.147]$$

Assuming

$$\begin{cases} X_1(p) = p^{-1}\left[\frac{K}{T_2T_3}U(p) - \frac{1}{T_2T_3}Y(p)\right] & \Rightarrow \dot{x}_1 = \frac{K}{T_2T_3}u - \frac{1}{T_2T_3}y \\ X_2(p) = p^{-1}\left[-\frac{KT_1}{T_2T_3}U(p) - \frac{T_2+T_3}{T_2T_3}Y(p) \right. & \Rightarrow \dot{x}_2 \\ \qquad \left. +X_1(p)\right] & = -\frac{KT_1}{T_2T_3}u - \frac{T_2+T_3}{T_2T_3}y + x_1 \\ Y(p) = X_2(p) & \Rightarrow y = x_2. \end{cases} \qquad [6.148]$$

or in the form of state equations:

$$\begin{bmatrix} \dot{x}_1 \\ \dot{x}_2 \end{bmatrix} = \begin{bmatrix} 0 & -\frac{1}{T_2T_3} \\ 1 & -\frac{T_2+T_3}{T_2T_3} \end{bmatrix} \begin{bmatrix} x_1 \\ x_2 \end{bmatrix} + \begin{bmatrix} \frac{K}{T_2T_3} \\ -\frac{KT_1}{T_2T_3} \end{bmatrix} u$$
$$y = \begin{bmatrix} 0 & 1 \end{bmatrix} \begin{bmatrix} x_1 \\ x_2 \end{bmatrix} \qquad [6.149]$$

$$O_{(A,C)} = \begin{bmatrix} C \\ CA \end{bmatrix} = \begin{pmatrix} 0 & 1 \\ 1 & -\frac{T_2+T_3}{T_2T_3} \end{pmatrix}. \qquad [6.150]$$

This representation is always observable; it is the canonical form of observability.

6.4.5.2. *Controllable canonical form*

$$H(p) = \frac{Y(p)}{U(p)} = \frac{K(1-T_1p)}{(1+T_2p)(1+T_3p)} = \frac{-\frac{KT_1}{T_2T_3}p + \frac{K}{T_2T_3}}{p^2 + \frac{T_2+T_3}{T_2T_3}p + \frac{1}{T_2T_3}}. \qquad [6.151]$$

Let:

$$H(p) = \frac{Y(p)}{U(p)} = \frac{N(p)}{D(p)} = \frac{N(p)}{V(p)}\frac{V(p)}{D(p)}. \qquad [6.152]$$

$$\begin{cases} Y(p) = N(p)V(p) - \left(\frac{KT_1}{T_2T_3}p + \frac{K}{T_2T_3}\right)V(p) & \Rightarrow y = -\frac{KT_1}{T_2T_3}\dot{v} + \frac{K}{T_2T_3}v \\ U(p) = N(p)V(p) = \left(p^2 + \frac{T_2+T_3}{T_2T_3}p + \frac{1}{T_2T_3}\right)V(p) & \Rightarrow u = \ddot{v} + \frac{T_2+T_3}{T_2T_3}\dot{v} + \frac{1}{T_2T_3}v \end{cases}$$
$$[6.153]$$

We take as state vector $x = \begin{bmatrix} v & \dot{v} \end{bmatrix}^T$ from which:

$$\begin{aligned} x_1 &= v, \\ x_2 &= \dot{v} = \dot{x}_1 \\ \dot{x}_2 &= \ddot{v} = -\frac{1}{T_2T_3}v - \frac{T_2+T_3}{T_2T_3}\dot{v} + u, \\ &= \ddot{v} = -\frac{1}{T_2T_3}x_1 - \frac{T_2+T_3}{T_2T_3}x_2 + u, \\ y &= \frac{K}{T_2T_3}x_1 - \frac{KT_1}{T_2T_3}x_2. \end{aligned} \qquad [6.154]$$

The state equations in matrix form:

$$\begin{bmatrix} \dot{x}_1 \\ \dot{x}_2 \end{bmatrix} = \begin{bmatrix} 0 & 1 \\ -\frac{1}{T_2T_3} & -\frac{T_2+T_3}{T_2T_3} \end{bmatrix} \begin{bmatrix} x_1 \\ x_2 \end{bmatrix} + \begin{bmatrix} 0 \\ 1 \end{bmatrix} u$$
$$y = \begin{bmatrix} \frac{K}{T_2T_3} & -\frac{KT_1}{T_2T_3} \end{bmatrix} \begin{bmatrix} x_1 \\ x_2 \end{bmatrix} \qquad [6.155]$$

$$C_{(A,B)} = \begin{bmatrix} B & AB \end{bmatrix} = \begin{pmatrix} 0 & 1 \\ 1 & -\frac{T_2+T_3}{T_2T_3} \end{pmatrix}. \qquad [6.156]$$

This representation is always controllable. This is the controllability canonical form.

6.4.6. *Exercise 6.13*

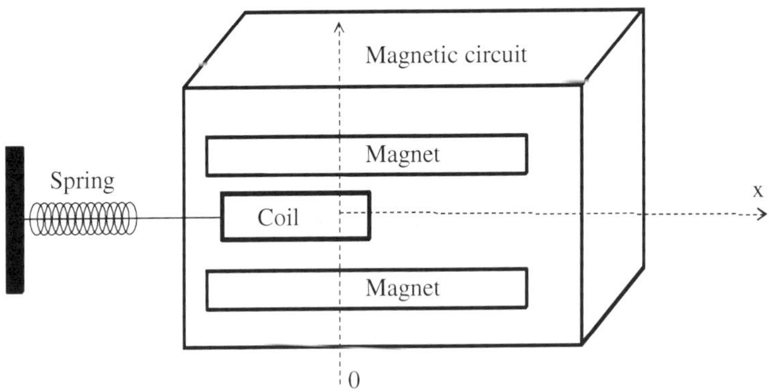

Figure 6.30. *Modeling of displacements*

– Electrical equation:

$$u(t) = Ri(t) + L\frac{di(t)}{dt} + e. \qquad [6.157]$$

– Mechanical equation:

$$m\frac{d^2x(t)}{dt^2} = \beta i(t) - kx(t) - f\frac{dx(t)}{dt}. \qquad [6.158]$$

Coupling equation:

$$e = \alpha \frac{dx(t)}{dt}. \qquad [6.159]$$

By applying the Laplace transform to equations [6.157]–[6.159], we achieve:

$$\begin{cases} U(p) = (R + Lp)I(p) + E(p) \\ F_a(p) = \beta I(p) = (Mp^2 + fp + k)X(p), \\ E(p) = \alpha p X(p). \end{cases} \qquad [6.160]$$

By rearranging this equation system, we get:

$$\begin{cases} I(p) = \frac{1}{R+Lp}(U(p) - E(p)) \\ E(p) = \alpha p X(p) \\ pX(p) = \frac{1}{Mp}(F_a(p) - fpX(p) - kX(p)). \end{cases} \qquad [6.161]$$

Hence, the block diagram shown in Figure 6.31.

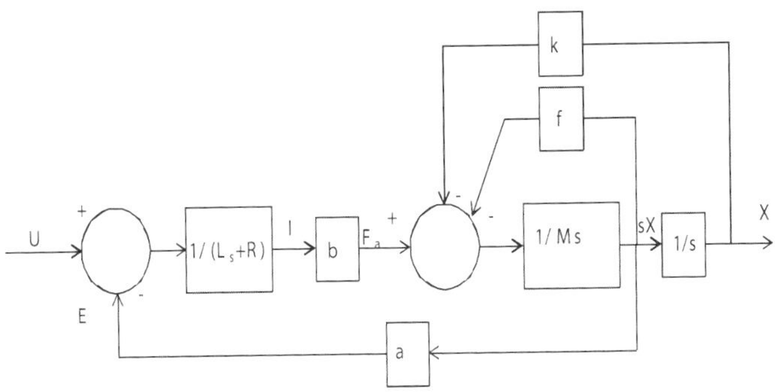

Figure 6.31. *Block diagram of the system*

– Equations [6.157]–[6.159] give us the following system:

$$\frac{di}{dt} = \frac{u}{L} - \frac{R}{L}i - \alpha \frac{dx}{dt}, \qquad [6.162]$$

$$\frac{d^2x}{dt^2} = \frac{\beta}{m}i - \frac{k}{m}x - \frac{f}{m}\frac{dx}{dt}.$$

Choosing the next state–space vector:

$$y = [i \;\; x \;\; \frac{dx}{dt}]^T. \qquad [6.163]$$

We obtain the following state–space representation:

$$\begin{aligned} \dot{y} &= Ay + Bu \quad \text{(state–space equation)}, \\ x &= Cy \quad \text{(output equation)} \end{aligned} \qquad [6.164]$$

$$\text{with } A = \begin{pmatrix} -\frac{R}{L} & 0 & -\frac{\alpha}{L} \\ 0 & 0 & 1 \\ \frac{\beta}{m} & -\frac{k}{m} & -\frac{f}{m} \end{pmatrix}; \quad B = \begin{pmatrix} \frac{1}{L} \\ 0 \\ 0 \end{pmatrix}; \quad C = \begin{pmatrix} 0 & 1 & 0 \end{pmatrix}. \qquad [6.165]$$

The characteristic polynomial is:

$$P(\lambda) = \det(\lambda I - A) = \lambda^3 + \lambda(\frac{f}{m} + \frac{k}{L}) + \lambda(\frac{k}{m} + \frac{Rf + \alpha\beta}{Lm}) + \frac{kR}{Lm}. \qquad [6.166]$$

Stability using the Routh criterion

– All terms are positive and present.

– All terms in the first column of the Routh table are positive. System of rank 3 ⇒ $a_1a_2 - a_3a_0 > 0$. In effect:

$$a_3p^3 + a_2p^2 + a_1p + a_0 = 0. \qquad [6.167]$$

$$\left| \begin{array}{ll} a_3 & a_1 \\ a_2 & a_0 \\ \frac{a_1a_2 - a_0a_3}{a_2} & \\ a_0 & \end{array} \right. \Rightarrow a_1a_2 - a_3a_0 > 0 \qquad [6.168]$$

$$\text{Let } \frac{f}{m}(\frac{k}{m} + \frac{Rf + \alpha\beta}{Lm}) + \frac{R}{L}\frac{Rf + \alpha\beta}{Lm} > 0. \qquad [6.169]$$

This term is always positive therefore the system is asymptotically stable.

Numerical application:

$$A = \begin{pmatrix} -1,000 & 0 & -625 \\ 0 & 0 & 1 \\ 25 & -250 & 0 \end{pmatrix}; \quad B = \begin{pmatrix} 125 \\ 0 \\ 0 \end{pmatrix}; \quad C = \begin{pmatrix} 0 & 1 & 0 \end{pmatrix}. \qquad [6.170]$$

The transfer function is given by:

$$H(p) = \frac{X(p)}{U(p)} = C(pI - A)^{-1}B. \qquad [6.171]$$

We first calculate the term $(sI - A)^{-1}$:

$$pI - A = \begin{pmatrix} p + 1,000 & 0 & 625 \\ 0 & p & -1 \\ -25 & 250 & p \end{pmatrix} \quad [6.172]$$

hence the determinant:

$$\begin{aligned} \Delta(p) &= (1,000 + p)(p^2 + 250) + 625(25p) \\ &= p^3 + 1,000p^2 + 15,875p + 250,000, \\ &= (p + 984.1271)(p^2 + 15.87292p + 254.0322), \end{aligned} \quad [6.173]$$

from which: $$(pI - A)^{-1} = \frac{1}{\Delta(p)} \times \begin{pmatrix} p^2 + 250 & 156,250 & -625p \\ 25 & p(p + 1,000) + 15,625 & p + 1,000 \\ 25p & -250(p + 1,000) & p(p + 1,000) \end{pmatrix} \quad [6.174]$$

then the transfer function is given by:

$$\begin{aligned} H(p) &= \frac{1}{\Delta(p)} \begin{pmatrix} 0 & 1 & 0 \end{pmatrix} \begin{pmatrix} \times & \times & \times \\ 25 & \times & \times \\ \times & \times & \times \end{pmatrix} \begin{pmatrix} 125 \\ 0 \\ 0 \end{pmatrix} \\ &= \frac{3,125}{\Delta(p)} = \frac{3,125}{(p + 984.1271)(p^2 + 15.87292p + 254.0322)}. \end{aligned} \quad [6.175]$$

The transfer function can also be written in the form:

$$H(p) = \frac{K\omega_n^2}{(1 + T_1 p)(p^2 + 2\xi\omega_n p + \omega_n^2)} \quad [6.176]$$

with $K = 3,125$, $T_1 = 10^{-3}$, $\omega_n = 16$ and $\xi = 0.5$. The pole $p = -984.12$ can be neglected because it is very far to the left in the complex plane. Its effect in the temporal domain disappears very quickly. The dynamic is thus described only by the dominant complex conjugate poles located near the imaginary axis.

6.4.7. *Exercise 6.14*

The stability of the system is inferred from the eigenvalues of A. These latter are given: $\lambda_1 = \lambda_2 = 0.5$. Since $|\lambda_1| < 1$ and $|\lambda_2| < 1$, the system is then stable. The transmittance in Z is given by:

$$H(z) = C[zI - A]^{-1}B \qquad \text{with} \qquad [zI - A]^{-1} = \begin{bmatrix} \frac{1}{z-0.5} & \frac{-1}{(z-0.5)^2} \\ 0 & \frac{1}{z-0.5}. \end{bmatrix}, \qquad [6.177]$$

$$\text{from which} \quad H(z) = -\frac{z+0.5}{(z-0.5)^2}. \qquad [6.178]$$

The recurrence equation:

$$H(z) = \frac{Y(z)}{U(z)} \qquad \Longrightarrow \qquad Y(z)(z^2 - z + 0.5) = -U(z)(z + 0.5), \qquad [6.179]$$

which yields:

$$\begin{aligned} y_{k+2} - y_{k+1} + 0.25y_k &= -u_{k+1} - 0.5u_k \quad \Longrightarrow \\ y_{k+2} &= y_{k+1} - 0.25y_k - u_{k+1} - 0.5u_k. \end{aligned} \qquad [6.180]$$

6.4.8. *Exercise 6.15*

Let the system be described as shown in Figure 6.32.

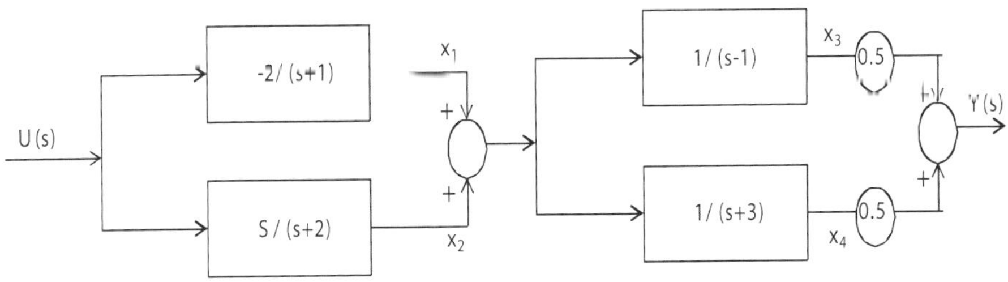

Figure 6.32. *Block diagram of the system*

6.4.8.1. *State–space representation of the system*

The state vector is given by

$$X = [x_1, x_2, x_3, x_4]^T. \quad [6.181]$$

The state–space representation:

$$\begin{cases} \dot{X} = AX + BU \\ Y = CX \end{cases} \quad [6.182]$$

$$\text{with } A = \begin{pmatrix} -1 & 0 & 0 & 0 \\ 0 & -2 & 0 & 0 \\ 1 & 1 & 1 & 0 \\ 1 & 1 & 0 & -3 \end{pmatrix}; \quad B = \begin{bmatrix} -2 \\ 3 \\ 0 \\ 0 \end{bmatrix}; \quad C = \begin{bmatrix} 0 & 0 & 0.5 & 0.5 \end{bmatrix}. \quad [6.183]$$

In the following sections, we give the definitions of controllability and observability.

6.4.8.2. *Controllability*

DEFINITION 6.1.– *The system is controllable if there is a finite control $u(t)$ that can bring the system from an initial state x_0 to a final state x_f in finite time.*

DEFINITION 6.2.– *The system is controllable if the controllability matrix*

$$Com = [B, AB, \ldots, A^{n-1}B] \text{ is of rank } n. \quad [6.184]$$

DEFINITION 6.3.– *In the case of a system expressed in diagonal form, it suffices that all the elements of vector B be non-zero.*

6.4.8.3. *Observability*

DEFINITION 6.4.– *The system is observable if, regardless of the initial state $x(0)$, it is possible in a finite time τ to determine $x(0)$ from $u(t)$ and $y(t)$ with $0 \leq t \leq \tau$.*

DEFINITION 6.5.– *The system is observable if the observability matrix*

$$Obs = [C^T, A^T C^T, \ldots, A^{T(n-1)} C^T] \text{ is of rank } n. \quad [6.185]$$

DEFINITION 6.6.– *In the case of a system expressed in the diagonal form, it suffices that all the elements of vector C be non-zero.*

Analyzing the controllability and the observability of this system:

$$Com = \begin{bmatrix} B & AB & A^2B & A^3B \end{bmatrix} = \begin{bmatrix} -2 & 2 & -2 & 2 \\ 3 & -6 & 12 & -24 \\ 0 & 1 & -3 & 7 \\ 0 & 1 & -7 & 31 \end{bmatrix}, \quad [6.186]$$

Com is of rank = 3 ⇒ non-controllable system (a non-controllable state).

$$Obs = \begin{bmatrix} C \\ CA \\ CA^2 \\ CA^3 \end{bmatrix} = \begin{bmatrix} 0 & 0 & 0.5 & 0.5 \\ 1 & 1 & 0.5 & -1.5 \\ -2 & -3 & 0.5 & 4.5 \\ 7 & 11 & 0.5 & -13.5 \end{bmatrix}, \quad [6.187]$$

Obs is of rank = 3 ⇒ non-observable system (a non-observable state).

Analyzing the controllability and the observability of this system making use of the diagonal form. After computation of the eigenvalues $\lambda_i (i = 1, \ldots, 4)$ and of the matrix P of the eigenvectors, we obtain the following state–space representation:

$$\begin{cases} \dot{Z} = P^{-1}APZ + P^{-1}BU \\ Y = CPZ \end{cases} \quad [6.188]$$

with

$$P^{-1}AP = \begin{bmatrix} -1 & 0 & 0 & 0 \\ 0 & -2 & 0 & 0 \\ 0 & 0 & 1 & 0 \\ 0 & 0 & 0 & -3 \end{bmatrix}; \; P^{-1}B = \begin{bmatrix} -2 \\ 3 \\ 0 \\ -2 \end{bmatrix}; \; CP = \begin{bmatrix} 0 & 2/6 & 0.5 & 0.5 \end{bmatrix} \quad [6.189]$$

The third state is non-controllable. The first state is non-observable.

The new block diagram is given hereafter.

The transfer function of the system is given by

$$H(p) = \frac{Y(p)}{U(p)} = \frac{1}{(p+3)(p+2)}. \quad [6.190]$$

In conclusion, the system is of the second order. The *minimal* representation of the process in the state form only introduces two states.

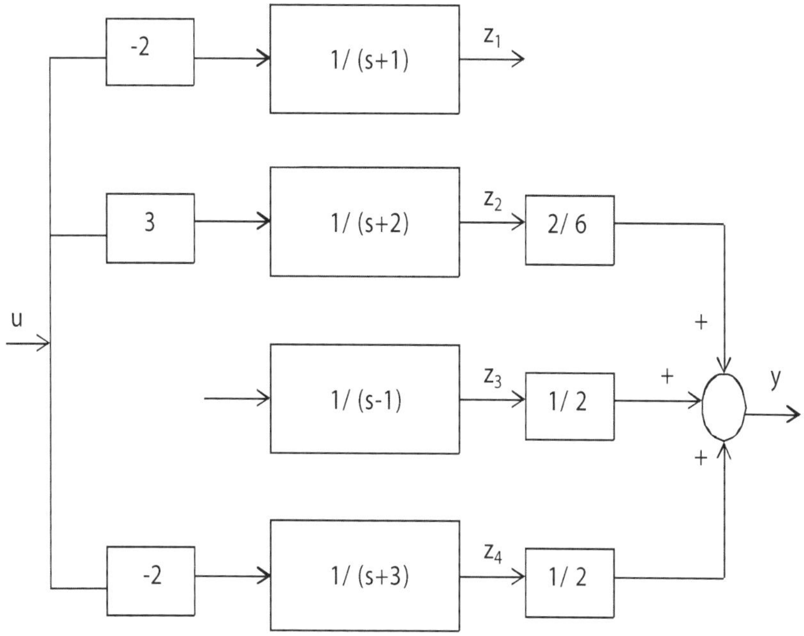

Figure 6.33. *Block diagram of the second-order system*

Caution: The real system is shown in Figure 6.33; we do not have the right to simplify the block diagram. The transfer function $H(p)$ does not actually reflect what is happening inside the process of the second-order system.

– Process delayed model (discrete model)

– Observability detectability

– Controllability reachability

– Stabilizability

– Linearization and separation theorem

– Observers

– Nonlinear systems control

Nonlinear systems observation and control

Consider nonlinear systems of the form:

$$\dot{x} = f(x) + g(x)u \; and \; \; y = h(x) \qquad [6.191]$$

$$\begin{aligned} \dot{x} &= f(x) + g(x)u \\ y &= h(x) \end{aligned} \qquad [6.192]$$

with $x \in$ U (U $\subset \mathcal{R}^n$), $u \in \mathcal{R}^m$ and $y \in \mathcal{R}^m$.

$$\begin{aligned} u &= col(u_1, \ldots, u_m) \qquad & g(x) &= [g_1(x), \ldots, g_m(x)] \\ y &= col(y_1, \ldots, y_m) \qquad & h(x) &= col(h_1(x), \ldots, h_m(x)) \end{aligned} \qquad [6.193]$$

$$\begin{array}{ccc} f : U \longrightarrow R^n & g_i : U \longrightarrow R^n & h_i : U \longrightarrow R \\ x \rightarrow f(x) & x \rightarrow g_i(x) & x \rightarrow h_i(x) \end{array} \qquad [6.194]$$

f, g_i and h_i of class C^∞.

Observation: Manipulative robots fall under this class.

– Loop and state-change linearization:

- relative degree (r);
- state change (normal form);
- linearization;
- zeros dynamics.

– State feedback stabilization;

– trajectory tracking;

– example.

The purpose of the linearization is to find a control u and a change i state such that:

$$\begin{aligned} u &= \alpha(x) + \beta(x)v \quad \text{new input signal,} \\ z &= \Phi(x), \end{aligned} \qquad [6.195]$$

which makes it possible to obtain the equivalent *linear* system in the CL system.

$$\begin{cases} \dot{x} = f(x) + g(x)\alpha(x) + \beta(x)v \\ y = h(x) \end{cases} \xrightarrow{z=\Phi(x)} \begin{cases} \dot{z} = Az + Bv \\ y = Cz. \end{cases} \qquad [6.196]$$

The system [6.191] has a (vector) relative degree $\{r_1, \ldots, r_m\}$ in x_0 if:

i) $L_{g_j} L_f^k h_i(x) = 0$
$\forall\, 1 \le j \le m, \forall\, 1 \le i \le m, \forall\, k < r_i - 1$ and $\forall\, x$ in the neighborhood of x_0.

ii) The following $m \times m$ matrix:

$$A(x) = \begin{bmatrix} L_{g_1} L_f^{r_1-1} h_1(x) & \ldots & L_{g_m} L_f^{r_1-1} h_1(x) \\ L_{g_1} L_f^{r_2-1} h_2(x) & \ldots & L_{g_m} L_f^{r_2-1} h_2(x) \\ \ldots & \ldots & \ldots \\ L_{g_1} L_f^{r_m-1} h_1(x) & \ldots & L_{g_m} L_f^{r_m-1} h_m(x) \end{bmatrix}, \qquad [6.197]$$

is non-singular in x_0.

Observation:

For any integer r_i associated with input i of the system, it can be said that $\forall\, k < r_i - 1$ the vector

$$\left[\, L_{g_1} L_f^k h_i(x) \;\; L_{g_2} L_f^k h_i(x) \;\ldots\; L_{g_m} L_f^k h_i(x) \,\right], \qquad [6.198]$$

is zero $\forall\, x$ in the neighborhood of x_0 and non-zero for $k = r_i - 1$.

Bibliography

[ARM 00] ARMSTRONG B., "Elastic-plastic friction model: contact compliance and stiction", *Proceedings of the 2000 American Control Conference*, Chicago, Illinois, June 28–30, 2000.

[ARQ 82] ARQUES, P.Y., *Décision en traitement du signal*, Masson, 1982.

[ASC 91] ASCH G. *et al.*, *Les capteurs en instrumentation industrielle*, Dunod, 1991.

[AST 97] ASTROM K., WINTTENMARK B., *Computer Controlled Systems*, Prentice-Hall, 1997.

[AST 89] ASTROM K.J., WITENMARK B., *Adaptive Control*, Addison-Wesley, 1989.

[AUB 84] AUBIN J.P. *et al.*, *Differential Inclusions, A Series of Comprehensive Studies in Mathematics*, Springer-Verlag, 1984.

[BAB 85] BABARY J.P., PELCZEWSKI W., *Commande optimale des systèmes continus déterministes*, Masson, 1985.

[BEL 12] BELANGER M., *Traitement numérique du Signal*, Dunod, 2012.

[BEN 87] BENVENIST A., METEVIER M., PRIOURET P., *Algorithmes adaptatifs et approximations stochastiques*: *Théorie et Applications*, Masson, 1987.

[BLA 81] BLANC A., PICINBONO B., *Fonctions aléatoires*, Masson, 1981.

[BLA 08] BLAU P.J., *Friction Science and Technologie: From Concepts to Applications*, 2nd ed., CRC Press T&F group, 2008.

[BOB 98] BOBROWS J.E. *et al.*, "Modeling identification and control of a pneumatically actuated, force controllable robot", *IEEE Trans. on Robotics and Automation*, no. 5, pp. 732–741, 1998.

[BOI 80] BOITE R., LEICH H., *Les Filtres Numériques*, Masson, 1980.

[BOU 92] BOUACHACHE B., *Time-Frequency Signal Analysis: Methods and Applications*, Longman-Cheshire, 1992.

[BOX 70] BOX G.E.P., JENKINS G.M., *Time Series Analysis: Forecasting and Control*, Holden-Day, 1970.

[CAN 86] Candy J.V., *Signal Processing: The Model Based Approach*, McGraw Hill, 1986.

[CAN 98] Canudas de Wit C., "Comments on a new model for control of systems with friction", *IEEE Transaction on Automatic Control*, vol. 43, no. 8, pp. 1189–1191, August 1998.

[CEL 96] Cellier F.E. *et al.*, "Modeling from Physical Principles", in Levine W.S. (ed.), *The Control Handbook*, CRC Press, 1996.

[COS 97] Costas T., "Adaptive impedance control applied to a pneumatic legged robot", *Int. Journal of Intelligent and Robotic Systems*, vol. 20, pp. 105–129, 1997.

[CRA 89] Craig J.J., *Introduction to Robotics*, Addison-Wesley, 1989.

[DE 84] De Coulon F., *Théroie et Traitement des Signaux*, Dunod, 1984.

[DE 92] De Larminat P., Thomas Y., *Automatique des systèmes linéaires*, Flammarion, 1992.

[DI 94] Di Stefano J.J., Stuberud A.R., Williams I.J., *Systèmes Asservis*, McGraw Hill, 1994.

[DOR 95] Dorf R., *Modern Control Systems*, Addison-Wesley, 1995.

[DUP 09] Dupont P., "Torsional kinematic model for concentric tube robots", *Conf. Proc. IEEE International Conference on Robotics and Automation*, pp. 3851–3858, 2009.

[FAE 81] Faes Y., *Commande de processus industriels par calculateurs*, Masson, 1981.

[FAR 86] Fargeon C., *Commande Numérique des Systèmes*, Masson, 1986.

[FEM 97a] Femmam S., Méthodes et algorithmes de segmentation des signaux par détection de ruptures, Report, Laboratoire de robotique de Paris, 1997.

[FEM 97b] Femmam S., Méthodes et algorithmes en temps-fréquence et temps-échelle, Report, Laboratoire de robotique de Paris, 1997.

[FEM 11] Femmam S., *Analyse, traitement et diagnostic des signaux non-stationnaires par détection de ruptures et temps-Fréquence: application à la perception et à la caractérisation*, Editions Universitaires Européennes, 2011.

[FLA 93] Flandrin P., *Temps-Fréquence*, Hermes, Paris, 1993.

[FUK 99] Fukuda T. *et al.*, "Recent topics on robotic actuation technologies", *Proc. of* 2nd *Int. Conference on Climbing and Walking Robots*, Portsmouth, vol. 3-15, 1999.

[GIL 67] Gilles J.C., Decaulne P., Pellegrin M., *Dynamique de la commande linéaire*, Dunod, 1967.

[GOO 84] Goodwin G.C., Sin K.S., *Adaptive Filtering: Prediction and Control*, Prentice Hall, 1984.

[HAY 01] Haykin S., *Communication Systems*, 4th ed., John Wiley & Sons, 2001.

[KAR 06] Karnopp C., *System Dynamics: Modeling and Simulation of Mechatronic Systems*, Wiley, 2006.

[KEL 87] Kellal A., Contribution à l'étude des asservissements électropneumatiques, Application à la réalisation d'un asservissement numérique d'un axe de robot, Thesis, INSA, Lyon, 1987.

[KHA 86] KHALIL W., KLEINFINGER W., GAUTIER J.F., "Reducing the computational burden of the dynamical model of robot", *Proc IEEE Conf. on Robotics and Automation*, San Fransisco, pp. 525–531, 1986.

[KUN 06] KUNT M., *Traitement Numérique des Signaux*, 8th ed., Dunod, 2006.

[KUO 75] KUO B.C., *Automatic Control Systems*, Prentice Hall, 1975.

[LAC 94] LACROUX G., *Les Actionneurs Électriques*, Technique et Documentation, 1994.

[LAN 79] LANDAU I.D., *Adaptive Control: the Model Reference Approach*, Dekker, 1979.

[LAN 88] LANDAU I.D., *Identification et Commande des Systèmes*, Hermes, 1988.

[LIF 79] LIFERMANN J., *Les Méthodes Rapides de Transformation du signal*, Masson, 1979.

[LIN 92] LIN X.F., Contribution à la modélisation et à la commande d'un vérin pneumatique de grande longueur, Thesis, INSA, Lyon, 1992.

[LIU 88] LIU R. *et al.*, "An analysis of pneumatic servo system and its application to a computer-controlled robot", *Transactions of the ASME Journal Dynamic Systems, Measurements and Control*, vol. 110, pp. 228–235, 1988.

[LJU 87] LJUNG L., *System Identification: Theory for the User*, Prentice Hall, 1987.

[LJU 83] LJUNG L., SODERSTROM T., *Theory and Practice of Recursive Identification*, MIT Press, 1983.

[LOP 86] LOPEZ P., *Commande des Systèmes Robotique*, Test, 1986.

[MAX 81] MAX J., *Méthodes et Techniques du Traitement du Signal*, Masson, 1981.

[MIC 92] MICHAUT F., *Méthodes adaptatives pour le signal. Outils mathématiques et mise en oeuvre*, Hermès, 1992.

[MID 90] MIDDLETON R.H., GOODWIN G.C., *Digital Estimation and Control: A Unified Approach*, Prentice Hall, 1990.

[MOR 90] MOREAU N., *Modélisation et estimation spectrale*, ENST, 1990.

[NAJ 82] NAJIM K., *Commande adaptative de processus industriel*, Masson, 1982.

[NAJ 83] NAJIM K., MURATET G., *Pratique de la régulation de processus industriels*, Masson, 1983.

[OGA 03] OGATA K., *Modern Control Engineering*, Prentice Hall, 2003.

[ORT 08] ORTEGA-SANCHEZ M. *et al.*, "A global model of a tidal jet including the effects of friction and bottom slope", *Journal of Hydraulic Research*, vol. 46, no. 1, pp. 80–86, 2008.

[PAP 77] PAPOULIS A., *Signal Analysis*, McGraw Hill, 1977.

[PAY 61] PAYNTER H.M., *Analysis and Design of Engineering Systems*, MIT Press, 1961.

[PIC 86] PICINBONO B., *Eléments de Théorie du Signal*, Dunod, 1986.

[POW 80] POWELL F., *Digital Control of Dynamic Systems*, Addison-Wesley, 1980.

[PU 96] PU J. *et al.*, "Acceleration characteristics of servo controlled pneumatic cylinder", *ASME, Fluid Power Syst Technol.*, vol. 3, pp. 119–130, 1996.

[RAB 75] RABINER L., GOLD B., *Theory and Applications of Digital Signal Processing*, Prentice Hall, 1975.

[RAG 82] RAGOT J., *Exercices et problèmes d'automatique*, Masson, 1982.

[RAG 90] RAGOT J., *Validation de données et diagnostic*, Hermes, 1990.

[RIC 96] RICHARD E., SCAVADA S., "Comparaison between linear and nonlinear control of an electropeumatic servodrive", *Trans. of ASME Journal of Dynamic Systems, Measurements and Control*, vol. 1118, pp. 245–252, 1996.

[ROS 82] ROSENBERG R.C., "General friction considerations for engine design", *Society of Automative Engineers*, Warrendale, Paper no. 821576, 1982.

[ROS 83] ROSENBERG R.C., KARNOPP D.C., *Introduction to Physical System Dynamics*, McGraw Hill, 1983.

[ROU 79] ROUBINE E., *Introduction à la théorie de la communication*, Masson, 1979.

[SCA 89] SCAVARDA A S., *Les asservissements électropneumatiques de position*, Hermes, 1989.

[SEV 69] SEVELY Y., *Systèmes linéaires et asservissements échantillonnés*, Dunod, 1969.

[SHE 56] SHEARER J.L., "Study of pneumatic process in the continous control of motion with compressed air I, II", *Trans ASME Journal Dynamic Systems, Measurements and Control*, pp. 233–249, 1956.

[SOD 83] SODERSTROM T., STOICA P., *The Instrumental Varaible Approach to System Identification*, Springer, Berlin, 1983

[SOD 89] SODERTROM T., STOICA P., *System Identification*, Prentice Hall 1989.

[SPA 73] SPATARU A., *Théorie de la transmission de l'information*, Masson, 1973.

[SPA 87] SPATARU A., *Fondements de la théorie de transmission de l'information*, Presses Polytechnique et Universitaires Romandes, 1987.

[THO 96] THOMASSET D., *et al.* "Non linear controller with perturbation estimation for electropneumatic actuators", 9th *Bath Fluidpower Workshop*, Bath, UK, pp. 72–86, 1996.

[TUS 47] TUSTIN W., Vibration and shock isolation trends and solutions, Equipment Reliability Institue, 1947.

[WAL 84] WALRATH C.D., "Adaptive bearing friction compensation based on recent knowledge of dynamic friction", *Automatica*, vol. 20, np. 6, pp. 717–727, 1984.

[WEL 79a] WELLSTEAD P.E., *Introduction to Physical System Modeling*, Academic Press, 1979.

[WEL 79b] WELLSTEAD P., ZARROP M., *Self-tuning Systems Control and Signal Processing*, Wiley-Blackwell, 1979.

[WID 75] WIDROW B., STEARNS S.D., *Adaptive Signal Processing*, Prentice Hall, 1975.

[WON 74] WONHAM, W.M., *Linear Multivariable Control: a Geometric Approach*, Springer Verlag, 1974.

http://www.aldebaran-robotics.com/en/

http://www.aldebaranrobotics.com/documentation/software/choregraphe/

http://robotdomotique.com/

http://fr.wikipedia.org/wiki/Robot_aspirateur,

http://www.maisonrobot.fr/comment-ca-marche/

http://www.mysti2d.net/polynesie//ETC/C041/32/capteursNAO/Documentstechniques6.html

https://community.aldebaran-robotics.com/doc/1-12/software/monitor/index.html

https://community.aldebaran-robotics.com/doc/1-12/software/naosim/index.html

http://www.aldebaran-robotics.com/en/Discover-NAO/Software/Webots-for-NAO.html

http://www.aldebaran-robotics.com/documentation/dev/matlab/index.html

http://www.cmake.org/cmake/resources/software.html

Index

A, C, D, E

F, H, I

N, P, S, W

Other titles from

ISTE

in

Digital Signal and Image Processing

2015

BLANCHET Gérard, CHARBIT Maurice
Digital Signal and Image Processing using MATLAB®
Volume 2 – Advances and Applications:The Deterministic Case – 2nd edition
Volume 3 – Advances and Applications: The Stochastic Case – 2nd edition

CLARYSSE Patrick, FRIBOULET Denis
Multi-modality Cardiac Imaging

DUBUISSON Séverine
Tracking with Particle Filter for High-dimensional Observation and State Spaces

GIOVANNELLI Jean-François, IDIER Jérôme
Regularization and Bayesian Methods for Inverse Problems in Signal and Image Processing

MAÎTRE Henri
From Photon to Pixel: The Digital Camera Handbook

2014

AUGER François
Signal Processing with Free Software: Practical Experiments

BLANCHET Gérard, CHARBIT Maurice
Digital Signal and Image Processing using MATLAB®
Volume 1 – Fundamentals – 2nd edition

DUBUISSON Séverine
Tracking with Particle Filter for High-dimensional observation and State Spaces

ELL Todd A., LE BIHAN Nicolas, SANGWINE Stephen J.
Quaternion Fourier Transforms for Signal and Image Processing

FANET Hervé
Medical Imaging Based on Magnetic Fields and Ultrasounds

MOUKADEM Ali, OULD Abdeslam Djaffar, DIETERLEN Alain
Time-Frequency Domain for Segmentation and Classification of Non-stationary Signals: The Stockwell Transform Applied on Bio-signals and Electric Signals

NDAGIJIMANA Fabien
Signal Integrity: From High Speed to Radiofrequency Applications

PINOLI Jean-Charles
Mathematical Foundations of Image Processing and Analysis
Volumes 1 and 2

TUPIN Florence, INGLADA Jordi, NICOLAS Jean-Marie
Remote Sensing Imagery

VLADEANU Calin, EL ASSAD Safwan
Nonlinear Digital Encoders for Data Communications

2013

GOVAERT Gérard, NADIF Mohamed
Co-Clustering

DAROLLES Serge, DUVAUT Patrick, JAY Emmanuelle
Multi-factor Models and Signal Processing Techniques: Application to Quantitative Finance

LUCAS Laurent, LOSCOS Céline, REMION Yannick
3D Video: From Capture to Diffusion

MOREAU Eric, ADALI Tulay
Blind Identification and Separation of Complex-valued Signals

PERRIN Vincent
MRI Techniques

WAGNER Kevin, DOROSLOVACKI Milos
Proportionate-type Normalized Least Mean Square Algorithms

FERNANDEZ Christine, MACAIRE Ludovic, ROBERT-INACIO Frédérique
Digital Color Imaging

FERNANDEZ Christine, MACAIRE Ludovic, ROBERT-INACIO Frédérique
Digital Color: Acquisition, Perception, Coding and Rendering

NAIT-ALI Amine, FOURNIER Régis
Signal and Image Processing for Biometrics

OUAHABI Abdeljalil
Signal and Image Multiresolution Analysis

2011

CASTANIÉ Francis
Digital Spectral Analysis: Parametric, Non-parametric and Advanced Methods

DESCOMBES Xavier
Stochastic Geometry for Image Analysis

FANET Hervé
Photon-based Medical Imagery

MOREAU Nicolas
Tools for Signal Compression

2010

NAJMAN Laurent, TALBOT Hugues
Mathematical Morphology

2009

BERTEIN Jean-Claude, CESCHI Roger
Discrete Stochastic Processes and Optimal Filtering / 2nd edition

CHANUSSOT Jocelyn *et al.*
Multivariate Image Processing

DHOME Michel
Visual Perception through Video Imagery

GOVAERT Gérard
Data Analysis

GRANGEAT Pierre
Tomography

MOHAMAD-DJAFARI Ali
Inverse Problems in Vision and 3D Tomography

SIARRY Patrick
Optimization in Signal and Image Processing

2008

ABRY Patrice *et al.*
Scaling, Fractals and Wavelets

GARELLO René
Two-dimensional Signal Analysis

HLAWATSCH Franz *et al.*
Time-Frequency Analysis

IDIER Jérôme
Bayesian Approach to Inverse Problems

MAITRE Henri
Processing of Synthetic Aperture Radar (SAR) Images

MAITRE Henri
Image Processing

NAIT-ALI Amine, CAVARO-MENARD Christine
Compression of Biomedical Images and Signals

NAJIM Mohamed
Modeling, Estimation and Optimal Filtration in Signal Processing

QUINQUIS André
Digital Signal Processing Using Matlab

2007

BLOCH Isabelle
Information Fusion in Signal and Image Processing

GLAVIEUX Alain
Channel Coding in Communication Networks

OPPENHEIM Georges *et al.*
Wavelets and their Applications

2006

CASTANIÉ Francis
Spectral Analysis

NAJIM Mohamed
Digital Filters Design for Signal and Image Processing

Printed and bound by CPI Group (UK) Ltd, Croydon, CR0 4YY